T0137241

Intelligent Systems Reference Library

Volume 159

Series Editors

Janusz Kacprzyk, Polish Academy of Sciences, Warsaw, Poland

Lakhmi C. Jain, Faculty of Engineering and Information Technology, Centre for
Artificial Intelligence, University of Technology, Sydney, NSW, Australia;
Faculty of Science, Technology and Mathematics, University of Canberra,
Canberra, ACT, Australia;
KES International, Shoreham-by-Sea, UK; Liverpool Hope University,
Liverpool, UK

The aim of this series is to publish a Reference Library, including novel advances and developments in all aspects of Intelligent Systems in an easily accessible and well structured form. The series includes reference works, handbooks, compendia, textbooks, well-structured monographs, dictionaries, and encyclopedias. It contains well integrated knowledge and current information in the field of Intelligent Systems. The series covers the theory, applications, and design methods of Intelligent Systems. Virtually all disciplines such as engineering, computer science, avionics, business, e-commerce, environment, healthcare, physics and life science are included. The list of topics spans all the areas of modern intelligent systems such as: Ambient intelligence, Computational intelligence, Social intelligence, Computational neuroscience, Artificial life, Virtual society, Cognitive systems, DNA and immunity-based systems, e-Learning and teaching, Human-centred computing and Machine ethics, Intelligent control, Intelligent data analysis, Knowledge-based paradigms, Knowledge management, Intelligent agents, Intelligent decision making, Intelligent network security, Interactive entertainment, Learning paradigms, Recommender systems, Robotics and Mechatronics including human-machine teaming, Self-organizing and adaptive systems, Soft computing including Neural systems, Fuzzy systems, Evolutionary computing and the Fusion of these paradigms, Perception and Vision, Web intelligence and Multimedia.

** Indexing: The books of this series are submitted to ISI Web of Science, SCOPUS, DBLP and Springerlink.

More information about this series at http://www.springer.com/series/8578

Anna Esposito · Antonietta M. Esposito ·
Lakhmi C. Jain
Editors

Innovations in Big Data Mining and Embedded Knowledge

Springer

Editors
Anna Esposito
Dipartimento di Psicologia and International
Institute for Advanced Scientific
Studies (IIASS)
Università degli Studi della Campania
"Luigi Vanvitelli"
Caserta, Italy

Antonietta M. Esposito
Sezione di Napoli, Osservatorio Vesuviano
Istituto Nazionale di Geofisica
e Vulcanologia
Napoli, Italy

Lakhmi C. Jain
University of Technology Sydney
Sydney, Australia

University of Canberra
Canberra, ACT, Australia

KES International
Shoreham-by-Sea, UK

Liverpool Hope University
Liverpool, UK

ISSN 1868-4394   ISSN 1868-4408 (electronic)
Intelligent Systems Reference Library
ISBN 978-3-030-15941-2   ISBN 978-3-030-15939-9 (eBook)
https://doi.org/10.1007/978-3-030-15939-9

Library of Congress Control Number: 2019935537

This Springer imprint is published by the registered company Springer Nature Switzerland AG
The registered company address is: Gewerbestrasse 11, 6330 Cham, Switzerland

Preface

Data Mining or Knowledge Discovery in Databases (KDD) had received a great interest among researchers and practitioners due to its focus on processes transforming big amount of data into novel, valid, useful, and structured knowledge by detecting concealed patterns and relationships in data.

The book reports new directions in knowledge discovery using data mining and knowledge embedding through models. Contributors have reported a number of schemes to explain how data mining or embedding knowledge can be beneficial to social organizations, domestic spheres, and ICT market.

The concept of knowledge is broad and speculative and had produced epistemological debates in western philosophies. The intensified interest in knowledge management and data mining stems from the difficulty in identifying computational models able to approximate to a certain degree human behaviors and abilities in resolving organizational, social, and physical problems. Current ICT interfaces are not adequate to be introduced into the domestic spheres in order to support and simulate abilities of medical doctors, teachers, assistants, housekeepers, and so on. The domestic world has been denied to the machines because, differently from industrial contexts where abilities are routinely applied, the domestic world is continuously changing and unpredictable. The questions posited in this field are very challenging. Does knowledge locked in conventions, rules of conducts, common senses, ethics, emotions, laws, cultures, and experiences be mined from data? It will suffice for socially believable and emotionally behaving automatic systems, to rule complex interactions only through the mining of big amount of data?

The themes afforded by this book are multidisciplinary in nature and suggest that computational models able to approximate human behaviors and abilities in resolving organizational, social, and physical problems.

Innovations in Big Data Mining and Embedded Knowledge will prove useful to:

a. The academic research community
b. The ICT market
c. Ph.D. Students and Early Stage Researchers

d. Schools, Hospitals, Rehabilitation, and Assisted-Living Centers
e. Representatives from multimedia industries and standardization bodies

The editors would like to thank the contributors for their rigorous and invaluable scientific contributions, dedication, and priceless selection process. We wish to acknowledge the time and expertise of the reviewers for their contributions. Thanks are also due to the Springer-Verlag for their excellent support during the developmental phase of this research book.

Caserta, Italy   Anna Esposito
Napoli, Italy   Antonietta M. Esposito
Canberra, Australia   Lakhmi C. Jain

Sponsoring Organizations

- Università della Campania "Luigi Vanvitelli", Dipartimento di Psicologia
- International Institute for Advanced Scientific Studies "E.R. Caianiello" (IIASS, www.iiassvietri.it/), Italy
- Società Italiana Reti Neuroniche (SIREN)

Contents

About the Editors

Anna Esposito received her "Laurea Degree" summa cum laude in Information Technology and Computer Science from the Università di Salerno in 1989 with a thesis on: The Behavior and Learning of a Deterministic Neural Net (published on Complex System, 6(6), 507–517, 1992). She received her Ph.D. in Applied Mathematics and Computer Science from Università di Napoli "Federico II" in 1995. Her Ph.D. thesis on: Vowel Height and Consonantal Voicing Effects: Data from Italian (published on Phonetica, 59 (4), 197–231, 2002) was developed at Massachusetts Institute of Technology (MIT), Research Laboratory of Electronics (RLE), under the supervision of professor Kenneth N. Stevens.

She has been Post Doc at the International Institute for Advanced Scientific Studies (IIASS), and Assistant Professor at Università di Salerno (Italy), where she taught Cybernetics, Neural Networks, and Speech Processing (1996–2000). She had a position as Research Professor (2000–2002) at the Department of Computer Science and Engineering at WSU, Dayton, OH, USA, where she is currently Research Affiliate. She has been the Italian MCM of COST ACTIONS: COST 277: www.cost.esf.org/domains_actions/ict/ Actions/277 (2001–2005); COST IC1002: www.co st.esf.org/domains_actions/ict/Actions/IC1002 (2010–2014); COST TIMELY: www.timely-cost.eu (2010–2014). She has been the proposer and chair of COST 2102: www.cost.esf.org/domains_actions/ict/Actions/ 2102 (2006–2010). Currently, she is the Italian MCM of COST CA15218: www.cost.eu/COST_Actions/ca/

CA15218?management, and the Italian MCM Substitute of COST Action IS1406: www.cost.eu/COST_Actions/isch/IS1406?management. Since 2006, she is a Member of the EuCognition, www.eucognition.org.

Her research interests are on contextual and interactional cross-modal analysis of typical and disordered (e.g., depressed or impaired individuals) human–machine interactional exchanges through speech, gestures, vocal, and facial emotional expressions in order to assess user's requirements and expectations to appropriately inform the implementation of emotional and social believable Human–Computer Interaction (HCI).

Antonietta M. Esposito was born and educated in Italy. She got the "Laurea" (Degree) from the University of Salerno (Italy) in 2001, with a Laurea thesis on the processing of speech signals (published in the IEEE Proceedings of the IEEE International Workshop on Circuits and Systems, R. L. Ewing et al. (Eds.), 2, 516–519, 2001).

From November 2001 to 2003, she participated at the Master in "Advanced Technologies of Information and Communication", supported by a 2-year fellowship of the "Ministero dell'Istruzione, dell'Università e della Ricerca (MIUR)". From November 2003 to April 2005, her research activity was on speech recognition, at the CIRTE SpA (a telecommunication company in Salerno), and jointly on the analysis of seismic signals, at the Department of Physics at Salerno University and the IIASS.

She joined the Istituto Nazionale di Geofisica e Vulcanologia, Sezione di Napoli Osservatorio Vesuviano, Napoli, Italy in May 2005 where she currently works as senior researcher for the monitoring and surveillance of Campanian Volcanoes (Vesuvio, Campi Flegrei and Ischia, Italy).

Her research interests are on seismic signal processing, speech, nonverbal communication features, neural networks, and mining of large amount of data with intelligent learning algorithms.

She is author of several contributes in peer-reviewed journals, books, and conferences and co-editors of three international books.

Dr. Lakhmi C. Jain, Ph.D., M.E., B.E.(Hons) Fellow (Engineers Australia) is with the University of Technology Sydney, Australia; University of Canberra, Australia and Liverpool Hope University, UK.

Professor Jain founded the KES International for providing a professional community, opportunities for publications, knowledge exchange, cooperation, and teaming. Involving around 5,000 researchers drawn from universities and companies world-wide, KES facilitates international cooperation and generate synergy in teaching and research. KES regularly provides networking opportunities for professional community through one of the largest conferences of its kind in the area of KES. www.kesinternational.org.

His interests focus on the artificial intelligence paradigms and their applications in complex systems, security, e-education, e-healthcare, unmanned air vehicles, and intelligent agents.

Keywords

Data mining
Embedded knowledge
Domestic spheres
Social agents and robots
Models
Assistive technologies
Complex autonomous systems
Socially believable interfaces
Computational complexity

Chapter 1
More Than Data Mining

Anna Esposito, Antonietta M. Esposito and Lakhmi C. Jain

Abstract Data Mining has received a great momentum of interest due to the automatic processes transforming big amount of data into novel, valid, useful, and structured knowledge by detecting concealed patterns and relationships in data. However, embedded knowledge has not been thoroughly considered in data mining. The chapters reported in this book discuss on several facets of embedded knowledge and propose solutions for data mining.

Keywords Data mining · Embedded knowledge · eHealth · Emotions · Socially and believable ICT interfaces

1.1 Introduction

Embedded knowledge is referred to the knowledge locked in conventions, rules of conducts, common senses, ethics, emotions, laws, cultures, experiences [18]. This embedding can have happened through the formalization of beneficial routines or was derived from experiential facts. Such knowledge is dynamically changing and able to rule complex interactions. It can be explicit, factual, tacit, or experiential. It may or may not evolves from one form (for example factual) to another (explicit)

A. Esposito (✉)
Dipartimento di Psicologia and International Institute for Advanced Sceintific Studies (IIASS), Università della Campania "Luigi Vanvitelli", Caserta, Italy
e-mail: iiass.annaesp@tin.it

A. M. Esposito
Istituto Nazionale di Geofisica e Vulcanologia, Sezione di Napoli Osservatorio Vesuviano, Naples, Italy
e-mail: antonietta.esposito@ingv.it

L. C. Jain
University of Technology Sydney, Sydney, Australia
e-mail: jainlc2002@yahoo.co.uk

University of Canberra, Canberra, ACT, Australia

Liverpool Hope University, Liverpool, UK

© Springer Nature Switzerland AG 2019
A. Esposito et al. (eds.), *Innovations in Big Data Mining and Embedded Knowledge*, Intelligent Systems Reference Library 159,
https://doi.org/10.1007/978-3-030-15939-9_1

and back again to the previous one [34], as well as, may vary along a continuum to the extent that categorization become hard to apply [9]. The intensified interest in knowledge management and data mining stems from the desire in defining computational models able to extract such embedded knowledge from data and provide support to humans in resolving organizational, social, and physical problems. The expectations are to develop intelligent ICT interfaces able to support and simulate abilities of medical doctors, teachers, assistants, housekeepers, and more, in order to enter the realm of the domestic world. This realm has been denied to machines right now, with the exception of very simple mechanical devices. This is because, differently from industrial contexts, where abilities are routinely applied, the domestic world is continuously changing and unpredictable, presenting problems of Non Polynomial (NP) solutions both in terms of computational time, and memory space. Data mining appears as the expedient to enter domestic spheres through massive data analyses, and its success is facilitated by the increased power of technology, both in terms of computational speed and memory capacity. It has become routine to collect large amount of data, and in particular heavy data such as videos, images, and speech waves, and detect from them any kind of medical, behavioral, and interactional information. However, successful knowledge management strongly depends on the mined data and the way such data are mined, in order to extract explicit, tacit, and embedded knowledge. Knowledge management Information Communication Technologies (ICT) interfaces work well with formalized/coded data, which can easily be stored, retrieved and modified but are less effective in handling unstructured thoughts, ideas, experiences, and behaviors, embedded in cultures and contexts. ICT knowledge management of unstructured data had to face technical (Does data mining will resolve NP-Complete and NP-hard computational problems?) and pragmatic questions (Does the knowledge extracted from the data will be adequate to allow the implementation of socially believable and emotionally behaving automatic systems?).

The chapters in the book aim to further progress in this direction providing examples of data mining philosophies from an embedded point of view. In particular, several data mining solutions are proposed to different problems exploiting specific embedded knowledge. Such problems are hard to be modeled, requiring sophisticated ICT tools for their computational handlings, and artificial intelligence solutions for extracting patterns and models from the mined data.

1.2 Discussion of Content

The mining of big amount of data requires the development of sophisticated computational instruments able to recognize, process, store, and synthesize relevant information. The relevance depends on the context of use, culture, social and ethical conducts, and effectiveness of the identified information.

In the current volume, Chap. 2 provides an example of the degree of sophistication to be handled by knowledge management tools. It proposes the development of an original application build upon a big data platform, i.e., a recommender system for touristic e-services ([6], this volume). More specifically, the authors organize their paper in a very appropriate way, introducing first relevant notions related to big data characteristics [31] in terms of volume (amount of data), velocity (speed at the which the information is generated), veracity (accuracy and quality of content), variety (distribution and heterogeneity of the data), and value (weight of the information content) of the data. Then they provide a background for recommender systems identified as systems equipped with a set of preprocessing algorithms devoted to collect information about user profiles and preferences in order to guide and facilitate their choices. The authors then discuss the evolving trends in tourism, and propose an e-service platform to support tourist operators in Italy.

Chapter 3 proposed by Kamath and colleagues ([25], this volume) affords a tacit knowledge management paradigm, where knowledge is extracted from sequence of data. This is done with relatively little a priori information to attenuate the dependence from experts' insights which are difficult to code and formalize. Along this line, the authors enfold concepts of machine learning and evolutionary computation in order to implement a transparent and scalable analysis able to extract features through evolutionary algorithms (the Evolutionary Features Generator, EFG) and scale them through a meta-learning algorithm (the Parallel Spatial Boosting Machine Learner, PSBML). The key aspects of both the proposed procedures are then combined into a unique framework called Evolutionary Machine Learning (EML). The authors proved, through validation experiments exploiting the worm and fly genomes datasets (http://www.wormbase.org) that their proposed EML framework compete in terms of accuracy and computation time with several state of art algorithms combining "*the best of both worlds*" (i.e. accuracy and computation time) "*in terms of finding discriminating features and carrying out parallel learning*" ([25], this volume).

Chapter 4 proposed by Kapros ([26], this volume) aims to assess how differences among societies and cultures may affect students' performances in readings, math, and several other science domains. The proposed methodology exploits the six cultural dimensions provided by Hofstede [24] and defined as (1) *Power Distance* (the extent to which a given culture consider acceptable social inequalities among people); (2) *Uncertainty/Avoidance* (the extent to which a culture accept unknown and ambiguous situations); (3) *Individualism versus Collectivism* (the extent to which a culture emphasizes autonomy versus individuals' integration into primary groups); (4) *Masculinity versus Femininity* (the extent to which a culture do not make primary differences between man and women); (5) *Long versus Short Term Orientation* (the extent to which a culture is tied with traditions and social obligations); (6) *Indulgence versus Restraint* (the extent to which a culture control personal needs and gratifications).

In order to detect information on how these cultural facets affect at a global scale students' performances the authors combine data from the International Large-Scale Assessments (ILSA; [19, 42]) and Cultural Dimensions [23] datasets and shows that high students' performances are correlated with *Long versus Short Term Orientation*, *Individualism versus Collectivism*, and *Uncertainty/Avoidance*. These cultural variables explain significant proportions of the variance in students' achievements, as well as, the effect of a given variable on the specific type of achievement. This research involves large and representative samples of students, teachers and schools, and requires the design of appropriate instruments to validate not only students' performance but also *"categories of background and explanatory variables"* ([21], p. 15). The data generated by these investigations are full of embedded and tacit knowledge and represent a huge challenge for data mining algorithms.

Chapter 5, proposed by Vinciarelli and colleagues ([39], this volume) provides a realistic view on perspective students' choices about high education domains, analyzing preferences, across different subjects, and differences in choices in terms of gender, and socio economic status. To this aims the authors administered to 4,885 perspective students a questionnaire aiming to assess whether there are science subjects more attractive than others as result, for example, of the prospect of a high salary once graduated, as well as, whether female and male participants differ significantly in the chosen disciplines, and in which disciplines differences are larger. Finally, the authors investigate whether participants' socio economic status produces inequalities among the selection of education's subjects. The authors show that some science domains are selected significantly more frequently than others ([39], Fig. 1.2, this volume) and interestingly, choices are not driven by financial and and/or occupational goals. Concerning gender differences, among the 109 proposed high education subjects, the authors identified, 19 male dominated and 14 female dominated subjects ([39], Fig. 1.4, this volume) along a *female to male* ratio of 1.55, while Education and Childhood Practice were shown to be definitively a female domain with a ratio of 14.1 and 31.0 respectively. Interestingly, the authors, as further explanation of the female—male divide, suggest considering emphatic and systemizing skills [4, 3] as being the first for the most part female qualities and the second for the most part male qualities. Finally, among socio economic factors, parental education seems to be the one that may generate significant inequalities among perspective students' choices. This chapter show how important can be to mine social data in order to improve educational offers and meet users' expectations.

Chapter 6, authored by Koutsombogera and Vogel provides a reflection on the difficulties of mining data derived from human interactional exchanges underlying challenges in modeling, understanding and automatically code heterogeneous multimodal signals resulting from such interactions. In this context, the verbal and nonverbal communication modes jointly cooperate in assigning semantic and pragmatic contents to the conveyed message. In order to understood, model, analyze, and automatize such behaviors converging competences from several scientific disciplines (such as social and cognitive psychology, linguistic, philosophy, and computer science) are needed. The exchange of (more or less conscious) information occurring during interactions builds up new shared knowledges. This often results in the need of storing it in order to allow efficient retrospectively use of data (for successive offline processing), but also in the need to appropriately support, assist, and dynamically optimize the interactional experience while it takes place. Currently the scientific efforts of the European community are devoted to implement context-aware multimodal systems able to infer the organizational, cultural, and physical context where interactions take place and support it in real-time and collaborative way. Such systems should act as co-workers, actively cooperating and contributing to the group's knowledge building, helping people to express their ideas, discuss them with others, differentiate their points of view from others, clarify disagreements and misunderstandings, negotiate common meanings and formulate a valid and useful knowledge and context of use [38]. To this end, the computational resources available should provide features showing abilities to automatically detecting cognitive connections, linking properly perceived visual/acoustic scenes to high-level descriptors of interactive situations driving interactional exchanges. In this context ([28], this volume) it is show that speech pauses play a fundamental role in providing a thorough understanding of human interaction mechanisms.

Chapter 7 authored by Navarretta and Oemig investigates on gesture and their role in multimodal communication. The goal is to review cross-modal features of verbal and nonverbal communication modes in order to identify models and procedures for implementing socially believable interactive dialog systems. Interactive dialog systems are computer interfaces represented by lifelike human characters capable of performing believable actions while reacting and interacting with human users. Such systems aim to allow a friendly human-machine interaction in the domains of health, housekeeping, assistance, security, front office, entertainment. Studies exploiting verbal and nonverbal communication features with the goal to improve the naturalness and effectiveness of interactive dialogue systems and enrich the way humans interact with machines are in their seminal stage, even though prototypes proving their usefulness and efficacy have already been envisioned [15–17]. The cross-modal analyses of audio and video recordings presented by Navarretta and Osteim ([33], this volume) suggests that the phasing of speech, gesture and facial expressions is critical for identifying the correct meaning of conveyed messages and solve message ambiguities. Taken singularly, the phasing of gestural motions can be identified through dynamic information related to gestural displacements with respect to a referential point. The phasing of individual emotional states can be captured through specific speech prosodic and facial expression analysis. Head motions can be captured through eyes tracking. What remain is the challenging work to combine these different communicative features and give meaning to the whole set of actions involved in the transmission of a message. This can be done only defining *"meta-entities involving mental processes more complex than those devoted to the simple peripheral pre-processing of the received signals"* [14]. These considerations, together with concerns about users' needs, cultural and individual differences, and ethical issues for processing personal data, are at the core of the research investigating the nature of knowledge that can be gained from multimodal data mining.

Chapter 8 authored by Moreau and colleagues [32] posits its emphasis on the demand for developing sophisticated and functional man-machine interfaces. This demand originates from emergent social needs, such as the world population aging and consequent difficulty to maintain an appropriate social welfare because of high associated human labor costs. The achievement of a human level of automaton intelligence raises the exigency of more precise approaches for advancing and validating cognitive architectures and pose imperative challenges in (a) identifying preprocessing algorithms able to capture invariant features from multimodal and social signals;

and (b) infer simple and fast computational algorithms for the detection, classification, and maintenance of hierarchically structured objects, such as feelings, personal traits, cultural conducts, all time dependent and reciprocally connected through complex relations. To solve these issues, Moreau and colleagues propose to popularize Artificial Intelligent (AI) tools, and in particular Natural Language Processing (NLP) technologies. According to the authors this is because NLP technology is first of all "*AI-complete*", involving intractable problems that can be solved through different approximation degrees. In addition, NLP is "*large ... [implicating] ... data that no one individual can ever hope to comprehend, ... have concrete applications, ... focuses on automatizing language-related tasks, ...[is] an experimental science, ... [and therefore] ... is intuitive and can be appealing to many people*" [32]. This popularization will be effortless considering that "*natural language processing is a function managed within the human cognitive architecture*", and therefore users' participation to its developments may help "*to assess the parameters of evolution in terms of how quickly stable (if ambiguous) conventions of use emerge*" [41].

Chapter 9 deals with knowledge discovery from biomedical databases. In particular, the work presented by Leonardi and colleagues ([29], this volume) aims to show how data mining and data knowledge extraction procedures are used in different context of use, proposing two case studies. The first case is devoted to detect food adulteration. In particular, through data obtained from the Fourier Transform Infrared (FTIR) spectroscopy [11], the authors tend to identify whether or not pure raspberry purees [27] have been contaminated by adding extra sugar or any other substance. In this case a dimension-reduced dataset of 1023 food instances (200 associated to pure raspberry samples—the negative class, and 823 adulterated samples) each described by 21 FTIR spectral features are given in input to a Support Vector and Random Forest learning algorithms [10, 36]. The authors show that machine learning algorithms are very useful in such case since clustering k-means and Expectation Minimization (EM) algorithms [13, 30] fail to provide an effective correct classification. The classification results obtained by such AI algorithms are very satisfying, clearly suggesting that knowledge discovery from data heavily depends on the exploited learning algorithm for that particular set of data. The second case study proposes a semantic process mining approach that exploits a knowledge-based abstraction mechanism in order to resort knowledge-based abstraction of event log traces in the medical field of stroke management. This last case is an example of how much complicate is the process of knowledge abstraction when timing is based on semantic processes. However, the semantic approach adopted produced "*significant advances with respect to the state of the art in process mining*" ([29], this volume).

Chapter 10, authored by Placidi and colleagues [35] describes web application devoted to detect and recognize spontaneous emotions through long EEG recording sessions. The general idea behind this proposal is to acquire long EEG signals during participants' daily activities and ask them to record instants when they experienced given emotional states, or otherwise, circumstances when relaxing situations had occurred. To implement such system, subjects participating to the data collection need to wear a portable EEG headset with huge storing capabilities, in order to record long brain activity periods. The annotations provided by participants must be processed in order to appropriately attach to portions of signals correct emotional labels which will be exploited for further processing and automatic emotion recognition. The challenge, in this context, is to manage in real time the huge amount of data coming from the EEG recordings, while dealing with unavoidable noise surrounding the emotional data. The authors propose cloud architecture as solution for storing and processing the large volume of EEG signals, a column-oriented database to store them, and a relational database with data sorting and sparse data managing capabilities to mine the data. The Hadoop framework [7] and the Map/Reduce paradigm [12] are used in order to parallelize the data management tasks, optimizing the computational time needed to process huge amount of data, and providing incremental capabilities in order to add new nodes, and increase computational resources. In addition to computational and memory space difficulties, this research need to deal with ethical and privacy issues which represent the "challenge" in the field of affective computing. However, ethical solutions are delicate and difficult to handle.

Anticipating user behaviors is the research core of Chap. 11 proposed by Böck and colleagues ([8], this volume). In this chapter, considerations are made on assigning meaning to statements advocating the development of ICT interfaces able to act intelligently in an environment, and being aware of the culture and community of its user. Such enculturation requires the development of systems intelligent enough to adjust their behavior and cognitive competencies with those of others they interacts with acquiring others' norms and culture. This become especially complex when empathy come to play, since the ability to understand ourselves by observing others, and understand others by imagining our own feelings in their shoes, is co-developed as we grow up, and the sense of ourselves and others is entwined. The authors argue that, in this context, the mining of big data alone would be not helpful in developing socially emphatic systems, since this knowledge is not in the data. To develop empathic systems it is necessary to enable them to make sense of themselves and their surrounding environments, equipping them with various sensors, actuators and other computation units, encouraging the collection of information from different sources such as social media, daily interactions, behavioral patterns, in order to acquire the ability to read situations and interpret people's behaviors in those situations, in terms of their possible intentions, emotional states, and proclivity to interact.

To fulfil this aim, the authors *"encourage the collection and use of enriched data by presenting methodologies for collecting data with rich disposition variety and predictable classifications based on a careful design and standardized psychological assessments"* ([8], this volume). The suggestion is that for future socially intelligent behaving systems to fluidly and easily interact with humans there is a need of a considerable period of enculturation. It will not be possible to program them off line and just place them into use. It will require fine tuning, sharing experiences and training over time.

Chapter 12 proposed by Amiriparian et colleagues ([2], this volume) is a first step toward the collection of *"enriched data"*. *"The aim of this chapter is to propose and assess the feasibility of a system which combines state-of-the-art big data knowledge extraction and annotation systems with machine learning algorithms for generation of real-world databases from multimedia data available on the Internet*" ([2], this volume).

The authors describe their CAS^{2}T toolkit system able to extract relevant data from online social media platforms [1]. Relevant data are identified through examples of provided data sources in order for CAST2T to identify similar data over social media. Successively, the collected data are annotated through the intelligent crowdsourcing platform iHEARu-PLAY (https://www.ihearu-play.eu). Such platform offers a large variety of annotation tasks for audio, video, images, and audio-visual data, exploiting machine learning algorithms to ensure the data quality management through several measures such as consistency, accuracy, and control questions to compute the intra-annotator and inter-annotator agreement [22]. Finally, the identified and annotated data are processed and classified through deep learning approaches. All the presented toolkits are open source and publicly available.

The final chapter of this book proposes a reflection on the use of data-driven approaches to extract patterns from dialogue corpora. The author consider inaccurate the assumption that "*automated machine learning analyses of large-scale dialogue corpora reveals rules to manage human-machine dialogue*" [20]. This is because, such data cannot provide knowledge on "how dialogues unfold" missing context-sensitive and context-appropriate information, These contexts incorporate concepts of appearance, verbal and nonverbal communication, acoustic, and other personal and interpersonal signals that humans process into an evaluative impression of a person to extract meaning from dialogues. These are complex concepts that cannot be ruled out from data. Factors such as attractiveness, credibility, dress, age, race, gender, socio-economic status influences the observer in making assumptions regarding what the sender is saying and which information is transferred fully objectively to the counterpart. This is also true for gestures, postures, gaze direction and vocal and facial emotional expressions. Individual voices and associated acoustic patterns affect how person judgements in terms of personality, emotional state, and competence. Embodied knowledge acquisition and embodied knowledge use cannot be gained through big data mining [5, 37, 40].

1.3 Conclusions

The readers of this book will get a taste of major efforts in big data mining and will be made aware of the importance of context for developing socially and emotionally believable ICT interfaces. Contexts drive situational awareness and affects interactional exchanges. Automatic behaving agents (and robots) need to evaluate the context (the organizational, physical, and social context) where interactions take places in order to understand how user behave and appropriately act in response of information exchanges. Future socially behaving systems should be designed towards context-sensitive and context-appropriate presence, and made aware of surrounding environments and social and behavioral rules that makes successful interactions with humans. Context is a multifaceted and multi-determined concept including several aspects of human's everyday life. Context has a physical dimension that includes objects and physical backgrounds surrounding interactions. Examples can be the weather of the day, room temperature, colors, places, furniture, and how it is arranged, etc. Socially believable ICT interfaces must be equipped of multiple sensors for audio, video, and other environmental inputs, in order to acquire contextually-accurate representations of their surroundings and appropriately integrate this information with feelings, thoughts, sensations, and emotions (the inner context) before to produce an interpretation of the exchanges. However context is also symbolic (meaning that previous events and interactions must be considered), as well as relational (relationships among people such as children and elders, father-son etc. must be accounted for), situational (referring to current situations, such as a lecture, a game, etc.), and cultural (social norms, social expectations, cultures of organizations). How big data mining can account of these facets remains an open research question.

References

1. Amiriparian, S., Pugachevskiy, S., Cummins, N., Hantke, S., Pohjalainen, J., Keren, G., Schuller, B.: CAST a database: rapid targeted large-scale big data acquisition via small-world modelling of social media platforms. In: Proceedings Biannual Conference on Affective Computing and Intelligent Interaction(ACII), San Antonio, TX, pp. 340–345 (2017)
2. Amiriparian, S., Schmitt, M., Hantke, S., Pandit, V., Schuller, B.: Humans inside: cooperative big multimedia data mining. This volume (2019)
3. Baron-Cohen, S., Wheelwright, S.: The empathy quotient: an investigation of adults with asperger syndrome or high functioning autism, and normal sex differences. J. Autism Dev. Disord. **34**(2), 163–175 (2004)
4. Baron-Cohen, S., Richler, J., Bisarya, D., Gurunathan, N., Wheelwright, S.: The systemizing quotient: an investigation of adults with asperger syndrome or high functioning autism, and normal sex differences. Philos. Trans. R. Soc. Lond. B Biol. Sci. **358**(1430), 361–374 (2003)
5. Barsalou, L.W., Niedenthal, P.M., Barbey, A.K., Ruppert, J.A.: Social embodiment. In: Ross, B.H. (ed.) The Psychology of Learning and Motivation, 43, 43–92. San Diego, Academic Press (2003)
6. Bellandi, V., Ceravolo, P., Damiani, E., Tacchini, E.: Designing a recommender system for touristic activities in a big data as a service platform. This volume (2019)
7. Berrada, G., van Keulen, M., Habib, M.: Hadoop for EEG storage and processing: a feasibility study. In: Brain Informatics and Health, 218–230 (2014)
8. Böck, R., Egorow, O., Höbel-Müller, J., Flores-Requardt, A., Siegert, I., Wendemuth, A.: Anticipating the user: acoustic disposition recognition in intelligent interactions. This volume (2019)
9. Botha, A., Kourie, D., Snyman, R.: Coping with Continuous Change in the Business Environment, Knowledge Management and Knowledge Management Technology. Chandice Publishing Ltd., London (2008)
10. Breiman, L.: Random forests. Mach. Learn. **45**(1), 5–32 (2001)
11. Davis, R., Mauer, L.J.: Fourier transform infrared (FTIR)-spectroscopy: a rapid tool for detection and analysis of foodborne pathogenic bacteria. Curr. Res. Technol. Educ. Top. Appl. Microbiol. Microb. Biotechnol. **2**, 1582–1594 (2010)
12. Dean, J., Ghemawat, S.: MapReduce: a flexible data processing tool. Commun. ACM **53**(1), 72 (2010)
13. Dempster, A.P., Laird, N.M., Rubin, D.B.: Maximum likelihood from incomplete data via the EM algorithm. J. R. Stat. Soc. Ser. B. **39**(1), 1–38 (1977). JSTOR 2984875. MR 0501537

14. Esposito, A.: COST 2102: cross-modal analysis of verbal and nonverbal communication (CAVeNC). In: Esposito, A. et al. (eds.) Verbal and nonverbal communication behaviours, LNCS, vol. 4775, 1–10, Springer, Basel, Switzerland (2007)
15. Esposito, A., Fortunati, L., Lugano, G.: Modeling emotion, behaviour and context in socially believable robots and ICT interfaces. Cogn. Comput. 6(4), 623–627 (2014)
16. Esposito, A., Esposito, A.M.: On the recognition of emotional vocal expressions: motivations for an holistic approach. Cogn. Process. 13(2), 541–550 (2012)
17. Fortunati, L., Esposito, A., Lugano, G.: Beyond Industrial robotics: social robots entering public and domestic spheres. Inf. Soc. 31(3), 229–23 (2015)
18. Gamble, P.R., Blackwell, J.: Knowledge Management: A State of the Art Guide. London, Kogan Page (2001)
19. Ganimian, A.J., Koretz, D.M.: Dataset of International Large-Scale Assessments. Cambridge, Harvard Graduate School of Education (2017). Last updated: 8 Feb 2017
20. Gnjatović, M.: Conversational agents and negative lessons from behaviourism. This volume (2019)
21. Gustafsson, J.-E.: Effects of international comparative studies on educational quality on the quality of educational research. Eur. Educ. Res. J. 7(1), 1–17 (2008). www.wwwords.eu/EERJ
22. Hantke, S., Appel, T., Schuller, B.: The inclusion of gamification solutions to enhance user enjoyment on crowdsourcing platforms. In: Proceedings of the 1st Asian Conference on Affective Computing and Intelligent Interaction (ACII Asia 2018), IEEE, Beijing, People's Republic of China (2018)
23. Hofstede, G.: Culture's Consequences: Comparing Values, Behaviors, Institutions and Organizations Across Nations. 2nd edn, Thousand Oaks, Sage (2001)
24. Hofstede, G.: Dimensionalizing cultures: the Hofstede model in context. Online Readings in Psychol. Cult. 2(1) (2011). https://doi.org/10.9707/2307-0919.1014
25. Kamath, U., Domeniconi, C., Shehu, A., De Jong, K.: EML: a scalable, transparent metalearning paradigm for big data applications. This volume (2019)
26. Kapros, E.: Towards addressing the limitations of educational policy based on international large-scale assessment data with Castoriadean magmas. This volume (2019)
27. Kemsley, E.K., Holland, J.K., Defernez, M., Wilson, R.H.: Detection of adulteration of raspberry purees using infrared spectroscopy and chemometrics. J. Agric. Food Chem. 44, 3864–3870 (1996)
28. Koutsombogera, M., Vogel, C.: Speech pause patterns in collaborative dialogs. This volume (2019)
29. Leonardi, G., Montani, S., Portinale, L., Quaglini, S., Striani, M.: Discovering knowledge embedded in bio-medical databases: experiences in food characterization and in medical process mining. This volume (2019)
30. Lloyd, S.P.: Least squares quantization in PCM. IEEE Trans. Inf. Theory 28(2), 129–137 (1982)
31. Madden, S.: From databases to big data. IEEE Internet Comput. 16, 4–6 (2012)
32. Moreau, E., Vogel, C., Barry, M.: A paradigm for democratizing artificial intelligence research. This volume (2019)
33. Navarretta, C., Oemig, L.: Big data and multimodal communication: a perspective view. This volume (2019)
34. Nonaka, I.: Theory of organizational knowledge creation. Organ. Sci. 5(1), 14–37 (1994)
35. Placidi, G., Cinque, L., Polsinelli, M.: A web application for characterizing spontaneous emotions using long EEG recording sessions. This volume (2019)
36. Platt, J.: Fast training of support vector machines using sequential minimal optimization. In: Schölkopf, B., Burges, C.J.C., Smola, A. (eds.) Advances in Kernel Methods, 185–208. Cambridge, MIT Press (1999)
37. Smith, E.R., Semin, G.R.: Socially situated cognition: cognition in its social context. Adv. Exp. Soc. Psychol. 36, 53–117 (2004)
38. Squartini, S., Esposito, A.: CO-worker: toward real-time and context-aware systems for human collaborative knowledge building. Cogn. Comput. 4(2), 157–171 (2012). https://doi.org/10.1007/s12559-012-9136-5

39. Vinciarelli, A., Riviera, W., Dalmasso, F., Raue, S., Abeyratna, C.: What do prospective students want? An observational study of preferences about subject of study in higher education. This volume (2019)
40. Vinciarelli, A., Esposito, A., André, E., Bonin, F., Chetouani, M., Cohn, J.F., Cristan, M., Fuhrmann, F., Gilmartin, E., Hammal, Z., Heylen, D., Kaiser, R., Koutsombogera, M., Potamianos, A., Renals, S., Riccardi, G., Salah, A.A.: Open challenges in modelling, analysis and synthesis of human behaviour in human-human and human-machine interactions. Cogn. Comput. **7**(4), 397–413 (2015)
41. Vogel, C., Esposito, A.: Advancing and validating models of cognitive architecture, unpublished manuscript (2017)
42. Wagemaker, H.: International large-scale assessments: from research to policy. In: Rutkowski, L. et al. (eds.) Handbook of International Large-Scale Assessment. Background, Technical Issues, and Methods of Data Analysis, 11–36. Boca Raton, CRC Press (2014). https://ilsa-gateway.org/ilsa-in-education

Chapter 2
Designing a Recommender System for Touristic Activities in a Big Data as a Service Platform

Valerio Bellandi, Paolo Ceravolo, Ernesto Damiani and Eugenio Tacchini

Abstract Designing e-services for tourist today implies to deal with a large amount of data and metadata that developers should be able to exploit for generating user perceived values. By integrating a Recommender System on a Big Data platform, we constructed the horizontal infrastructure for managing these services in an application-neutral layer. In this chapter, we revise the design choices followed to implement this service layer, highlighting the data processing and architectural patterns we selected. More specifically, we first introduce the relevant notions related to Big Data technologies, we discussed the evolving trends in Tourism, and we introduce fundaments for designing Recommender Systems. This part provides us with a set of requirements to be fulfilled in order to integrate these different components. We then propose an architecture and a set of algorithms to support these requirements. This design process guided the implementation of an innovative e-service platform for tourist operators in Italy.

2.1 Introduction

In this chapter, we focus on recommender systems for the Tourism Industry. We, in particular, discuss the design choices we followed to implement a horizontal infrastructure for managing a multi-agent Recommender System on top of Big Data technologies.

V. Bellandi (✉) · P. Ceravolo · E. Damiani
Dipartimento di Informatica, Università degli Studi di Milano, via bramante 65, 26013 Crema, Italy
e-mail: valerio.bellandi@unimi.it

P. Ceravolo
e-mail: paolo.ceravolo@unimi.it

E. Damiani
e-mail: ernesto.damiani@unimi.it

E. Tacchini
Universitá Cattolica di Piacenza-Cremona, Via Emilia Parmense 84, 29122 Piacenza, Italy
e-mail: eugenio.tacchini@unicatt.it

© Springer Nature Switzerland AG 2019
A. Esposito et al. (eds.), *Innovations in Big Data Mining and Embedded Knowledge*, Intelligent Systems Reference Library 159,
https://doi.org/10.1007/978-3-030-15939-9_2

The development of Web 2.0 has led to a rapid growth in the quantity and variety of data available on the Internet. These data include the published contents as well as a large quantity of metadata: for example, data about content usage, data on interactions between users and user preferences.

In e-commerce Internet, largely overcoming the limitation of physical spaces, expanded the range of available items and its possible buyers. To grasp the roots of this trend, it is possible to use the concept of "long tail", an expression coined by Chris Anderson in 2004. Through this definition, he tried to describe and explain new economic and commercial models related to the world of media and entertainment industry. Starting from the analysis of some literary phenomena connected to each other in that period, Anderson realizes how important was the existence of a potentially infinite catalog of consumer choices. The fact that people could do much more in-depth research of all available titles, without having to limit themselves to the circumscribed field of best-sellers or current trends, has led to a digital entertainment economy completely different from the current mass market.

Within this vast and constantly growing environment, one of the key problems that emerges is linked to information overload [1]. For this reason, Recommender Systems have been introduced in the last decades, and more and more research and studies are dedicated to its development and improvement. Marcia J. Bates argues that the Web and the online research databases (together with a series of technologies developed between the nineteenth and twentieth centuries, such as different classification methods, specific catalogs in alphabetical or thematic order, etc.) represent an important effort to make accessible a small part of information within ever-larger collections. This happens above all because of the human tendency to make the least effort in research and to be rather passive partly due to the fact that most of the information we need comes automatically from the social context of most people in the story of humanity [2].

Recommender Systems are tools based on the use of one or more algorithms, which through the collection and processing of information related to user preferences, build profiles and propose a series of customized items based on the profile that most clearly can be assigned to a user [3].

Data can be acquired by the system implicitly or explicitly. In the first case, assessments of appreciation/refusal carried out by the user are usually collected; in the second, instead, the behavior of the users is monitored, through the information recorded by a system, for example, the analysis of the navigation history, the downloads made, the time spent on the page and more.

When the information processed is mainly demographic, it is entered directly by the user through questions usually proposed at the time of registration or placed within their profile, for example, age, nationality, or gender.

It should also be noted that in recent years, more and more websites and applications have access directly to information relating to the state of the physical world through geolocation systems, RFID, real-time health signals.

Opposite to the non-customized systems, Recommender Systems have many advantages both from the user's point of view and from the point of view of the service provider. In the latter case, a recommender system allows to understand the needs of the users and improve their experience, or increase the sale of items, even very different from each other. The benefits enjoyed by the user are, for example, the possibility of finding all the items of possible interest or a specific one, of getting the suggestion of always new articles through the memory of the systems, being warned of the possible combination of several elements or of a sequence of objects. In addition, the user through his own opinion is able to help and influence other users [3].

It is possible to identify a set of criteria that determine the effectiveness of a recommendation system, regardless of the algorithm used. Firstly, it is important that items as individual as possible are proposed, not simply based on a general and generic judgment (for example an average evaluation of objects), but on a careful analysis of the user object of the suggestion. Secondly, since most of the Recommender Systems operate within a commercial context, the speed with which suggestions are made is very important. Closely related to this aspect there is a problem of scalability of the algorithm: despite the huge amount of data and assessments already existing, it is essential that the response time is not overly influenced by the size of the data.

Finally, in the face of the constant insertion of new reviews by users in the recommendation datasets, algorithms must rapidly manage continuous updating flows.

2.2 The Big Data Era

In the last five years, the concept of Big Data has become increasingly important, especially in business contexts where information is seen as one of the fundamental resources. A Big Data set can be defined, in a generic way, as a gigantic collection of information, and the concept can be extended by incorporating the technologies related to their management. These technologies are aimed at managing huge amounts of data, whose dimensions are in the order of zettabyte.

In particular, it is important to underline how a Big Data platform differs from a classic system based on relational databases. If in a relational database it is possible to define a global schema that expresses the structural organisation of a collection, i.e., following specific logical models of representation, Big Data volumes are of many orders higher than in classic Databases, and that is why typically data from a Big Data platform are distributed over a large number of computers, not depending on any pre-ordered schema [4].

Moreover, the explosion of social networks, the unprecedented acceleration of technological development, the digital revolution, represent some of the aspects that in recent years have brought to the increasing adoption of Big Data technologies. Data, produced by the most diverse sources, such as social networks, video and digital images, sensor networks and many other means, have recently been the protagonists of an exponential growth that does not intend to stop. Today's world is therefore submerged by Big Data, produced with considerable speed and in the most diverse formats, whose processing requires certain resources and technologies different from conventional systems of storage and management of data.

In 2010, the then CEO of Google, Eric Schmidt, said that *"within two days we produce the same amount of data generated since the dawn of civilization until 2003"*.

In such a context, it immediately became clear to companies that the analysis of this large amount of data could have been a useful treasure to forecast, analyze and solve the problems of thousands of people. Combined with sophisticated business analysis, Big Data has the potential to give businesses insights into market conditions, customer behavior, making decision-making more effective and faster than the competitors. The World Economic Forum, in a report published in 2011 by the Washington Post, has included them in a new category of economic resources, as a raw material [5].

2.2.1   Dimensional Model of the Big Data

Over the years, many analysts have developed dimensional models to define the concept of Big Data. In particular, the constituent characteristics of Big Data are included in the "Vs storylines". Three dimensions were initially identified as crucial:

- **Volume**: It is the volume that represents one of the main characteristics of the data. The adjective "Big" with which this new phenomenon is described, is significant as it refers to the ability to store and access large amounts of data.
- **Velocity**: The information that makes up a Big Data set is generated more and more rapidly. These informational piers must be stored and analyzed with equal promptness and velocity, almost in real time. Identifying a trend, an opportunity is the key to gain competitive and economic advantages.
- **Veracity**: With many forms of Big Data, quality and accuracy are less controllable (just think of Twitter posts with hashtags, abbreviations, typos and colloquial speech as well as the reliability and accuracy of content).

In [6] two more dimensions was proposed.

- **Variety**: Distributed and heterogeneous data can no longer be organized by a schema because they depend on multiple formats, semantics and data structures types.
- **Value**: Having so much information about a particular object or subject does not necessarily imply that the information is useful. It is necessary to eliminate the "noise" contained in it.

So you can safely argue that "value" is the most important V of Big Data.

2.2.2   Tourism in the Future

Within this context, it arises that a new generation of tourism applications will be available. In the recent years, a constant trend has been established: users have evolved from simple observers to direct purchasers of tourist services, thanks to the direct accessibility offered by the Web and to the recommendations that can be provided by online marketplaces or sharing platforms.

This constitutes a significant advantage for the consumer, to whom research is simplified and speeded up. Regarding the owners of tourist services, as hoteliers, owners of transport companies, etc., they must be able to emerge within a very wide and often very uniform commercial landscape. The new trend gets realization through four main actions available to the user:

- Inspiration: recommendations play a significant role in proposing destinations that could reflect the preferences of a user. For example, knowing the preference for maritime destinations, the system offers tourists a series of locations that are very popular at that time of year.
- Research: once the destination has been established, a search process about all the aspects necessary for the development of the holiday begins, starting from transport, to the accommodation and activities on the spot. In this case, the proposals are sorted according to a series of filters selected by the user, for example, the price or the review score, and, of course, according to availability in the selected period.
- Booking: once the most appropriate solution is identified, the system should support booking in short and easy steps. Often, many users, after carrying out informative research, prefer to contact directly the physical agencies for saving or because a face to face interaction increases confidence.
  Note that the booking step can be split into two sub-phases: (i) before the departure (usually reserved for the purchase of packages, transport, and accommodation) and (ii) after arrival to the destination (for additional services, entertainment or cultural activities, prolonging the stay, booking restaurants, car hire, etc. ...).
- Sharing: it concerns the sharing of opinions, thoughts, photos, videos or texts through Social Media. The user's friends can get a lot of information or advice on a particular place that might be of interest to him in the near. A relevant role is also played by ratings, reviews and votes that the tourist makes through special sites consulted directly by other travelers who are looking for tourist experiences.

In other words, this new way of managing and sharing information around tourist services permits to merge information coming from tour operators with information directly generated by the users. This largely increases the data available and open the doors to a new generation of Tourism Recommender Systems.

2.3 Recommender System

Recommender Systems are "personalized information agents that provide recommendations: suggestions for items likely to be of use to a user" [7].

The "item" is the object of the recommendation, it can be, for example, a product to buy, a piece of content to consume or a person to get in touch with.

Recommender systems are used for different kind of tasks; according to the summary provided by [8], based on the analysis in [3, 9], here are some of the most common uses:

- Find some good items: for example, recommending to a user, according to his profile, the 10 movies he is expected to like more.
- Find all good items: for example, a prior art search application, which allows users to find all the patents, publications and other contents related to a specific topic.
- Bundle recommendation: for example, an e-commerce Web site providing suggestions about accessories that fit with the product the customer is buying/viewing.
- Sequence recommendation: for example, a music playlist recommender system, where the order of the items suggested also matters.
- Group recommendation: for example, a recommender system suggesting destinations to a group of people, taking into consideration the profiles of all the group's members.

2.3.1 Basic Techniques

Recommender Systems require keeping track of the items the users liked in the past. This can be implemented through an explicit rating system or through implicit feedback techniques, e.g. if a user repeatedly listens to a song, the system assumes they like the song [3].

Recommender systems have been traditionally classified into three main categories, according to the technique used to provide recommendations: content-based recommender systems, collaborative filtering recommender systems, and hybrid recommender systems [10].

Content-based (CB) recommender systems base the suggestions on the analysis of the items: each item is analyzed and described using a set of features, e.g. a movie could be represented using as features the director's name, the genre, the year of production and the country of production and the system normally suggests to the users' items similar to the ones they liked in the past, e.g. a movie having the same genre or director.

Collaborative filtering (CF) is "the process of filtering or evaluating items through the opinions of other people" [11]: CF recommender systems don't require the analysis of the items' content, they use the feedback provided by users as the features of the item. The basic intuition behind CF recommender systems is that if two items have been rated similarly by the same users, the items are similar (item-based) and if two users rated similarly the same items the two users are similar.

Computing the similarities among users allow suggesting to a user the (unrated) items that other similar users (called neighbors) rated positively (known as the user-based approach).

Computing the similarities among items allow suggesting to a user the (unrated) items that are similar to items the user liked in the past (known as the item-based approach).

Moreover, in memory-based CF recommender systems, all the ratings in the system are directly used to predict ratings not yet available, while model-based CF techniques use the ratings to produce a predictive model (based, for example, on Bayesian Clustering, Latent Semantic Analysis or Singular Value Decomposition) and use the model to predict the ratings for unrated user-item pairs not available [12].

An additional technique is the demographic approach: in this case the system suggests items to users according to their demographic characteristics, e.g. age class or country; it is not a personalized suggestion, because of the preferences of a user can differ from the majority of the people sharing the same demographic profile, however, this approach can the sole solution in a *cold-start* situation, when you don't know anything about the user except some demographic data.

2.3.2 Evaluation Metrics

The metric most used to evaluate recommender systems is *accuracy*: to measure how accurate are the suggestions provided by a system, several indexes have been used in literature, most of them are quite popular and also used in other disciplines; we mention here the Root Mean Square Error (RMSE), which measures the average error made in a ratings prediction task and the Mean Average Precision (MAE) which measures the accuracy in a TOP-n recommendation task, also considering the order of the suggested items.

The accuracy measures, however, don't take into consideration other aspects of the recommendations such as diversity, novelty, and serendipity: a suggestion, while accurate, can be in fact too obvious or too similar to other, already suggested, items. The literature has largely discussed this topic, for examples [13] introduced metrics such as *r-unexpectedness* and *r-serendipity*, [8] introduced the notion of *serendipity cost*. It has been demonstrated that adopting these metrics in calculating ranking scores, relevant improvements in long-term users satisfaction can be achieved [14, 15].

2.3.3 Tourism Recommender Systems

While recommender systems are applied in many different domains, tourism is a domain where they can be particularly useful: considering the amount of information available, users can use tourism recommender systems to plan the visit to a city or to receive suggestions for activities, hotels, restaurants related to a particular destination. At the same time, Tourism Recommender Systems impose novel functions and requirements.

Borras et al. [16] provide a survey of intelligent tourism recommender systems and describe these systems according to (among the others) two aspects: interface and functionalities.

Considering the interface, about 50% of the systems included in the survey offered a standard Web interface only, 20% a Web interface plus a mobile interface and about 30% only a mobile interface. For the system, we are describing in this work, we are not focusing on a specific interface: our engine exposes API endpoints that can be used by an end-user application to get recommendations, the interface used by such application doesn't have any impact on the recommendations.

Considering the functionalities, [16] considers four different types of functionality provided: *travel destination and tourist packs*, *ranked list of suggested attractions*, *planning a route* and *social aspects*.

While some systems only provide recommendations for a *destination* (according to the user's preferences), most of the systems suggest *attractions* (events, hotels, restaurants, activities, ...) once the user has decided the trip's destination. Among these systems, the most sophisticated ones also take into account some contextual aspects, for example, [17, 18].

Some systems suggest attractions but also aim at creating an optimal route through the attractions. The aspects these systems take into consideration can be multiple: the amount of time the user can spend for the visit, the opening time of the attractions, the distances or the walking speed are some of them [19].

Some systems include a focus on social features which allow users to share their experiences and interact with each other. The recommendation to groups of users also falls into this category. A recent example is proposed in [20].

2.4   Architecture

The underlying technology of our environment should be flexible enough to support data acquisition, storage, and processing. Below we describe the Big Data logical architecture we adopted.

2.4.1   Lambda Architecture

The Lambda Architecture, shown in Fig. 2.1, consists of three layers: (i) speed, (ii) batch and (iii) serving layer.

The batch layer merges incoming data with historical data and reiterates the procedural workflows on all the combined data input, in order to construct results. The accuracy of batch views comes at the cost of high latency, therefore Lambda Architecture must also be able to compute incremental updates via speed layer, in order to guarantee a good level of responsiveness.

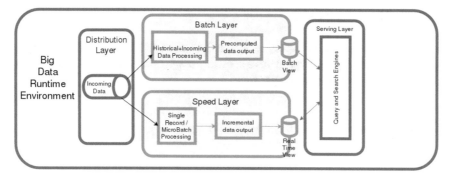

Fig. 2.1 Lambda architecture

The speed layer takes in input the new data in the form of either micro-batches (small chunks of data ingested on a regular basis) or single records, and the last update of batch data output, which implies that a mandatory condition for algorithms to run in the speed layer is being capable of processing data incrementally, providing a short time model updates influenced by fresh data delivered.

One of the most prominent hurdles to overcome, in order to deal with batch and speed views, is synchronizing them to prevent data redundancy or data loss.

In order to address this problem, [21] proposes the solution of tagging data as soon as it enters the system, to keep track of its delivery time and thus keeping track of what corresponding information can be removed from speed data views, activating batch procedures.

The serving layer arranges some ad-hoc low-latencies queries to indexed data which is stored to the output scalable storages for batch and real-time views in order to provide an aggregate view of both batch and speed layer data, resulting as a suitable interface layer for reporting and visualization tasks.

Selecting the correct architectural layout takes into account several specifications and it depends on the application use cases.

Application use cases involving Recommender Systems make large use of machine learning algorithms, like collaborative filtering via Matrix Factorization (MF).

In order to compute the whole MF matrix over all the historical data, algorithms like ALS must perform a reiteration over all the historical data whenever new data chunks are delivered, while other algorithms exist to incrementally update the MF models, but these provide only an approximation of the former MF matrix calculated with ALS [22].

2.4.1.1   Distribution Layer

Every time raw data is delivered to a Big Data application distribution layer, the process of ingesting data may involve several storage options depending on the size of data batched to be processed per time.

2.4.1.2   Storage and Data Representation

One key factor to consider when the batches of raw data to be processed are demanding a lot of storage resources, and the retainment of long-term historical data is a requirement, is optimizing the data format for splittable compression.

Importing GZip files to HDFS, which is designed to scale out when the storage consumption increases, does not prevent per se Map Reduce tasks to start processing but does not allow the file to be split at the map and reduce stages, thus resulting in lower performances.

A better solution is to wrap the source file in a container file format such Avro or SequenceFiles to support splittable compression. Other aspects to discuss are how data will be updated when imported and how it is accessed.

HDFS implements the "write once read many" paradigm, allowing no further update after the file has been created. Data is usually appended into "delta files" containing the incremental updates of last data batches imported. Having many small delta files inside the directories could involve a compaction process running in order to reduce the overhead at files reading, merging the HDFS files. Compaction jobs on HDFS will also detect records with the same key across the delta files and keep only the updated version of the record to the merged file.

HDFS along with compaction is a valid alternative when processing involves large scans, on the other hand, HBase [23] manages to fire off the compaction job in background after a put command occurs but it's optimized for random access rather than large scans.

Regarding ways to access stored data, if data are mostly accessed with sequential scans, like when data transformation steps require several iterations over the whole dataset, then HDFS is the best option to store data, conversely, if data is randomly accessed (e.g. frequent updates by key) then HBase is the most convenient choice, finally, Solr [24] provides some distributed search capabilities, it uses shards to logically partition data indexes in the cluster.

Amongst data formats, CSV files, and more generally, structured tabular formats with fixed length columns, can be stored as HDFS files, then imported to a Hive table where SQL queries are required to process it. RDBMS-wise, conversely, sometimes data come natively unstructured or semi-structured, using key-value, or columnar layouts, thereby demanding a flexible representation. HBase is a suitable solution in terms of metadata flexibility, since it allows to define a column key which rules how to read the record values, [25] addresses how to efficiently and economically achieve semi-structured spatio-temporal data storage and query in HBase.

A very common use case in sensor analytics is time-series data, solutions include using HBase or Cassandra, aiming to retain big volumes of time-related data, regardless of data-fields, with a focus to fast update response time [26].

A usual pattern to use HBase as a time-series DB is setting a composite row-key which includes the record key and the start time (defined as a reverse epoch number in order to sort the records from the newest to the oldest) and saving the stop-time in the timeStamp, this way the row is able to track both start and stop time.

In order to update the last entry for a given record with a new row, this approach requires a simple-record scan, and two updates to set the stop time for the most current entry (the first one), and storing the new record on top with a null value for the stopping time [27].

2.4.1.3 Ingestion

Ingestion tools are chosen according to many factors, one of which is data ingestion timeliness, namely to the time lag from when data is available for ingestion to when its accessible to tools in the Hadoop ecosystem: [28] a common classification for timeliness distinguishing:

- *Macro Batch*. Processes ingesting data for 15 min to hours.
- *Micro Batch*. Micro batch ingestion covers data delivery time ranges of 2–15 min, and they are often configured to continuously supply a small amount of single events of data to a stream topic per-time, to be asynchronously consumed and processed by short living processes.
- *Near Real-time Decision Support*. This is considered to be immediately actionable by the recipient of the information, with data delivered in less than 2 min but greater than 2 s.
- *Near Real-time Event-processing*. This is usually referring to single events per time which are processed in a time range going from 100 ms to 2 s.
- *Real Time*. Real-time ingestion imports and processes single events per time and usually occurs in a period no longer than 100 ms.

Along with ingestion timeliness, there are several factors to take into account to choose a suitable ingestion tool, we are referring to some of them in Table 2.1.

2.4.1.4 Data Preparation

Common preparation tasks include:

- Clean selected data for better quality.
- Treat missing values.
- Identify or remove outliers.
- Resolve redundancy caused by data integration.
- Correct inconsistent data.

Table 2.1 Ingestion tools

Tool	Fault tolerance	Parallel ingestion	Transformation	Sources	Security
File transfer	None	Single-threaded	Only after landing in HDFS	External file system	SSL
Sqoop [29]	Map tasks commit transactions periodically, resulting in a partial import/export up to the last commit	Using MapReduce mappers to write a portion of the table to HDFS, loading multiple tables in parallel	Only after landing in HDFS	RDBMS	SSL
Kafka [30]	Kafka replicates data to partitions, the level of fault tolerance is highly customizable, tuning in favour of availability or replicas consistency	Partitioned topics supply multiple consumers tasks in parallel Only as data source for a streaming engine like spark streaming or storm	Only as data source for a streaming engine like Spark streaming or storm	No native source nor sink implementation provided, but extreme pluggability thanks to pub/sub interfaces	SSL and Kerberos
Flume [31]	Supports splitting data on ingestion to feed a backup cluster	Flume supports multithreading and allows to tune up the fan-in with multi-agent ingestion	Supports low latency processing with interceptors	Spool directory, HTTP, JMS, AVRO and many others	SSL

- Flatten nested data formats to table formats.
- Ordering.

There is no silver bullet to accomplish these tasks without any ad-hoc transformation tailored to the data format and the nature of the discrepancies found.

Discrepancies in a data set occur for many reasons, including failures during the ingestion process, data integration of different sources, and could result in anomalies not so easy to find.

The adoption of machine learning or sophisticated data quality analysis tools [32] allows to detect automatically discrepancies, like typos in a date or schematic heterogeneity across the columns, these are fixed with custom ETL batch tasks, running over the whole dataset analyzed.

Apache Pig is the standard Hadoop script interpreter to execute ETL instructions using map reduce tasks.

Anomalies detection and transformations are usually reiterated until no other discrepancy is found in the dataset.

Assuming that a reiteration of these steps could take minutes to hours in a large data set, and may involve the human intervention, a better approach called Potter?s Wheel Framework proposed in [32] slims down the human-in-the-loop interaction steps applying the discrepancy detection incrementally on the records visible to display as the user scrolls down the dataset, thereby he can quickly respond to the discrepancies found and apply a fixing transformation simulation to the same subset.

The same process goes on until no more anomalies are discovered, and results in a custom procedure ready to be executed on the whole dataset performing all the simulated transformations.

A list of transform operations proposed is illustrated in [32] where it is shown how almost all the composite transformations applied to fix discrepancies can be optimally performed as a constituent sequence of these operations.

2.4.2 A Multi-agent Tourism Recommender System

The recommender system we are proposing is a multi-agent recommender that can provide suggestions for several kinds of tourism-related items: activities, hotels, restaurants, and others.

In the following chapters, we will focus on a specific category of items: *activities*. An activity is an experience a tourist can have in a specific location, for example, a guided tour of the Louvre museum in Paris or a weekend in a beauty farm.

We have a set of m registered users U and a set of n activities A, the goal of our recommendation engine is to suggest relevant activities to users when they ask for recommendations. Furthermore, the system should be able to suggest activities to users who are not registered and to groups.

To achieve its goals, our system uses a combination of three different techniques: *collaborative filtering*, *content-based*, and *demographic*.

2.4.2.1 Collaborative Filtering

The collaborative filtering approach is used to compute the similarities among activities based on the users who liked them. We have in input a matrix $F(m, n)$ containing, for each of the $m \times n$ combinations, the rating the user gave to the activity, if the user consumed the activity and submitted a feedback. We have a missing value for a specific combination of (users, activity) when the user didn't consume the activity or didn't submit a feedback. The ratings are in a one-to-five range.

To compute the similarities among activities, we consider each activity as a vector of user ratings and we compute the Pearson correlation between all the couples of activities, obtaining a similarity matrix $S_{cf}(n, n)$. Considering a couple of vectors X and Y, representing the ratings two activities have received by N users who rated both the activities, the Pearson correlation r is computed as follows:

$$r = \frac{N \sum XY - (\sum X \sum Y)}{\sqrt{[N \sum x^2 - (\sum x)^2][N \sum y^2 - (\sum y)^2]}} \tag{2.1}$$

The similarity matrix S_{cf} allows us to provide relevant suggestions for users if we know their history: we consider the activity the user liked the most in the past and we provide as a suggestion the fifty most similar activities the user has not consumed yet.

The collaborative filtering approach assumes to have a rich history of consumptions and feedbacks. This is not always the case, especially in a cold-start situation; when the system has just been implemented, the matrix F could be very sparse, therefore, at least for some activity couples, the Person correlation is computed on a small sample size, not capturing in its entirely the underlying similarity between the activities. This, in turn, could make the suggestions provided to the users not relevant. In these situations, a content-based approach can help.

2.4.2.2   Content-Based

We use a content-based approach to compute the similarities among activities (as we did for collaborating filtering) but in this case, the similarity is based on a set of features describing the activities and not on the users' feedbacks. Since these features are something we can always provide, we are able to build a similarity matrix S_{cb} even without any activity consumption or feedback data.

Our content model describes the activities using categorical data; each feature can have a finite set of values: some of the features are binary, e.g. "Activity suitable for kids" can be true or false, other features can have a number of possible values greater than two, for example "geographic area". All the values assigned to the features are integer values: for binary features we use 0 to denote false and 1 to denote true, for non-binary features we use a finite set of integer values that represent each possible category; for example if we decide to use the "Geographic area" feature to represent the country where the activity takes place, we can assign a different integer value i for $i = 1, 2, \ldots, n$ to denote the n different countries we want to represent.

A subset of the features, in particular, is used to categorize the actual experience the tourist using binary assignments, some features could be, for example: "Sport", "Museums and Arts", "Food and Wine" or "Shows and Concerts". This allows to describe an activity based on multiple categories: for example, a Gallery tour that combines Wine tasting will have both the "Museums and Arts" and "Food and Wine" features set to 1.

We have, this way, in input a matrix $AF(m, k)$ where, for each activity m, we describe the activity assigning a value to each of the k features. To compute the similarities among activities, we consider each activity as a vector of features values and we compute a similarity matrix $S_{cb}(m, m)$ based on the Jaccard index, which measures, for each couple of activities, the proportion of feature whose values differ.[1]

Since we are using categorical data, the Jaccard index captures better than the Pearson correlation the distance (similarities) between items: if two activities are described, for a specific feature, by values 1 and 10 respectively, they differ exactly as two activities that for the same feature have values 1 and 2.

Once we have the similarity matrix S_{cb} we can use the same approach we used for the collaborative filtering technique: to suggest relevant recommendations to a user: we select the activity they like more in the past and we suggest the fifty most similar activities according to S_{cb}, excluding the activities the user already consumed.

Note that, even if the computation of the similarity matrix does not suffer from the cold-start problem, we are still unable to provide meaningful suggestions to users we do not have relevant information about their history.

2.4.2.3 Demographic Approach

To solve the cold-start problem, we mentioned in the previous section, we use demographic data.

The idea is to compute the similarities among users and to suggest to a user U_i activities based on what the user's neighbors (the users most similar to U_i) liked. The approach is similar to a classic collaborative filtering user-based approach, but instead of using consumption/feedback data we used demographic data, assuming we can always have access to some demographic data about the user. Some of the features we use are, for example: "Age class", "Gender", "Has children", "Speaks English".

The representation level is an input matrix $UD(n, z)$, where, for each user n, we describe the user assigning a value to each of the z demographic features. To compute the similarities between a user and all the others, we consider each user as a vector of demographic feature values and we compute the Jaccard index between them; we can then suggest to a user U_i the activities most liked by their neighbors: the users most similar to U_i.

Please, note that features do not need to be strictly demographic: we can use any kind of categorical feature including user preferences, if available; for example "Interested in watching sports". This allows mitigating the cold start problem using information that can be extracted from social media: for example, using Facebook login[2] it would be possible to get the pages liked by a user. The model used to

---

[1] This represents the distance between two activities, not the similarity, but we can still easily get, for each activity, the most similar ones by sorting according to the distance, ascending.

[2] https://developers.facebook.com/docs/facebook-login.

represent a user can be extended adding or removing features without any impact on the implementation.

This approach has been introduced, in particular, to suggest relevant activities to *prospects*, i.e. users who browse the Website but are not customers yet, and want instant recommendations. Since the process of computing similarities among users can take time if the number of users is very high, we implemented a model based on *fingerprints* (explained in Sect. 2.4.3.1), representing, the feature values of each user, allowing the fast computation of the neighbours and therefore the generation of a list of favourites activities for a typical customer having the same characteristics of the prospect who is querying the Recommender System.

2.4.2.4   Groups

All the techniques we have just explained aim at serving an individual customer or prospect. We also have a specific module which provides suggestions for groups, i.e. two or more people who want to consume an activity together, for example, a group of friends or a family.

This modular implementation is very similar to the one we used for prospects: we model different groups according to an extensible set of features, e.g. average age of the group, group with children, etc. We then suggest to a prospect group the activities most liked by the groups having similar characteristics.

2.4.3   RS Architecture

Our system is composed by two main components: the recommendation engine, which pre-computes the recommendations and the API, which, among the other things, provides the actual recommendations for a user, a prospect, or a group once queried.

The recommendation engine is implemented in Python, for several tasks we took advantage of the features provided by the SciPy library.[3]

The data is stored in a MongoDB[4] database; in particular we store in MongoDB both the input data and the pre-computed recommendation data.

2.4.3.1   Input Data

The input data consist of the following MongoDB collections:

---

[3]https://www.scipy.org/.

[4]https://www.mongodb.com.

```
activities (id, val_feat_1, val_feat_2,...val_feat_n)

customers (id, val_feat_1, val_feat_2,...val_feat_n,

fingerprint)

customers_activities (id_customer,id_activity,rating,date),
```

these collections represent the information related to the activities, the customers and the consumption/feedback history of the customers; in particular, the `customers_activities` collection represents, for each customer, the activities they consumed, the date of consumption and the rating.

It is worth to describe how the fingerprint field works: it is a list whose elements contain (one element for each feature) the code of the feature and the feature's value, for a customer. For example, let's assume that in our model we only have two features for customers: "Age class" (possible values: $1, 2, 3, 4$) and "Country" (possible values: $1–206$, each value represents a specific country); let's also assume that we denote "Age class" with the feature code "001" and "Country" with the feature code "002"; for an Italian customer, 25 years old, the fingerprint will be ["001002", "002034"], assuming that 2 is the value corresponding to the 18–25 age class and that Italy is the country 34.

The fingerprint field allows the computation of the neighbors of a prospect in one MongoDB query; for example, to get the neighbors of an Italian prospect, 25 years old, we can execute the following query:

```
neighbors = db.customers.aggregate([{ "$project":

{"id": 1, "count": {    '$size':

{'$setIntersection': ["$features",['001002','002034']]},

  "_id": 0 } },{ "$sort" : { "count" : -1} },

{ "$limit" : 50 }]),
```

which returns the 50 most similar neighbors.

This approach also allows to change the model without changing the Python implementation code of the demographic technique: if you add or remove a customer's feature from the model you just need to update the schema of the customers' collection and the related records.

The information represented in the field *fingerprint* is redundant: the same information, in fact, is represented through the `value_feature_1,value_feature`

_2, ...value_feature_n fields; however, we are keeping those fields in the schema because they work as fast middleware between the recommendation engine and other systems which might expect a schema where each field is represented by a feature.

Some additional collections are used to represent information related to groups:

```
groups (id, val_feat_1, val_feat_2,...val_feat_n,

fingerprint)

group_activities (id_group,id_activity,rating,date)
```

The first collection, group, represents the information related to each group, the second one is similar to customers_activities and represents the consumption/feedback history of a group. Note that a record of the collection groups represents all the groups having the same characteristics while in customers the corresponding collection represents a single customer.

2.4.3.2   Output Data

Once the recommendations are computed, they are stored in some MongoDB collections. In particular, for each recommendation technique we explained in the previous chapter, we have a collection that stores the IDs of the activities suggested for a particular target, i.e., a customer, a prospect, or a group.

For each row, we store the id of the customer and an array containing the list of 50 activity IDs to suggest to the user, sorted by relevance.

Storing the pre-computed recommendations as an array in MongoDB, one per user, is a very fast and computationally inexpensive way to provide recommendations, even in a Big Data context, where both the total number of users n and the total number of items m (activities, hotels, or any other kind of item that needs to be recommended) are very high and also the recommendation requests concurrency is high. A typical Tourism Recommender System may expect to operate with millions of users and items, largely below the volumes addressed by Big Data applications in other domains.

2.4.4   *Multi-agent Approach*

While we have discussed in detail about the recommendation of *activities*, adding the support for a new typology of item is straightforward. Most of the work consists in modeling a new set of features for this new type of item, in order to use the content-based technique. We also need additional data structures (MongoDB collections) to store all the data related to the new category of items.

For example, to add the support for hotels, we could describe each hotel using some categorical features such as the hotel luxury rating, the suitability for families or the price category. We also need to add all the MongoDB collections needed to represent the information related to hotels and to the ratings the hotels can receive by customers.

While *hotels* and *activities* will have a different feature sets, `customers_hotels` and `customers_activities` will have the same structure.

The collaborative filtering approach normally does not need changes: a user can, in fact, rate an activity, a hotel, a restaurant or any other kind of item and we always store just the rating and the date (if available). The demographic approach also does not need change: for both customers and groups, we still need to compute their neighbors (according to their features) and then the suggestions (according to the hotels most liked by the neighborhood).

For each type of item we can use a different recommendation technique, according to the specific needs and data availability: for example if we have a rich consumption/feedback dataset for hotels but we don't have much data for activities, we could use a collaborative filtering approach for hotels and a content-based one for activities.

2.4.5 Future Work

The recommendation engine proposed is a work-in-progress project and we have planned several improvements.

First, we want to work on cross-domain recommendations: our engine can recommend different types of items, belonging to different domains (e.g. activities and hotels); we might know very well the preferences of a user for a particular domain but not for *all* the domains. The adoption of a cross-domain strategy allows to exploit the knowledge available in a domain to recommend items belonging to another domain, for example, we could recommend, to a user that normally books luxury hotels, activities in the same price range, even if we do not have any consumption history, for that user, in the *activities* domain. This can be implemented in several ways, for example by defining a subset of features that work for all the domains or by exploiting, in order to suggest items to a user for a domain, the knowledge we have on that domain for the user's neighbors.

On the implementation side, we want to adopt Apache Spark[5] to measure the similarities among users and the similarities among activities (or items in general) we use in the collaborative filtering and the content-based approach. In particular, we want to use the MongoDB connector for Spark, which provides integration between MongoDB and Apache Spark. This would allow using the Spark library directly on

---

[5]https://spark.apache.org/.

the MongoDB datasets, resulting in a more efficient approach, especially considering the Big Data scenario.

Acknowledgements This work was partly supported by the "eTravel project" funded by the "Provincia di Trento", and by the program "Piano sostegno alla ricerca 2015–17" funded by Università degli Studi di Milano.

References

1. Ojokoh, B.A., Isinkaye, F.O., Folajimi, Y.O.: Recommendation systems: principles, methods and evaluation. Egypt. Informatics J.-Cairo Univ. **2**, 75–82 (2015)
2. Bates, M.J.: Toward an integrated model of information seeking and searching. New Rev. Inf. Behav. Res. **3**, 1–15 (2002)
3. Ricci, F., Rokach, L., Shapira, B.: Introduction to recommender systems handbook. In: Recommender Systems Handbook, pp. 1–35. Springer (2011)
4. Ardagna, C.A., Bellandi, V., Ceravolo, P., Damiani, E., Bezzi, M., Hebert, C.: A Model-Driven Methodology for Big Data Analytics-as-a-Service, pp. 105–112 (2017)
5. Ardagna, C.A., Bellandi, V., Bezzi, M., Ceravolo, P., Damiani, E., Hebert, C.: Model-based big data analytics-as-a-service: take big data to the next level. IEEE Trans. Serv. Comput. (2018)
6. Demchenko, Y., Grosso, P., de Laat, C., Membrey, P.: Addressing big data issues in scientific data infrastructure. In: 2013 International Conference on Collaboration Technologies and Systems (CTS), pp. 48–55 (2013)
7. Burke, R.: Hybrid web recommender systems. In: The Adaptive Web, pp. 377–408. Springer (2007)
8. Tacchini, E.: Serendipitous Mentorship in Music Recommender Systems. Ph.D. thesis. PhD thesis, Computer Science Ph.D. School—Università degli Studi di Milano (2012)
9. Herlocker, J.L., Konstan, J.A., Terveen, L.G., Riedl, J.T.: Evaluating collaborative filtering recommender systems. ACM Trans. Inf. Syst. (TOIS) **22**(1), 5–53 (2004)
10. Adomavicius, G., Tuzhilin, A.: Toward the next generation of recommender systems: a survey of the state-of-the-art and possible extensions. IEEE Trans. Knowl. Data Eng. **17**(6), 734–749 (2005)
11. Schafer, J.B., Frankowski, D., Herlocker, J., Sen, S.: Collaborative filtering recommender systems. In: The Adaptive Web, pp. 291–324. Springer (2007)
12. Desrosiers, C., Karypis, G.: A comprehensive survey of neighborhood-based recommendation methods. In: Recommender Systems Handbook, pp. 107–144. Springer (2011)
13. Oku, K., Hattori, F.: Fusion-based recommender system for improving serendipity. DiveRS@ RecSys **816**, 19–26 (2011)
14. Damiani, E., Ceravolo, P., Frati, F., Bellandi, V., Maier, R., Seeber, I., Waldhart, G.: Applying recommender systems in collaboration environments. Comput. Hum. Behav. **51**, 1124–1133 (2015)
15. Ge, M., Delgado-Battenfeld, C., Jannach, D.: Beyond accuracy: evaluating recommender systems by coverage and serendipity. In: Proceedings of the Fourth ACM Conference on Recommender Systems, pp. 257–260. ACM (2010)
16. Borràs, J., Moreno, A., Valls, A.: Intelligent tourism recommender systems: a survey. Expert Syst. Appl. **41**(16), 7370–7389 (2014)
17. Adomavicius, G., Tuzhilin, A.: Context-aware recommender systems. In: Recommender Systems Handbook, pp. 191–226. Springer (2015)
18. Bahramian, Z., Ali Abbaspour, R., Claramunt, C.: A context-aware tourism recommender system based on a spreading activation method. Int. Arch. Photogr. Remote Sens. Spat. Inf. Sci. **42** (2017)

19. Gavalas, D., Kasapakis, V., Konstantopoulos, C., Pantziou, G., Vathis, N., Zaroliagis, C.: The ecompass multimodal tourist tour planner. Expert Syst. Appl. **42**(21), 7303–7316 (2015)
20. Christensen, I., Schiaffino, S., Armentano, M.: Social group recommendation in the tourism domain. J. Intell. Inf. Syst. **47**(2), 209–231 (2016)
21. van Seghbroeck, G., Vanhove, T., De Turck, F.: Managing the synchronization in the lambda architecture for optimized big data analysis. IEICE Trans. **2**, 297–306 (2016)
22. Bell, R., Koren, Y., Volinsky, C.: Matrix factorization techniques for recommender systems. Computer **42**(8), 30–37 (2009)
23. Apache Foundation. HBase (2018)
24. Apache Foundation. Solr (2018)
25. Zhang, X.F.C., Chen, X., Ge, B.: Storing and querying semi-structured spatio-temporal data in hbase. WAIM Workshops (2016)
26. Raveendran, V., Kalakanti, A.K., Sudhakaran, V., Menon, N.: A comprehensive evaluation of nosql datastores in the context of historians and sensor data analysis. In: IEEE International Conference on Big Data (Big Data), pp. 1797–1806 (2015)
27. Seidman, J., Grover, M., Malaska, T., Shapira, G.: Pattern: Hadoop Application Architectures: Designing Real-World Big Data Applications. O'Reilly Media, Inc. (2015)
28. Seidman, J., Grover, M., Malaska, T., Shapira, G.: Hadoop Application Architectures: Designing Real-World Big Data Applications. O'Reilly Media Inc. (2005)
29. Apache Foundation. Sqoop (2018)
30. Apache Foundation. Kafka (2018)
31. MultiMedia LLC. Flume (2018)
32. Raman, V., Hellerstein, J.M.: An interactive framework for data cleaning. Comput. Sci. Div. (2000)

Chapter 3
EML: A Scalable, Transparent Meta-Learning Paradigm for Big Data Applications

Uday Kamath, Carlotta Domeniconi, Amarda Shehu and Kenneth De Jong

Abstract The work presented in this chapter is motivated by two important challenges that arise when applying ML techniques to big data applications: the scalability of an ML technique as the training data increases significantly in size, and the transparency (understandability) of the induced models. To address these issues we describe and analyze a meta-learning paradigm, EML, that combines techniques from evolutionary computation and supervised learning to produce a powerful approach for inducing transparent models for big data ML applications.

3.1 Introduction

An important factor in the success of machine learning (ML) applications is the choice of the features used to create the feature space used by an ML technique on which the induced models are built. For many applications the feature space is a fairly obvious part of the application, or provided by domain experts. However, an increasing number of exceptions to this are showing up in real-world machine learning problems that involve extracting knowledge from sequence data. Some examples of this include (1) communication, such as speech, handwriting, music, language, and text; (2) temporal point processes, such as stock prices, weather readings, and web events; and (3) molecular biology, such as predicting function or specific properties of DNA, mRNA, proteins, and other biological molecules from biological sequence

U. Kamath (✉)
Digital Reasoning, Franklin, USA
e-mail: uday.kamath@digitalreasoning.com

C. Domeniconi · A. Shehu · K. De Jong
George Mason University, Fairfax, USA
e-mail: cdomenic@gmu.edu

A. Shehu
e-mail: ashehu@gmu.edu

K. De Jong
e-mail: kdejong@gmu.edu

© Springer Nature Switzerland AG 2019
A. Esposito et al. (eds.), *Innovations in Big Data Mining and Embedded Knowledge*,
Intelligent Systems Reference Library 159,
https://doi.org/10.1007/978-3-030-15939-9_3

data. Invariably, in all these application domains, the goal is to detect interactions among various elements in sequence data that encode the presence or absence of a specific property or signal. This is a challenging task, particularly when little or no a-priori information is available on what types of interactions among the elements denote the presence of the sought signal. Increasingly, insights from domain experts have proven to be difficult to translate into *features* that encode local or distal interactions. Prominent examples from molecular biology include the prediction of enzymatic activity, hypersensitive sites in DNA, or splice sites in DNA/RNA sequence data [38, 39, 43–45, 50, 67, 98].

The complexity of how the elements of a sequence interact with one another to give rise to a global property illustrates the need for machine learning methods that are not limited by the lack or unreliability of domain-expert insight, the types or complexity of features needed to encode possible interactions, or the ability to enumerate such features. Currently, two complementary approaches are proposed in machine learning to attenuate the dependence on domain experts. The first operates under the umbrella of deep learning (DL) and utilizes Convolution Neural Networks (CNN) or Long Short-Term Memory (LSTM) networks to automatically extract features. DL-based methods have been shown highly effective in diverse application domains, such as computer vision, text classification, and protein fold recognition in molecular biology [20, 41, 101]. While effective at automating the process of feature generation, DL generates, black-box complex models that are uninterpretable and so cannot be readily used by domain scientists to extract insight [19].

On the other hand, methods that focus on specific feature types and enumerate features result in transparent models but are typically challenged by the size and dimensionality of the feature space. This suggests the need for feature generation algorithms that are automated and effective at assisting the underlying ML technique to attain good prediction power. Large data sets present another challenge to existing supervised learning algorithms. This is particularly evident with sequence-based datasets which are often in the millions or billions in size. Most machine learning methods are ineffective in such cases because they require the entire training dataset to be in memory, or because the learning time during model induction grows significantly with the size of the training dataset size, or both [12]. Basic solutions like reducing the size of the training datasets via sampling can be used at the cost of introducing sampling errors [2]. Another approach is to employ complex changes specific to the desired machine learning algorithm for running on parallel and distributed architectures [26], but the flexibility and range of applicability is lost.

In this chapter we demonstrate how one can leverage and operationalize concepts and techniques from the evolutionary computation (EC) community to assist with both challenges: transparency and scalability. Specifically, we show that evolutionary algorithms (EAs) have much to offer regarding effective automation of feature construction even in vast and high-dimensional spaces of complex features. In addition, we show how concepts from spatially-structured EAs can be utilized to obtain scalable machine learning methods. Together, they comprise a powerful *meta-learning* paradigm that significantly improves the state of the art in big data ML applications.

In the following sections we first summarize relevant concepts and related work, and then devote the rest of this chapter to describing and evaluating our meta-learning paradigm, EML, with respect to its capability for handling complex feature spaces as well as its capability for learning from big data. We then demonstrate its effectiveness on a class of large and difficult molecular biology sequence data applications.

3.2 Background

In this section we briefly review supervised learning methods. Since we will be demonstrating our meta-learning paradigm on large biological sequence data sets, we focus mainly on how ML techniques have been used for classification of sequence data. Typically, the process involves first transforming sequence data into vectors on which the ML technique operates to produce a classification model using a selected subset of training data. The learned model is then applied to novel sequences to make label predictions and thus *detect* or *recognize* the presence of the sought signal. Classification methods can be categorized into statistical-based or feature-based, though most methods are often a combination of the two categories. Below we summarize each category.

3.2.1 Statistical Learning

In statistical learning, the focus is on inducing an underlying statistical model for the classification, which can be generative or discriminative. The transformation of sequence data into numeric data is conducted a priori through a kernel function or a feature-based method that explicitly extracts features of relevance for the transformation. Once provided with numeric data, generative models learn the joint probability $P(x, y)$ of inputs $x \in X$ with labels $y \in Y$. Bayes rule is used to calculate the posterior $p(y|x)$ and predict the most likely label for an unlabeled input. Generative models are able to exploit unlabeled data, in contrast to discriminative models which learn the posterior directly but are limited to a supervised setting that demands labeled training data. Discriminative models are preferred in many classification settings, as they provide a more direct way to model the posterior [65]; for instance, Bayesian discriminative algorithms, such as the Maximum Supervised Posterior (MSP) algorithm, have yielded superior results on biological sequence classification problems [33].

Various hybrid methods have been proposed that combine discriminative and generative models [9] to address the fact that generative methods lose their ability to exploit unlabeled data when trained discriminatively [8]. Hybrid methods have shown superior performance in a variety of application settings [48]. Representative methods include the position-specific scoring matrix (PSSM) method [32, 86], the weight array model (WAM) [88], higher-order Markov models and Markov Random

Fields (MRFs) [7, 49, 96, 97], Bayesian networks [5, 16], a mixture of Bayesian trees and PSSMs [3], smooth interpolations of PSSMs, and empirical distributions [52].

3.2.1.1 Kernel-Based Methods

Support Vector Machines (SVMs) are probably the most popular discriminative learning method due to their ease of implementation and solid grounding in statistical theory [66, 93]. They have been employed in many sequence classification problems, including prediction of transcription start sites on DNA [81], translation initiation sites [89], gene finding [77], transcription factor-binding sites [40], and DNA regulatory regions [76]. The predictive power of SVMs greatly depends on the chosen kernel function [75]. This function maps input (sequence, here) data onto an usually higher-dimensional feature space, where provided samples of the two classes can be linearly separated by a hyperplane. The kernel function can be viewed as providing a similarity measure between any two sequences. Many kernels are designed for sequence classification, of which the most relevant and state-of-the-art are Weighted Degree (WD) and Weighted Degree with Shift (WDS) devised for recognition of DNA splice sites [82]. In these kernels, limited-range dependencies between neighboring nucleotides are considered to encode features for the SVM, thus blurring the boundaries between kernel- and feature-based methods. Concepts from evolutionary computation have also been proposed to learn effective, possibly more complex, kernels for a given classification problem at hand [44, 46].

3.2.2 Feature-Based Learning

Here the primary focus is on finding effective features to be used to transform sequence data into feature vectors in order to use standard ML techniques. Feature-based methods offer *transparent learning*, in that they provide constructed features for inspection and further analysis by domain experts. Feature construction, however, is a non-trivial task. The predominant approach is to use enumeration to list all considered features but limit the focus to (*spectrum*) features, *k-mers*, which are strings of k symbols over the alphabet of building blocks in the considered sequence data [59]. Again, the essential idea is to transform given sequences into numeric vectors recording the frequency or occurrence of the listed k-mers and then employ supervised learning techniques, such as SVMs, to separate the training data in the resulting vector space [67]. Spectrum features have been shown useful in many classification problems, including prediction of DNA promoter regions, cis sites, HS sites, splice sites, and more [1, 30, 59, 102].

The tendency to enumerate spectrum features is typically the result of a lack of biological and/or expert insight into which are the important and effective features, resulting in significant inefficiency during model induction, in that the majority of spectrum features are seldom useful and are removed by feature selection algorithms [22]. Reducing the size and complexity of the features to minimize this inefficiency seldom works either, in that simple spectrum features generally do not provide sufficient discriminatory power [37, 39, 43, 44, 46, 50, 67, 98].

To improve efficiency and accuracy, other non-spectrum features have been used. For example, using features that encode correlations or simultaneous occurrences of particular k-mers at different positions in a sequence has been demonstrated to be important for classification accuracy [39, 45, 47]. In general, expanding the types of features considered, as in [39, 51], has resulted in improvement of accuracy, but the reliance on enumeration as a mechanism to explore a vast feature space ultimately imposes limits on the possible types of features considered, as well as their complexity.

3.2.3 EAs for Exploration of Vast, Complex Features Spaces

The previous sections serve in part to emphasize the importance of and the need for effective methods for constructing and selecting feature sets in the absence of a priori knowledge or insight. It is our view that evolutionary algorithms (EAs) present an appealing approach for efficiently exploring vast and high-dimensional spaces of complex features to automate the process of feature construction [25]. Their utility for feature construction was recognized early [79], and many following studies have shown them effective in various domains [14, 36, 43, 44, 46, 47, 50, 55, 58, 68, 72]. Recent work has shown that genetic algorithms (GAs), a particular class of EAs, improve classification accuracy when used to replace feature enumeration techniques in predicting promoter regions, HS sites, and splice sites in DNA, and even enzymatic activity in proteins [43–47, 50].

Another class of EAs, Genetic Programming (GP), evolves individuals represented as variable-length trees composed of functions and variables. Since their introduction, GP algorithms have seen an increase in their usage in diverse problems in bioinformatics [24, 50, 63, 71, 94, 99]. Application areas include drug design, cancer classification from gene expression data, classification of genetically-modified organisms, and classification of cognitive states from fMRI data [24, 27, 35, 56, 62, 63, 71, 94, 99].

Our recent work has introduced a GP-based method for feature construction in the context of DNA splice site recognition [47]. In this chapter, we present a more general EA-based approach that makes use of a GP algorithm to explore complex feature spaces and generate predictive features from sequence data.

3.2.4 Big Data Machine Learning

As noted in Sect. 3.1, challenging problems for machine learning are characterized not only by vast, complex feature spaces, but also by vast training datasets. Such data, now referred to as Big Data, pose scalability challenges and require ML improvements.

In statistical learning theory, a formal relationship between the notion of *margin* and the generalization classification error has been established [93]. Classifiers that converge to a large margin have better performance in terms of generalization accuracy. The classification boundary provided by an SVM has the maximum-minimum distance from the closest training points. SVMs have training times of $O(n^3)$ and space complexity of $O(n^2)$, where n is the size of the training set [91]. SVMs have been modified to scale to large datasets [13, 29, 31, 42, 92], but many of these modifications introduce a bias caused by approximations due to sub-sampling of the data or assuming a linear model; both can lead to a loss in generalization.

Another approach to handle big data relies on using distributed architectures and network computing via algorithm-specific parallelizations [18, 21, 28, 95]. These modifications are carried out on decision trees, rule inductions, and boosting algorithms [78, 87]. Most of these algorithms are changed to parallelize computations like matrix transforms in SVMs or tree node learning, and many of them use a message passing interface (MPI) for exchanges. MapReduce, which gives a generic framework for a divide-and-conquer-based approach, has also been used in conjunction with learning algorithms to scale to large datasets [21]. Ensemble-based learning on parallel networks has also been employed on various tree-based algorithms for learning from large-scale datasets [87].

3.2.4.1 Boosting and Large-Margin Classifiers

The large-scale learning approach we are presenting in this chapter is based on the ML concepts of boosting and ensemble learning. The AdaBoost technique, and boosting in general, is an example of a learning methodology known as ensemble learning [69], in which multiple classifiers are generated and combined to make a final prediction. Ensemble learning has been shown to be effective with unstable classifiers, by improving upon the performance of the base learners. AdaBoost, for instance, induces a classification model by estimating the hard-to-learn instances in the training data. A formal analysis of the AdaBoost technique has derived theoretical bounds on the margin distribution to which the approach converges [74]. Through the consolidation of different predictors, ensemble learning can lead to significant reductions in generalization error [6].

3.2.4.2 Spatially-Structured EAs (SSEAs)

Our meta-learning paradigm achieves scalability through parallelism. But, rather than modify existing ML techniques to run on parallel architectures, we adopt another well-established EA technique, Spatially-structured EAs (SSEAs), which use topologically-distributed populations and local neighborhood selection. SSEAs have been well analyzed in the EC literature [73]. SSEAs have been shown to main-

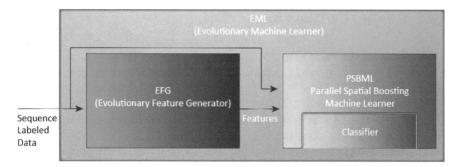

Fig. 3.1 Overall schematic of the transparent, feature-based, distributed learning approach we highlight in this chapter

tain a diverse set of fit individuals longer, resulting in improved performance in many applications [90]. However, the key feature that SSEAs provide is their "embarrassingly parallel" architecture; at each topological grid point, a local algorithm is running that has only local interactions with immediate neighbors.

3.2.5 Summary

The goal of this section has been to motivate the need for improved ML approaches for large data applications. This is particularly true with the growing number of applications involving sequence data for which improved feature generation methods are required in addition to improved scalability. We describe our approach to addressing these issues in the next section.

3.3 Methodology

The schematic of the overall approach is shown in Fig. 3.1. An EA-based approach is employed to construct complex features, and feature-based representations of a big training dataset are then fed to a distributed machine learning algorithm for inducing a scalable model. We now describe each of the components of this framework in detail.

3.3.1   Feature Construction via EFG

The Evolutionary Feature Generator (EFG) approach uses a GP algorithm to explore a large and complex space of potentially useful features from the given training dataset. Features are represented as standard GP trees, and a population of features is evolved over time using standard GP mechanisms of mutation and crossover. EFG uses a surrogate filter-based fitness function to estimate the usefulness of the GP-generated features, since the wrapper methodology to find effectiveness of the features is costly. A *hall of fame* mechanism incrementally collects the best-estimated features for subsequent use with a classifier.

3.3.1.1   Feature Representation

Various researchers, as highlighted in Sect. 3.2, have individually discovered many building blocks that can be very effective in finding the patterns in sequence classification. The novelty of the EFG algorithm is that it not only defines many new building blocks, but it also gives a structure through strongly-typed GP evolution, combining various building blocks in an effective and human-understandable manner. This structured way of searching a vast feature space involves building a complex structure given the constraints defined from simpler ones. The strongly-typed GP plays the role of giving structure and guidance to the vast search space of features. Next, we highlight the building blocks from the simplest short subsequence known as a motif, which becomes the common building block to the complex higher-order signals that can be constructed through the algorithm. We have arranged the explanation at various levels of complexity starting from level 1 (the simplest) to level 3 (the most complex) (Table 3.1).

Level 1: Motif

The most common building block is the presence of a short subsequence of strings of a given length, which are constructed as parse trees from the given alphabets. These motifs are used as a building block in all the second level constructs.

Level 2: Compositional Features

The `matches` operator allows constructing simple compositional features. The nucleotides that make up the motif serve as leaves. The evaluation involves obtaining the occurrence of the motif in a given sequence.

Level 2: Regional Features

The `regional` operator can be considered an extension of compositional features, restricted to certain bounds given by the domain experts. Region-specific features were found to be important functional signals in many sequence classification problems [39, 70].

Level 2: Positional Features

Table 3.1 A table of the non-terminals and terminal nodes employed by EFG

Name	Args	Return type
matches	motif	Boolean
matchesAtPosition	motif, position	Boolean
positionalShift	motif, position, Shift	Boolean
correlational	motif, motif, position, close	Boolean
regional	motif, region	Boolean
and	2 non-terminal boolean	Boolean
or	2 non-terminal boolean	Boolean
not	2 non-terminal boolean	Boolean
motif-*	ERC-chars	Motif
position	ERC-int	Integer
shift	ERC-int	Integer
close	ERC-int	Integer
region	ERC-int, ERC-int	Integer
ERC-char	{Symbols}	Character
ERC-int	$\{1, \ldots, length\}$	Integer

The matchesAtPosition operator allows constructing simple positional features from the motifs at a given position. The positional features correspond to local features often employed in classification of sequences. In these features, the goal is to find a specific motif at a specific position in the sequence.

Level 2: Positional Shift Features

The positionalShift operator allows constructing positional features that may be displaced in either direction by a small shift given as a parameter. The positional shift features were discovered to be very effective in complex sequence/series classification problems [82].

Level 2: Correlational Features

The correlational operator captures the presence of positional features adjacent to each other, within a distance. Correlational features are generated from two simpler motifs: position in the sequence and closeness capturing the adjacency.

Level 3: Complex Higher Order Signals

Many statistical learning approaches, such as Bayesian networks and Markov models, rely on higher order elements formed from lower order signals as the discriminative features [57]. The approach of having logical combinatorial operators like and, or, and not acts in a similar way to construct more complex features combining the simple Level 1 motifs or Level 2 elements, such as positional, compositional, correlational, shift-positional, and regional features, or even the Level 3 features to form any level of complex chains, from simpler conjuncts or disjuncts.

3.3.1.2   Genetic Operators

As in most EAs, individuals have to undergo some modifications through genetic breeding operators to generate new individuals from the existing population. Studies have shown robust EAs incorporate both mutation and crossover as the breeding operators [85].

Mutation

The role of mutation in EAs is to make small, incremental change to an individual to form a new individual with a small change. GP-styled evolution normally has a more disruptive mutation operator that randomly generates a subtree and replaces a node in the given individual with that subtree [54]. Instead, EFG employs problem-specific mutations as small, incremental operators. These mutation operators are motif mutation, positional mutation, shift mutation, and adjacency mutation.

Crossover

EFG employs the standard subtree crossover, one of the most common genetic recombination operators used in GP [54]. Subtree crossovers have been very effective in GP, as they form complex trees and explore vast search spaces more effectively.

3.3.1.3   Fitness Function

EFG employs a surrogate fitness function, or a "filter" approach, which is considered to be a fast and effective way for feature evaluation [53]. Since most sequence classification data are imbalanced and have very few positives and a large number of negatives, the goal is to improve precision while managing the discriminating power of features. We formulate the fitness function as follows:

$$\text{Fitness}(f) = \frac{C_{+,f}}{C_+} \times \left| \frac{\frac{C_{+,f}}{C_+} - \frac{C_{-,f}}{C_-}}{C_+ + C_-} \right|$$

In this equation, f refers to a feature, $C_{+,f}$ and $C_{+,f}$ are the number of positive and negative training sequences that contain the feature f, respectively. C_+ and C_- are the total number of positive training sequences. This fitness function tracks the occurrence of a feature in positive sequences, as negative sequences may not have any common features or signals, while penalizing non-discriminating features equally found in positive and negative sequences.

The goal is to maximize the fitness function, but the Koza fitness in the standard GP formulation is sought to be minimized [54]. Specifically, the Koza fitness of a feature is defined by f as $\text{Koza}(f) = 1/(\text{Fitness}(f))$. EFG converts the Koza fitness back into the GP-adjusted fitness $1/(1 + \text{Koza}(f))$ to select fit individuals. Note that the GP-adjusted fitness takes values in $[0, 1]$.

3.3.1.4 Hall of Fame

Since GP is a generational EA (that is, the parents die after producing the offspring), there can be "genetic drift" and convergence to a local optimum [25]. This can result in the loss of some of the best individuals, which can be useful discriminating features for classification. Introducing elitism, which is the ability to keep some of the best individuals in the population, helps to overcome this at the expense of introducing strong selection pressure. The EFG algorithm uses an external storage of features known as the *hall of fame*. At the end of the EA execution, these highly-fit features in the hall of fame constitute the solutions that are employed to transform sequence data into numeric data for the PSBM algorithm described next.

3.3.2 The PSBML Algorithm

Our approach (called Parallel Spatial Boosting Machine Learner, or PSBML) introduces a *new paradigm* to perform scalable classification with massive datasets. PSBML can be described as a *meta-learning* algorithm, because it can use any classifier capable of producing confidence measures on predictions. To achieve this goal, it uses concepts from stochastic optimization and ensemble learning. The specific stochastic optimization technique it uses is the SSEA summarized in Sect. 3.2.

By leveraging ideas from both stochastic optimization and ensemble learning, PSBML makes interesting connections between two otherwise disconnected fields. At the same time, PSBML represents a new approach to perform large-scale machine learning. Through the emerging behavior of a grid of classifiers, PSBML enables an effective processing of large amounts of data that culminates with the discovery of the fraction of data that is crucial for the problem at hand. The emerging behavior only requires local interaction among the learners, thus enabling a high degree of parallelism. Via a multi-thread implementation (one thread per classifier), PSBML is able to efficiently analyze the entire collection of data. As such, PSBML does not sacrifice accuracy like sampling does, and at the same time it achieves a general scalable solution that does not need to be customized for each specific classification algorithm. As a consequence, the PSBML framework is *scalable and general*. In the following, we describe the different phases of the algorithm. The pseudo-code of PSBML is given in Algorithm 1.

Initialization

Given a collection of labeled data, an independent, fixed validation set is created, and the rest is used for training. PSBML uses a wrap-around toroidal grid to distribute the training data and the specified classifier to each node in the grid. The training data is distributed across all the nodes using stratified uniform sampling of the class labels (Line 1 of Algorithm 1). The parameters for the grid configuration, i.e., width and height of the grid, the replacement probability, and the maximum number of iterations, are all included in *GridParam*.

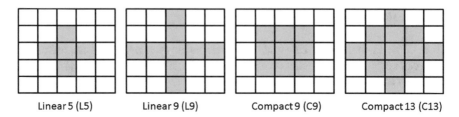

| Linear 5 (L5) | Linear 9 (L9) | Compact 9 (C9) | Compact 13 (C13) |

Fig. 3.2 Two-dimensional, toroidal grid with various neighborhood structures

Node Behavior at Each Epoch

The algorithm runs for a pre-determined number of iterations (or epochs) (Line 4 of Algorithm 1). The behavior of a node at each epoch can be divided into two phases, training and testing. During training, a node performs a standard training process using its local data (Line 5). For testing, a node's training data is combined with the instances assigned to the neighboring nodes (Lines 7 and 8). Each node in the grid interacts only with its neighbors, based on commonly used neighborhood structures as shown in Fig. 3.2. Each classifier outputs a confidence value for the prediction of each test instance, which is then used for weighting the corresponding instance. Every node updates its local training data for the successive training epoch by probabilistically selecting instances based on the assigned weights (Line 9). The confidence values are used as a measure of how difficult it is to classify a given instance, allowing a node to select, during each iteration, the most difficult instances from all its neighbors. Since each instance is a member of the neighborhood of multiple nodes, an ensemble assessment of difficulty is performed, similar to the boosting of the margin in AdaBoost [74]. Specifically, in PSBML the confidence cs_i of an instance i is set equal to the smallest confidence value obtained from any node and for any class: $cs_i = \min_{n \in N_i} c_{ni}$, where N_i is a set of indices defined over the neighborhoods to which instance i belongs, and c_{ni} is the confidence credited to instance i by the learner corresponding to neighborhood n. These confidence values are then normalized through linear re-scaling: $cs_i^{norm} = (cs_i - cs_{min})/(cs_{max} - cs_{min})$, where cs_{min} and cs_{max} are the smallest and the largest confidence values obtained across all the nodes, respectively. The weight $w_i = (1 - tcs_i^{norm})$ is assigned to instance i to indicate its relative degree of classification difficulty. The w_is are used to define a probability distribution over the set of instances i, and used by a node to perform a stochastic sampling technique (i.e. weighted sampling with replacement) to update its local set of training instances. The net effect is that, the smaller the confidence credited to an instance i is (i.e. the harder it is to learn instance i), the larger the probability will be for instance i to be selected. Instead of deterministically replacing the whole training data at a node with new instances, a user defined replacement probability P_r is used. Due to the weighted sampling procedure, and to the constant training data size at each node, copies of highly weighted instances will be generated, and low weighted instances will be removed with high probability during each epoch.

Grid Behavior at Each Epoch

At each iteration, once all nodes have performed the local training, testing, re-weighting, and have generated a new training dataset, a global assessment of the grid is performed to track the "best" classifier throughout the entire iterative process. The unique instances from all the nodes are collected and used to train a new classifier (Lines 10 and 11). The independent validation set created during initialization is then used to test the classifier (Line 12). This procedure resembles the "pocket algorithm" used in neural networks, which has been shown to converge to the optimal solution [64]. The estimated best classifier is given in output and used to make predictions for unseen test instances (Line 17).

Iterative Process

The weighted sampling process and the interaction of neighboring nodes enable the hard instances to migrate throughout the various nodes, due to the wrap-around nature of the grid. The rate at which the instances migrate depends on the grid structure, and more importantly on the neighborhood size and shape. Thus, the grid topology of classifiers and the data distribution across the nodes provides the parallel execution, while the interaction between neighboring nodes and the confidence-based instance selection give the ensemble and boosting effects.

Algorithm 1 PSBML*(Train, Validation, GridParam)*

1: INITIALIZEGRID(Train, GridParam)   ▷ *Distribute the instances over the nodes in grid*
2: currentMin ← 100
3: Pr ← GridParam.pr   ▷ *Probability of replacement*
4: **for** i ← 0 **to** GridParam.iter **do**   ▷ *Train all nodes*
5:    TRAINNODES(GridParam)
6:    TESTANDWEIGHNODES(GridParam)   ▷ *Collect neighborhood data and assign weights*
7:    PrunedData ← { }
8:    **for** j ← 0 **to** GridParam.nodes **do**
9:      NeighborData ← COLLECTNEIGHBORDATA(j)
10:     NodeData ← NodeData ∪ NeighborData
11:     ReplaceData ← WEIGHSAMPLING(NodeData, Pr)
12:     PrunedData ← UNIQUE(ReplaceData)   ▷ *Unique keeps one copy of instances in set*
13:    ValClassifier ← createNew(GridParam.classifier) ▷ *New classifier for validation*
14:    error ← VALIDATE(PrunedData, Validation, ValClassifier) ▷ *Use validation set to track model learning*
15:    currentMin ← error
16:    bestClassifier ← ValClassifier
17:    marginData ← PrunedData
18: **return** bestClassifier, marginData

3.3.3 EML Framework

In the previous sections, two functionally-orthogonal algorithms, EFG for feature generation in sequences and PSBML for large scale learning, are described. One can combine key aspects of these algorithms in a single framework for large-scale learning. From now on, we refer to this framework as Evolutionary Machine Learning (EML) (Fig. 3.3).

3.3.3.1 Data Parallelization in EML Framework

The first process in the EML framework is to split the training data using stratified folds, i.e., the same balance of labels are divided equally amongst all the nodes. The splitting of data into equal parts is used both in the feature generation and parallel classification. The design choice on the upper and lower bounds on this is affected by processor/cores and by the complexity of the learner, respectively. The grid configuration like the number of nodes running further depends on the processors/cores supported by the underlying hardware running the EML framework. For example, on a 256 POWER7 4.25 GHz IBM machine, a 256 nodes in 16×16

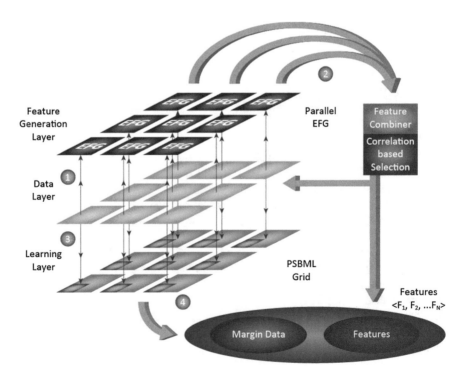

Fig. 3.3 Schematic of the EML framework

arrangement, will be dividing the large datasets in stratified manner into 256 samples. If the training data is not really too big, then dividing it by large number of nodes can lead to a configuration where each classifier/node in the learning mode has less data than the minimum requirement as given by the complexity of that classifier [10]. Thus, the balance between parallel execution and complexity of the classifier has to be achieved for effective learning and parallelization.

3.3.3.2   Parallel EFG

Parallel EFG in the grid consists of each node running the EFG algorithm on its stratified data, partitioned as discussed above, in parallel. Each node in the EFG grid generates the features in parallel in its own hall of fame storage described in the EFG section. Note that there is no communication between the EFG running in the nodes with other nodes in the grid. However, in the future, highly fit individuals migrating among neighboring nodes or breeding individuals from neighbors may be incorporated to give more robust features. Many island model-based EAs often employ migrations or neighborhood interactions for an effective evolutionary behavior [80].

3.3.3.3   Feature Combination and Selection

Since each node runs the EFG algorithm, there can be many disjoint sets of features generated by the whole grid, equal to the number of the nodes in the grid. These independent sets of features have to be combined to form a single set of features for use in the PSBML algorithm. They are combined to find a unique set by forming a union among all the disjoint sets generated by nodes in the grid. One can easily contemplate more sophisticated mechanisms of feature weighting in the future. If a certain feature is truly representative of an over-represented pattern, then it will emerge in more EFG nodes, and just reducing it to a single feature without taking into the account the normalized count of its appearance may seem bit contrived. However, imposing the feature weights puts a constraint on the classifier chosen by the PSBML algorithm to be one that handles weighted features and is a limited set. Also, complexity of feature weighting and instance weighting during PSBML iterations may lead to a positive feedback loop, introducing a biased induction and may perform badly on noisy data that has over-represented features.

Since there can presumably be a large number of features even after removing duplicates, a feature selection algorithm that further selects only relevant and non-redundant features becomes a mandatory dimensionality reduction step. We have investigated various feature selection algorithms such as information gain with ranking, correlational feature selection with GAs, and fast correlation feature selection with parallel greedy forward search were researched. Each of these methodologies have a trade-off between accuracy and the speed of searching for quality features. In the current EML framework, fast correlation based forward (FCBF) search, which was the fastest and had very good accuracy in general, has been employed as the dimensionality reduction technique [100].

3.3.3.4 Parallel Feature Interpretation

The feature set generated by the above feature reduction step is passed to each node. Each node in the grid has a feature interpreter that parses the features and maps it to the numerical values 0/1. Thus, parallel feature interpretation is accomplished and the raw sequence data gets transformed to high dimensional vector space in this process.

3.3.3.5 PSBML

There are no changes made to the PSBML algorithm to use it in the EML framework, as it is already designed as a parallel learning algorithm that works on the stratified samples of large training data on a grid configuration with some known features. The parallel data stratified and interpreted in the above step with the parallel features generated by feature selection respectively will be used in the PSBML algorithm. The choice of the classifier is generally a design decision based on the complexity of the resulting model, number of features, noise level in the data, and speed of the classifier in learning the model to name a few. Using the feature selection and reduction techniques, the features available to PSBML have more independence, lower correlation between them, and higher correlation with the label. Also because of the fast training speed and robustness to noise as compared to other classic classifiers analyzed in the PSBML analysis, Naïve Bayes is chosen as a default classifier in PSBML training nodes. The PSBML algorithm reduces the training data to substantially smaller but relevant margin data. These reduced margin data with EFG features become the inputs for the classifier for learning the model to predict the future unseen testing/validation data.

3.4 Experimental Validation

Here we demonstrate the effectiveness of the EML framework on a difficult, well studied problem that allows for rigorous comparisons. Specifically, we tackle the problem of genome-wide sequence splice site recognition, where training data sizes are large, the data is imbalanced, and finding interesting patterns is very difficult.

3.4.1 Evaluation Setup: Genome-Wide Sequence Classification

Whole-genome sequencing has yielded important insights into genetic variations, explaining the evolutionary changes and mapping the gene functions to the genome sequence regions of organisms [15]. These genomic sequences from various organisms are carefully annotated into various important regions such as regulatory, promoters, and splice site regions. Instead of using costly wet-laboratory experiments, the idea is to use computational methods to find interesting functional signals that become the predictor of these complex regions in the organisms. The computational biology community is avidly seeking cost-efficient, high-throughput, and accurate prediction tools for genome annotation [4, 60]. Training the model based on large data available in genome sequencing lends itself to the Big Data problem. Various learning algorithms, such as Markov chain-based and kernel-based ones, have performed relatively well on this problem [82].

For evaluation of the EML framework, genome-wide recognition of splice site locations from two different organisms, namely *Caenorhabditis elegans* ("worm") and *Drosophila melanogaster* ("fly") are used. The framework is compared to three other state-of-the-art algorithms in automated whole-genome sequencing, two kernel-based algorithms, and one Markov chain-based algorithm.

3.4.2 Experimental Setup: Hardware and Software

All the scalability experiments (in which the running times were measured) were run on a dual, 3.33 GHz, 6 processor Intel Xeon 5670 processor, accounting for 12 hardware threads. The EFG algorithm running on each node has basic settings of using motifs between length 2–8, IUPAC code with ambiguity symbols for the motif, population size of 5000, external storage of 250 features, crossover rate at 0.7 and three different mutations at 0.1 each.

PSBML is implemented as a multithreaded standalone Java program that can run on any JVM version above 1.5. All the experiments with PSBML are run using a maximum heap size of 8GB and a number of threads equal to the number of nodes in the grid. Since the hardware had only 12 threads, all the experiments are run using a 3 × 3 grid with a C9 configuration for the neighborhood.

The kernel-based algorithms, the weighted degree positional kernel (WD), and the weighted positional kernel with shift (WDS) algorithms are run using the publicly-available Shogun toolkit [84] along with the publicly-available LibSVM as the SVM implementation [17]. The SVM with an RBF kernel is run using WEKA [34], and the Markov chain (MC) method is run using the JSTACS software [33].

Table 3.2 Whole-organism genome training datasets size and true positives fractions in the total set

Evaluation	Worm dataset		Fly dataset	
	Acceptor	Donor	Acceptor	Donor
Training data size	1,105,886	1,744,733	1,289,427	2,484,854
True positive fraction	3.6%	2.3%	1.4%	0.7%

3.4.3 Datasets

The datasets are extracted from the worm and fly genomes and prepared as in [82]. Briefly, the genome is aligned through blat with all known cDNA sequences available at http://www.wormbase.org and all known EST sequences in [11] to reveal splicing sites confirming the introns and exons. Sequencing errors were corrected by removing minor insertions and deletions. Clustering of the alignments is then performed, initializing both cDNA and EST sequences to be in different clusters. Clusters are joined iteratively: if any two sequences from distinct clusters match to the same genomic location, this indicates alternate splicing. From the clustered alignments, a compact splicing graph representation is obtained, which can be easily used to generate a list of positions of true acceptor and donor splice sites.

Within the boundaries of the alignments, all positions exhibiting the AG, GT, or GC dimer, and which were not in the list of confirmed splice sites, became the decoys (false acceptor and donor sites). The sequence length in all sets is 141 nt; for acceptor splice sequences, the consensus dimer AG is at position 61, and for donor GT/GC, the consensus is at position 81. Details of the dataset and the distribution of true positives is given in Table 3.2.

3.4.4 Comparative Evaluation on the Splice Site Recognition Problem

The evaluation of classification models uses several criteria, including area under the Receiver Operator Characteristic Curve (auROC) and area under the Precision Recall Curve (auPRC). These are based on the basic notions of the number of true positives (TP), false positives (FP), true negatives (TN), and false negatives (FN). To briefly summarize what these measures capture, consider that predicted instances (sequences assigned a label by the classification model) can be ordered from most to least confident. Then, given a particular confidence threshold, the data above the threshold can be considered correctly labeled. The true positive rate (TPR = TP/TP + FN) and the false negative rate (FNR = FN/FN+TN) can then be computed as one varies this threshold from 0.0 to 1.0. In an ROC, one typically plots TPR as a function of FNR. We note that the auROC is a summary measure indicating whether

Table 3.3 Area under ROC (auROC) and area under PRC (auPRC) comparisons of various state-of-the-art algorithms with the EML framework on the worm and fly datasets for both acceptor and donor splice sites

Algorithms	Evaluation on worm dataset				Evaluation on fly dataset			
	Acceptor		Donor		Acceptor		Donor	
	auROC	auPRC	auROC	auPRC	auROC	auPRC	auROC	auPRC
MC	99.6	90.2	99.4	90.1	98.7	80.27	99.12	78.4
WD	99.36	91.7	99.5	88.2	99.02	84.82	99.49	86.40
WD-S	99.2	91.1	99.8	90.1	99.12	86.4	99.5	87.4
EML	97.9	93.2	96.7	93.3	96.9	89.9	96.8	90.3

prediction performance is close to random (0.5) or perfect (1.0). Further details can be found in [61].

The auROC can be a misleading indicator of prediction for unbalanced datasets, as it is independent of class size ratios; for instance, large auROC values may not necessarily indicate good performance. The auPRC is a better performance measure when the class distribution is heavily unbalanced [23]. The PRC measures the fraction of negative points misclassified as positives; as such, it plots the precision (TP/TP+FP) versus the recall ratio (i.e., TPR, sometimes referred to as sensitivity). Again, as one varies the threshold, precision is calculated at the threshold that achieves that recall ratio. auPRC is a less forgiving measure, and a high value indicates that the classification model makes very few mistakes. Thus, the higher the auPRC value, the better is the performance.

Training and testing of each of the models (algorithms under comparison) for each of the two organisms on the acceptor and donor recognition task was performed using 5-fold cross-validation. The reported auROC and auPRC are averaged scores over the five cross-validation unbiased splits. Each of the five training cycles contains 4/5 of training data going through the EML process and 1/5 for testing. MC, WD, and WD-S are compared with EML in terms of the auROC and auPRC metrics. The comparison is carried out by running EML 10 times, and a paired-t test is used to measure statistical significance. The results are related in Table 3.3.

The average time taken by WD and WD-S was 6.5 and 7.0 h, respectively, as measured by wall-clock time, while the EML framework took 3.5 h on average on each of the datasets. The MC algorithm took more than 10 h on average on each of the datasets. For large training datasets, these results confirm the hypothesis that the EML framework, by using EFG and PSBML, combines the best of both worlds in terms of finding discriminating features and carrying out parallel learning.

Upon closer inspection, for the worm dataset, the features generated by EML framework confirm the presence of 7mer motifs **GGTAAGT**, **AGGTAAG**, **GGTAGGT** around −43 nt, matching the donor consensus motif **AGGTAAGT**. Another important positional feature, **TAAT**, is found in the region [−18, −14] nt; this feature is a well-known branch site signal [83]. Shift-positional features around

−3 nt consisting of motifs **TTTCAGG** and **TTTCAGA** matching the acceptor consensus **TTTCAG(A/G)** motif are also present.

For the fly datasets, overrepresented features, such as compositional patterns of **GGTAA**, **GGTGAG**, and **GGTGAGT**, which are close to the consensus donor motifs, are top features in the EML framework. Acceptor features contain compositional features, such as **ATTTCAG**, **TTACAGA**, and **CTTGCAGA**, which are very similar to the consensus acceptor motifs. Interesting regional patterns, such as the downstream presence of **GTAAGT** and the upstream presence of **GTA**, also reproduce results obtained by past research [83].

3.5 Conclusion

In summary, the EML framework presented here conducts distributed learning over big sequence data by combining features of spatially-structured evolutionary algorithms with the well-known machine learning techniques of ensemble learning and boosting. It is worth noting that the framework does so in a way that does not require changes to the underlying machine learning methods, maintains or improves classification accuracy, and can achieve significant speedup in running times via a straightforward mapping of multithreaded shared-memory architectures.

An important take-away from the presented evaluation is the precision with which the constructed features characterize complex discriminatory patterns. It is worth emphasizing that this result would have taken a significant number of k-mers; moreover, the positional, correlational, and compositional context would not have been captured. This would not only result in lower information gain at the cost of a higher number of features, but it would also generate a large number of false positives. Markov models and positional matrix-based algorithms would have captured more of the patterns outlined, but not the complex combinations that the EFG algorithm in the presented EML framework captures. Moreover, the complex features can be interpreted in meaningful ways by domain experts, providing additional insights into the learning problem at hand.

The presented work opens up several directions of future work, as described in Sect. 3.3, including migration and/or breeding of individuals among neighboring nodes in the EFG algorithm to obtain more robust features, and more sophisticated mechanisms of feature weighting.

References

1. Anwar, F., Baker, S.M., Jabid, T., Hasan, M.M., Shoyaib, M., Khan, H., Walshe, R.: Pol II promoter prediction using characteristic 4-mer motifs: a machine learning approach. BMC Bioinform. **9**, 414–421 (2008)
2. Bacardit, J., Llorà, X.: Large scale data mining using genetics-based machine learning. In: Proceedings of the GECCO-2012, pp. 1171–1196 (2012)

3. Barash, Y., Elidan, G., Friedman, N., Kaplan, T.: Modeling dependencies in protein-DNA binding sites. In: Istrail, S., Pevzner, P., Waterman, M. (eds.) International Conference on Research in Computational Molecular Biology (RECOMB), pp. 1–8. ACM Press, New York, NY (2003)
4. Batzoglou, S., Jaffe, D.B., Stanley, K., Butler, J., Gnerre, S., Mauceli, E., Berger, B., Mesirov, J.P., Lander, E.S.: ARACHNE: a whole-genome shotgun assembler. Genome Res. **12**(1), 177–189 (2002)
5. Ben-Gal, I., Shani, A., Gohr, A., Grau, J., Arviv, S., Shmilovici, A., Posch, S., Grosse, I.: Identification of transcription factor binding sites with variable-order bayesian networks. Bioinformatics **21**(11), 2657–2666 (2005)
6. Bennett, K.P., Demiriz, A., Maclin, R.: Exploiting unlabeled data in ensemble methods. In: Proceedings of the KDD, pp. 289–296. ACM Press (2002)
7. Bernal, A., Crammer, K., Hatzigeorgiou, A., Pereira, F.: Global discriminative learning for higher-accuracy computational gene prediction. PLoS Comp. Biol. **3**(3), e54 (2003)
8. Bishop, C.M.: Pattern Recognition and Machine Learning. Springer, Singapore (2006)
9. Bishop, C.M., Lasserre, J.: Generative or discriminative? getting the best of both worlds. Bayesian Stat. **8**, 3–24 (2007)
10. Blumer, A., Ehrenfeucht, A., Haussler, D., Warmuth, M.K.: Learnability and the vapnik-chervonenkis dimension. J. ACM (JACM) **36**(4), 929–965 (1989)
11. Boguski, M.S., Lowe, T.M., Tolstoshev, C.M.: dbest-database for "expressed sequence tags". Nat. Genet. **4**(4), 332–333 (1993)
12. Bordes, A., Bottou, L., Gallinari, P.: Sgd-qn: careful quasi-Newton stochastic gradient descent. J. Mach. Learn. Res. **10**, 1737–1754 (2009)
13. Bottou, L., Bousquet, O.: The tradeoffs of large scale learning. In: Platt, J., Koller, D., Singer, Y., Roweis, S. (eds.) Advances in Neural Information Processing Systems, pp. 161–168. MIT Press (2008)
14. Brill, F.A., Brown, D.E., Martin, W.N.: Fast genetic selection of features for neural networks. IEEE Trans. Neural Netw. **3**(2), 324–328 (1992)
15. Burton, P.R., Clayton, D.G., Cardon, L.R., Craddock, N., Deloukas, P., Duncanson, A., Kwiatkowski, D.P., McCarthy, M.I., Ouwehand, W.H., Samani, N.J., et al.: Genome-wide association study of 14,000 cases of seven common diseases and 3,000 shared controls. Nature **447**, 661–678 (2007). https://doi.org/10.1038/nature05911
16. Cai, D., Delcher, A., Kao, B., Kasif, S.: Modeling splice sites with bayes networks. Bioinformatics **16**(2), 152–158 (2000)
17. Chang, C.C., Lin, C.J.: LIBSVM: A Library for Support Vector Machines. Online (2001)
18. Chang, E.Y., Zhu, K., Wang, H., Bai, H., Li, J., Qiu, Z., Cui, H.: Parallelizing support vector machines on distributed computers. In: Neural Information Processing Systems (NIPS) (2007). http://books.nips.cc/papers/files/nips20/NIPS2007_0435.pdf
19. Chen, X., Lin, X.: Big data deep learning: challenges and perspectives. IEEE Access **2**, 514–525 (2014). https://doi.org/10.1109/ACCESS.2014.2325029
20. Chen, Y., Jiang, H., Li, C., Jia, X., Ghamisi, P.: Deep feature extraction and classification of hyperspectral images based on convolutional neural networks. IEEE Trans. Geosci. Remote Sens. **54**(10), 6232–6251 (2016). https://doi.org/10.1109/TGRS.2016.2584107
21. Chu, C.T., Kim, S.K., Lin, Y.A., Yu, Y., Bradski, G., Ng, A.Y., Olukotun, K.: Map-reduce for machine learning on multicore. In: Advances in Neural Information Processing Systems, pp. 281–288. MIT Press (2007)
22. Chuzhanova, N.A., Jones, A.J., Margetts, S.: Feature selection for genetic sequence classification. Bioinformatics **14**, 139–143 (1998)
23. Davis, J., Goadrich, M.: The relationship between precision-recall and ROC curves. In: Proceedings of the 23rd International Conference on Machine Learning, pp. 233–240. ACM (2006)
24. Davis, R.A., Chariton, A.J., Oehlschlager, S., Wilson, J.C.: Novel feature selection method for genetic programming using metabolomic ^{1}h NMR data. Chemom. Intell. Lab. Syst. **81**(1), 50–59 (2005)

25. De Jong, K.A.: Evolutionary Computation: A Unified Approach. MIT Press, Cambridge, MA (2006)
26. Domingos, P., Hulten, G., Edu, P.C.W., Edu, C.H.G.W.: A general method for scaling up machine learning algorithms and its application to clustering. In: Proceedings of the Eighteenth ICML, pp. 106–113 (2001)
27. Driscoll, J.A., Worzel, B., MacLean, D.: Classification of gene expression data with genetic programming. In: chap Genetic Programming: Theory and Practice. Kluwer (2003)
28. Drost, I., Dunning, T., Eastman, J., Gospodnetic, O., Ingersoll, G., Mannix, J., Owen, S., Wettin, K.: Apache Mahout. Apache Software Foundation (2010). http://mloss.org/software/view/144/
29. Fan, R.E., Chang, K.W., Hsieh, C.J., Wang, X.R., Lin, C.J.: LIBLINEAR: a library for large linear classification. J. Mach. Learn. Res. **9**, 1871–1874 (2008)
30. Fletez-Brant, C., Lee, D., McCallion, A.S., Beer, M.A.: Kmer-SVM: a web server for identifying predictive regulatory sequence features in genomic data sets. Nucl. Acids Res. **41**, W544–W556 (2013)
31. Fung, G., Mangasarian, O.L.: A feature selection Newton method for support vector machine classification. Tech. Rep. 02-03, Data Mining Institute, Computer Sciences Department, University of Wisconsin, Madison, Wisconsin (2002)
32. Gershenzon, N.I., Stormo, G.D., Ioshikhes, I.P.: Computational technique for improvement of the position-weight matrices for the DNA/protein binding sites. Nucl. Acids Res. **33**(7), 2290–2301 (2005)
33. Grau, J., Keilwagen, J., Gohr, A., Haldemann, B., Posch, S., Grosse, I.: A java framework for statistical analysis and classification of biological sequences. J. Mach. Learn. Res. **13**, 1967–1971 (2012)
34. Hall, M., Frank, E., Holmes, G., Pfahringer, B., Reutemann, P., Witten, I.H.: The WEKA data mining software: an update. SIGKDD Explor. Newsl. **11**(1), 10–18 (2009). https://doi.org/10.1145/1656274.1656278
35. Hong, J.H., Cho, S.B.: Lymphoma cancer classification using genetic programming. In: Seventh European Conference (EuroGP), pp. 78–88 (2004)
36. Huang, J., Cai, Y., Xu, X.: A hubrid genetic algorithm for feature selection wrapper based on mutual information. J. Pattern Recognit. Lett. **28**, 1825–1844 (2007)
37. Islamaj-Dogan, R., Getoor, L., Wilbur, W.J.: A feature generation algorithm for sequences with application to splice-site prediction. In: Lecture Notes in Computer Science: Knowledge Discovery in Databases, vol. 4213, pp. 553–560. Springer (2006)
38. Islamaj-Dogan, R., Getoor, L., Wilbur, W.J.: A feature generation algorithm with applications to biological sequence classification. In: Liu, H., Motoda, H. (eds.) Computational Methods of Feature Selection. Springer, Berlin, Heidelberg (2007)
39. Islamaj-Dogan, R., Getoor, L., Wilbur, W.J., Mount, S.M.: Features generated for computational splice-site prediction correspond to functional elements. BMC Bioinform. **8**, 410–416 (2007)
40. Jiang, B., Zhang, M.Q., Zhang, X.: OSCAR: one-class SVM for accurate recognition of cis-elements. Bioinformatics **23**(21), 2823–2838 (2007)
41. Jo, T., Hou, J., Eickholt, J., Cheng, J.: Improving protein fold recognition by deep learning networks. Sci. Rep. **5**(17), 573 (2015)
42. Joachims, T.: Making large-scale support vector machine learning practical. In: Schölkopf, B., Burges, C.J.C., Smola, A.J. (eds.) Advances in Kernel Methods, pp. 169–184. MIT Press (1999). http://dl.acm.org/citation.cfm?id=299094.299104
43. Kamath, U., De Jong, K.A., Shehu, A.: Selecting predictive features for recognition of hypersensitive sites of regulatory genomic sequences with an evolutionary algorithm. In: GECCO, ACM, pp. 179–186 (2010). https://doi.org/10.1145/1830483.1830516
44. Kamath, U., Shehu, A., De Jong, K.A.: Feature and kernel evolution for recognition of hypersensitive sites in dna sequences. In: Suzuki, J., Nakano, T. (eds.) BIONETICS: International Conference on Bio-inspired Models of Network, Information, and Computing Systems, pp. 213–238. Springer, Boston, MA (2010)

45. Kamath, U., Shehu, A., De Jong, K.A.: Using evolutionary computation to improve SVM classification. In: WCCI: IEEE World Congress on Computational Intelligence, Barcelona, Spain (2010)
46. Kamath, U., Shehu, A., De Jong, K.A.: A two-stage evolutionary approach for effective classification of hypersensitive dna sequences. J. Bioinform. Comput. Biol. **9**(3), 399–413 (2011)
47. Kamath, U., Compton, J., Islamaj-Dogan, R., De Jong, K.A., Shehu, A.: An evolutionary algorithm approach for feature generation from sequence data and its application to DNA splice-site prediction. IEEE/ACM Trans. Comput. Biol. Bioinform. **9**(5), 1387–1398 (2012)
48. Keilwagen, J., Grau, J., Posch, S., Strickert, M., Grosse, I.: Unifying generative and discriminative learning principles. BMC Bioinform. **11**(98), 1–9 (2010)
49. Keilwagen, J., Grau, J., Paponov, I.A., Posch, S., Strickert, M., Grosse, I.: De-novo discovery of differentially abundant transcription factor binding sites including their positional preference. PLoS Comput. Biol. **7**(2), e1001,070 (2011)
50. Kernytsky, A., Rost, B.: Using genetic algorithms to select most predictive protein features. Proteins: Struct. Funct. Bioinf. **75**(1), 75–88 (2009)
51. Kim, W., Wilbur, W.J.: DNA splice site detection: a comparison of specific and general methods. In: AMIA Symposium, pp. 390–394 (2002)
52. King, O.D., Roth, F.P.: A non-parametric model for transcription factor binding sites. Nucl. Acids Res. **31**(19), e116 (2003)
53. Kohavi, R., John, G.H.: Wrappers for feature subset selection. Artif. Intell. **97**(1–2), 273–324 (1997)
54. Koza, J.: On the Programming of Computers by Means of Natural Selection. MIT Press, Boston, MA (1992)
55. Kuncheva, L.I., Jain, L.C.: Nearest neighbor classifier: simultaneous editing and feature selection. Pattern Recognit. Lett. **20**(11–13), 1149–1156 (1999)
56. Langdon, W., Buxton, B.: Genetic programming for mining DNA chip data from cancer patients. Genet. Program. Evol. Mach. **5**(3), 251–257 (2004)
57. Lasserre, J., Bishop, C.M.: Generative or discriminative? getting the best of both worlds. Bayesian Stat. **8**, 3–24 (2007). http://research.microsoft.com/~cmbishop/downloads/Bishop-WCCI-2008.pdf
58. Leardi, R., Boggia, R., Terrile, M.: Genetic algorithms as a strategy for feature selection. J. Chemom. **6**(5), 267–281 (2005)
59. Leslie, C., Noble, W.S., Eskin, E.: The spectrum kernel: a string kernel for SVM protein classification. In: Pacific Symposium on Biocomputing, Baoding, China, vol. 7, pp. 564–575 (2002)
60. Li, R., Zhu, H., Ruan, J., Qian, W., Fang, X., Shi, Z., Li, Y., Li, S., Shan, G., Kristiansen, K., Li, S., Yang, H., Wang, J., Wang, J.: De novo assembly of human genomes with massively parallel short read sequencing. Genome Res. **20**(2) (2010). https://doi.org/10.1101/gr.097261. 109
61. Mitchell, T.M.: Machine Learning, 1st edn. Mc-Graw Hill Companies Inc., Boston, MA (1997)
62. Moore, J.H., Parker, J.S., Olsen, N.J., Aune, T.M.: Symbolic discriminant analysis of microarray data in autoimmune disease. Genet. Epidemiol. **23**(1), 57–69 (2002)
63. Muni, D.P., Pal, N.R., Das, J.: Genetic programming for simultaneous feature selection and classifier design. Annu. Rev. Genomics Hum. Genet. **36**(1), 106–117 (2006)
64. Muselli, M.: On convergence properties of pocket algorithm. IEEE Trans. Neural Netw. **8**(3), 623–629 (1997)
65. Ng, A.Y., Jordan, M.I.: On discriminative vs. generative classifiers: a comparison of logistic regression and naive bayes. Neural Information Processing Systems, pp. 1–8 (2002)
66. Noble, W.S.: Support vector machine applications in computational biology. In: Schölkopf, B., Tsuda, K., Vert, J.P. (eds.) Kernel Methods in Computational Biology. MIT Press, Cambridge, MA (2004)

67. Noble, W.S., Kuehn, S., Thurman, R., Yu, M., Stamatoyannopoulos, J.A.: Predicting the in vivo signature of human gene regulatory sequences. Bioinformatics **21**(Suppl 1), i338–i343 (2005)
68. Oh, I.S., Lee, J.S., Moon, B.R.: Hybrid genetic algorithms for feature selection. IEEE Trans. Pattern Anal. Mach. Learn. **26**(11), 1424–1437 (2004)
69. Opitz, D., Maclin, R.: Popular ensemble methods: an empirical study. J. Artif. Intell. Res. **11**, 169–198 (1999)
70. Pertea, M., Lin, X., Salzberg, S.L.: Genesplicer: a new computational method for splice site prediction. Nucleic Acids Res. **29**(5), 1185–1190 (2001)
71. Ramirez, R., Puiggros, M.: A genetic programming approach to feature selection and classification of instantaneous cognitive states. Lect. Notes Comput. Sci. Appl. Evol. Comput. **4448**, 311–319 (2007)
72. Raymer, M.L., Punch, W.F., Goodman, E.D., Kuhn, L.A., Jain, A.K.: Dimensionality reduction using genetic algorithms. IEE Trans. Evol. Comput. **4**(2), 164–171 (2000)
73. Sarma, J., De Jong, K.: An analysis of the effects of neighborhood size and shape on local selection algorithms. In: Parallel Problem Solving from Nature (PPSN), pp. 236–244. Springer (1996)
74. Schapire, R.E., Freund, Y., Bartlett, P., Lee, W.S.: Boosting the margin: a new explanation for the effectiveness of voting methods. Ann. Stat. **26**(5), 1651–1686 (1998)
75. Schölkopf, B., Tsuda, K., Vert, J.P.: Kernel Methods in Computational Biology. Computational Molecular Biology. MIT Press, Cambridge, MA, USA (2004)
76. Schultheiss, S.J.: Kernel-based identification of regulatory modules. In: Computational Biology of Transcription Factor Binding Sites, Methods Mol Biol, vol. 674, pp. 213–223. Springer (2010)
77. Schweikert, G., Zien, A., Zeller, G., Behr, J., Dieterich, C., Ong, C.S., Philips, P., De Bona, F., Hartmann, L., Bohlen, A., Krüger, N., Sonnenburg, S., Rätsch, G.: mGene: accurate SVM-based gene finding with an application to nematode genomes. Genome Res. **19**(11), 2133–2143 (2009)
78. Shafer, J., Agrawal, R., Mehta, M.: Sprint: a scalable parallel classifier for data mining. In: Proceedings of the 22nd International Conference on Very Large Databases, Morgan Kaufmann, pp. 544–555 (1996)
79. Siedlecki, W., Sklansky, J.: A note on genetic algorithms for large-scale feature selection. Pattern Recognit. Lett. **10**(5), 335–347 (1989)
80. Skolicki, Z.: An analysis of island models in evolutionary computation. In: Proceedings of the 2005 Workshops on Genetic and Evolutionary Computation, ACM, pp. 386–389 (2005)
81. Sonnenburg, S., Zien, A., Rätsch, G.: ARTS: accurate recognition of transcription starts in human. Bioinformatics **22**(14), e472–480 (2006)
82. Sonnenburg, S., Schweikert, G., Philips, P., Behr, J., Rätsch, G.: Accurate splice site prediction using support vector machines. BMC Bioinform. **8**(10), S7 (2007)
83. Sonnenburg, S., Zien, A., Philips, P., Rätsch, G.: POIMs: positional oligomer importance matrices—understanding support vector machine based signal detectors. Bioinformatics **24**(13), i6–i14 (2008)
84. Sonnenburg, S., Rätsch, G., Henschel, S., Widmer, C., Behr, J., Zien, A., de Bona, F.: The SHOGUN machine learning toolbox. J. Mach. Learn. Res. **11**, 1799–1802 (2010). http://www.shogun-toolbox.org
85. Spears, W.M.: Crossover or mutation. In: FOGA, pp. 221–237 (1992)
86. Staden, R.: Methods to locate signals in nucleic acid sequences. Nucleic Acids Res. **12**(1), 505–519 (1984)
87. Svore, K., Burges, C.: Large-Scale Learning to Rank using Boosted Decision Trees. Cambridge University Press (2011)
88. Taher, L., Meinicke, P., Morgensten, B.: On splice site prediction using weight array models: a comparison of smoothing techniques. J. Phys. Conf. Ser. **90**(1), 012,004 (2007)
89. Tech, M., Pfeifer, N., Morgenstein, B., Meinicke, P.: TICO: a tool for improving predictions of prokaryotic translation initiation sites. Bioinformatics **21**(17), 3568–3569 (2005)

90. Tomassini, M.: Spatially structured evolutionary algorithms: artificial evolution in space and time. Natural Computing series. Springer (2005). http://books.google.com/books?id=Tkj2nWddNdUC
91. Tsang, I.W., Kwok, J.T., Cheung, P.M.: Core vector machines: fast SVM training on very large data sets. J. Mach. Learn. Res. **6**, 363–392 (2005). http://dl.acm.org/citation.cfm?id=1046920.1058114
92. Tsang, I.W., Kocsor, A., Kwok J.T.: Simpler core vector machines with enclosing balls. In: Proceedings of ICML, ACM, pp. 911–918 (2007). https://doi.org/10.1145/1273496.1273611. URL http://doi.acm.org/10.1145/1273496.1273611
93. Vapnik, V.N.: The Nature of Statistical Learning Theory. Springer, New York, NY (1995)
94. Venkatraman, V., Dalby, A.R., Yang, Z.R.: Evaluation of mutual information and genetic programming for feature selection in QSAR. J. Chem. Inf. Comput. Sci. **44**(5), 1686–1692 (2004)
95. Woodsend, K., Gondzio, J.: Hybrid MPI/OpenMP parallel linear support vector machine training. J. Mach. Learn. Res. **10**, 1937–1953 (2009)
96. Xing, E.P., Jordan, M.I., Karp, R.M., Russell, S.: A hierarchical Bayesian Markovian model for motifs in biopolymer sequences. In: Becker, S., Thrun, S., Obermayer, K. (eds.) Advances in Neural Information Processing Systems, pp. 200–207 (2002)
97. Yakhnenko, O., Silvescu A., Honavar, V.: Discriminatively trained Markov model for sequence classification. In: IEEE International Conference on Data Mining (ICDM), pp. 1–8 (2005)
98. Yeo, G.: Maximum entropy modeling of short sequence motifs with applications to RNA splicing signals. J. Comput. Biol. **11**(2), 377–394 (2004)
99. Yu, J., Yu, J., Almal, A.A., Dhanasekaran, S.M., Ghosh, D., Worzel, W.P., Chinnaaiyan, A.M.: Feature selection and molecular classification of cancer using genetic programming. Neoplasia **9**(4), 292–303 (2007)
100. Yu, L., Liu, H.: Feature selection for high-dimensional data: a fast correlation-based filter solution. In: International Conference on Machine Learning, pp. 856–863. AAAI Press (2003)
101. Zhang, X., Zhao, J., LeCun, Y.: Character-level convolutional networks for text classification. In: International Conference on Neural Information Processing Systems (NIPS), pp. 649–657. MIT Press (2015)
102. Zhou, X., Ruan, J., Wang, G., Zhang, W.: Characterization and identification of microrna core promoters in four model species. PLoS Comput. Biol. **3**(3), e37 (2007)

Chapter 4
Towards Addressing the Limitations of Educational Policy Based on International Large-Scale Assessment Data with Castoriadean Magmas

Evangelos Kapros

Abstract This chapter discusses the limitations of educational policy that is based on data and reports from International Large-scale Assessments (ILSAs) and ways to overcome them. Firstly, an analysis is performed on data from several ILSAs from 1963 to 2015, and correlated with cultural dimensions statistical data for the same countries. Expanding the analysis to several ILSAs for a large timeframe shows differences in success predictors from literature review. In particular, long-term orientation is a better predictor than individualism overall. Secondly, cultural factors are rarely factored as a policy aspect due to the disagreement of a definition of it. This chapter proposes the mathematical construct of *magmas* as defined by Castoriadis to be a suitable working tool for such definitions. Finally, the chapter proceeds to compose the above into principles for designing technological systems of educational assessment that overcome the aforementioned limitations.

4.1 Introduction

Global educational assessment surveys, such as the popular PISA, rely on aggregating large-scale data from individual students. In parallel, student autonomy is also often understood as individual autonomy. However, no individual can exist without their social context, without their cultural background with which and through which they organise the world [5].

This chapter discusses (a) the relation of achievement in International Large-Scale Assessments (ILSA) and culture, and (b) the implications for designing culturally-appropriate educational technology. We include data from more assessments than the influential PISA [21, 23], which we correlate with cultural data measured for each country by [13].

E. Kapros (✉)
Science Gallery, Trinity College Dublin, Dublin, Ireland
e-mail: ekapros@tcd.ie

© Springer Nature Switzerland AG 2019   61
A. Esposito et al. (eds.), *Innovations in Big Data Mining and Embedded Knowledge*,
Intelligent Systems Reference Library 159,
https://doi.org/10.1007/978-3-030-15939-9_4

In addition, in order to discuss the meaning of these findings regarding culture, the mathematical construct of *magmas* is introduced from [5], as well as its relation to 'soft' concepts such as culture, and is applied in the context of this chapter.

Then, in light of our previous work on Educational Assessment Technology (EdAT) [17, 18], some design principles are presented for designing culturally-appropriate EdAT. While these principles are practical, they are not specific guidelines, in order to allow for the necessary flexibility required for cultural differentiation.

4.1.1 Previous Work on ILSA and Culture

While there is much previous work on cultural aspects on ILSAs, this section will focus on the pieces of work mostly relevant in this study. For example, there are many studies on individual countries (for example see [9] or [11]), but here the interest is in a global comparison.

One particularly related work to this one is the work of [12]. In that paper, PISA scores have been correlated with the cultural dimensions of [13] using simple regression models. *Individualism* and *long-term orientation* are shown as equally good predictors for achievement in PISA. Moreover, this work shows agreement with previous, older results, thus proving that the approach of simple regression models is sufficient to indicate meaningful achievement predictors.

Other work has investigated the relation of PISA performance and cultural dimensions with education expenditure [7]. This is interesting, especially since while [13] has noted an effect on cultural dimensions by investment, [7] has shown that some cultural dimensions are good predictors for education expenditure; moreover, a relation between expenditure and performance is shown.

Finally, some previous work has compared clusters of countries (Confucian vs. Anglo, and Confucian vs. Europe) using either self-reported cultural perceptions or assessments [14, 20]. Self-reporting makes these studies difficult to compare to other studies, since either the achievement data or the cultural data are not standardised. Using data such as ILSA data and the Cultural Dimensions data of [13] (see more below) solves this issue.

While the previous work mentioned above has strong merits and has greatly impacted this analysis, it also has a shortcoming: it has focused almost only on PISA. While many have tried to improve these results by improving the statistical methods performed on PISA data [29], it is possible to enrich the data in other ways: namely, by including data from more ILSAs (such as TIMSS and others) in addition to PISA data.

How all ILSAs, PISA being one of them, can inform us about the relation between aspects of culture and assessment at a global scale is described below.

4.2   Data Analysis

This data analysis uses two datasets: the ILSA dataset from [10] and the Cultural Dimensions dataset from [13].

The Cultural Dimension dataset (6CD) includes a score for each country for 6 cultural dimensions measured in [13].[1] These are:

1. *Power Distance*, related to the different solutions to the basic problem of human inequality;
2. *Uncertainty Avoidance*, related to the level of stress in a society in the face of an unknown future;
3. *Individualism versus Collectivism*, related to the integration of individuals into primary groups;
4. *Masculinity versus Femininity*, related to the division of emotional roles between women and men;
5. *Long Term versus Short Term Orientation*, related to the choice of focus for people's efforts: the future or the present and past;
6. *Indulgence versus Restraint*, related to the gratification versus control of basic human desires related to enjoying life.

In this context, we have treated countries as follows. To match the ILSA set, all scores have been averages for countries: regions have been removed from the dataset, while scores from different communities in a country have been averaged to a single value (Table 4.1).

The ILSA dataset, as described by its authors, "includes system-level data for 37 International Large-Scale Assessments (ILSAs) for 328 sub-national and national school systems, from 1963 to 2015" [10]. The included countries in this paper and the scores can be seen at Table 4.2.

In order to accommodate for the different scoring styles the following were done. Firstly, it was observed in the ILSA dataset that two scoring styles were reported: raw percentage (PC), and item response theory (IRT) scoring. The raw percentage is unweighted: if a student has replied correctly in $X\%$ of the questions, their score is X. In contrast, with IRT a student who has replied correctly in $X\%$ of difficult questions, which carry a high weight k, will score $X + k, k > 0$, while another student who will reply in $X\%$ of easy questions which carry a low weight m will achieve a lower score $X + m, m > 0, m < k$. Since $m < k$, it follows that $X + k > X + m$. The organisations that coordinate the specific ILSAs set the values for k and m, and the students taking the tests do not necessarily have access to this weighting.

Secondly, a bivariate plot was drawn for each country's IRT and PC ILSA scores, and the linear relation between them was examined (see Fig. 4.1). The model used was a simple linear regression.

---

[1]Hereinafter these specific dimensions will be capitalised as Cultural Dimensions; in contrast, when discussing dimensions of culture in general, which may or may not be part of this framework, the lowercase cultural dimensions will be used.

Table 4.1 Countries included in the 6CD dataset that have never undertaken an ILSA

Countries

Andorra, Bangladesh, Belarus, Bosnia, Ecuador, Ethiopia, Guatemala, Iraq, Jamaica, Mali,
Moldova, Pakistan, Panama, Puerto Rico, Rwanda, South Africa, El Salvador, Serbia, Suriname,
Tanzania, Vietnam, Zambia, Zimbabwe

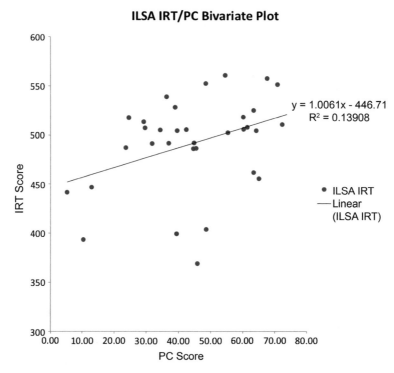

Fig. 4.1 Bivariate plot of each country's IRT and PC ILSA scores. A moderate linear relation can
be observed

Calculating the coefficient of determination yields a relatively low value $R^2 = 0.13908$, which means that the model explains approximately 14% of the cases; in other words, the IRT and PC scores for the same country may have a strong correlation for most countries, but are perfectly explained by the model for almost 14% of the countries.

As a result of the above, the ILSA set was broken down into two sets, depending on the scoring method: the ILSA IRT score, and the ILSA PC score datasets. Each of these is a subset of the initial dataset and includes the countries that participate in the specific type of ILSA and the score they received for each and every of these ILSAs.

Table 4.2 Countries that participated in ILSAs and their average scores

Country	IRT	PC	Country (cont/d.)	IRT	PC
Korea South	561	54.52	U.S.A.	442	
Hong Kong	558	67.57	Iran	442	5.30
Japan	552	48.49	Georgia	426	
Taiwan	551	70.78	Turkey	425	
Nigeria	540		Costa Rica	423	
Finland	539	36.33	Azerbaijan	421	
Canada	525	63.38	Colombia	421	
Estonia	521		Slovenia	420	
Ireland	518	60.08	Mexico	416	
Germany	518	24.57	Argentina	405	
Venezuela	516		Jordan	404	48.50
Denmark	514		Trinidad and Tobago	400	
Australia	513	29.31	Brazil	399	39.40
Sweden	510	72.25	Albania	398	
Singapore	509		Peru	394	10.50
Netherlands	508	61.35	Indonesia	394	
Belgium	507	29.73	Egypt	382	
Czech Rep	507		Macedonia Rep	369	45.82
Slovak Rep	506	60.23	Montenegro	368	
Romania	506		Algeria	362	
Hungary	505	42.36	Dominican Rep	339	
Great Britain	505	34.36	Ghana	319	
Luxembourg	505		Armenia	443	
Norway	504	64.20	Chile	447	13.07
Saudi Arabia	504	39.57	Thailand	455	
Kyrgyz Rep	504		Cyprus	455	
Poland	502	55.30	Burkina Faso		43.88
Austria	501		China		73.70
Lithuania	498		Portugal	497	
Croatia	492		India		5.73
Russia	492	44.85	Philippines	497	
Spain	491	31.94	Ukraine	491	
Iceland	489		Morocco		28.30
Italy	487	23.73	New Zealand		62.50
Latvia	487		Uganda	490	
France	486	45.39	Israel	486	44.60
Uruguay	471		Greece	466	
Switzerland	465		Malaysia	462	
Malta	462	63.30	Bulgaria	461	

The ILSA PC set includes 38 countries, the ILSA IRT includes 76 countries, and the overall ILSA set includes 81 countries. It can be observed that $38 + 76 > 81$, and specifically that $38 + 76 = 81 + 33$, which means that 33 countries are included in both ILSA IRT and ILSA PC datasets, as these countries participated in ILSAs each of which had a different scoring mechanism.

The ILSA PC dataset includes 11 ILSAs from 1963 to 2010, among them FISS and PASEC. The IRT set includes 26 ILSAs from 1988 to 2015, among them TIMSS and PISA.

For each country the average score was calculated. The set was not controlled for missing data at this stage with imputation, but rather with *available case analysis* (namely by pairwise deletion); while there exist methods for imputation [29], previous work has shown that simple regression has been enough to show patterns and trends at a global scale [12]. However, if one wanted to look into a specific country and policies, then it might be ideal to control for missing data by imputation. What is implied here that explains this distinction, is that many countries have values that are difficult to explain: while a model might predict correctly which values are missing, matching the imputed values to specific countries is non-trivial [7]. Studying a specific country removes this statistical pain-point, which remains in the case of a global analysis like this one, thus making available case analysis a viable option for controlling for missing data.

Then, linear regression models were calculated and plotted for ILSA PC and ILSA IRT ranked country scores.[2]

Both linear regressions show a very good fit of the data to the model. The coefficient of determination is very high in both sets: $R^2 = 0.91903$ for IRT, which means that the model explains approximately 92% of the IRT ILSA cases, and $R^2 = 0.96423$ for PC, which means that the model explains approximately 96% of the PC ILSA cases.

Finally, correlation values were calculated and aggregated in correlation matrices for the ILSA scores and the Cultural Dimensions. As above, the correlations were calculated for the available cases. A visualisation of the correlation matrix is presented at Fig. 4.2. Since ILSA IRT and PC rows have different degrees of freedom ($df = N - 2$: $df_{IRT} = 74$, $df_{PC} = 36$), this needed to be accounted for in terms of statistical significance.

In this correlation matrix we can examine the following. Firstly, we can examine if there are any statistically significant correlations between ILSA scores and Cultural Dimensions, for the purpose of investigating the relation between culture and assessment. Secondly, we can observe if IRT and PC scores produce different correlations, so as to control the results for the assessment method.

---

[2]The plots are not printed here due to limitations in space and are available upon request. The same applies to descriptive statistics tables.

country	ILSA IRT	Power Distance	Individualism	Masculinity	Uncertainty Avoidance	Long-term Orientation	Indulgence vs Restraint	ILSA PC
ILSA IRT		-0.18886	0.36092	0.07380	-0.19291	0.42397	0.00654	0.37293
Power Distance	-0.18886		-0.59841	0.11467	0.22864	0.00003	-0.28422	-0.12499
Individualism	0.36092	-0.59841		0.08299	-0.16516	0.12343	0.13690	0.05821
Masculinity	0.07380	0.11467	0.08299		-0.06087	0.03145	0.06704	-0.04931
Uncertainty Avoidance	-0.19291	0.22864	-0.16516	-0.06087		-0.01215	-0.07408	-0.30698
Long-term Orientation	0.42397	0.00003	0.12343	0.03145	-0.01215		-0.45545	0.32124
Indulgence vs Restraint	0.00654	-0.28422	0.13690	0.06704	-0.07408	-0.45545		0.11200
ILSA PC	0.37293	-0.12499	0.05821	-0.04931	-0.30698	0.32124	0.11200	

Fig. 4.2 Correlation matrix for Hofstede's cultural traits and ILSA scores. Positive correlation is coloured in blue, negative in red. Statistically significant values have a red stroke around their coloured box. The intensity of blue and red also represents the value, where darker means values closer to statistical significance

4.3   Findings

The findings of the correlation described above produced several **statistically significant** results.

The first observation is that there is a positive statistically significant correlation (Pearson Coefficient $r = 0.37293$) between IRT and PC results. On the one hand both scoring systems showed that they follow a linear pattern, but since their bivariate plot did not show a very strong relation, this correlation was not a given. This is an important finding, given that previous research has focused on IRT, and specifically on PISA.

With regard to correlations between 6CD and ILSA data, the following Cultural Dimensions are correlated with high performance in assessment:

- *Long Term versus Short Term Orientation*: Has the highest significant correlation with either ILSA, and is positively correlated with both ($r_{IRT} = 0.42397, r_{PC} = 0.32124$);
- *Individualism versus Collectivism*: Statistically significantly positively correlated to IRT, but not to PC at all ($r_{IRT} = 0.36092, r_{PC} = 0.05821$);
- *Uncertainty Avoidance*: Moderate negative correlation with PC but not with IRT ($r_{IRT} = -0.19291, r_{PC} = -0.30698$).

The following Cultural Dimensions are *not* correlated with high performance in assessment:

- *Masculinity versus Femininity*: The trait with values closest to zero ($r_{IRT} = 0.07380, r_{PC} = -0.04931$);
- *Power Distance*: ($r_{IRT} = -0.18886, r_{PC} = -0.12499$);
- *Indulgence versus Restraint*: ($r_{IRT} = 0.00654, r_{PC} = 0.11200$).

These correlations can be validated, to an extent, internally in the model. For example, one can see that Power Distance has a very significant negative correlation with Individualism ($r = -0.59841$). Therefore, it is not surprising that IRT has a positive correlation with Individualism but a moderately negative one with Power Distance. The same can be said for Long-term Orientation and Indulgence versus Restraint. However, this validation is not absolute, since PC has a slightly negative correlation with Power Distance, even though the positive correlation with Individualism is absent.

Having said that, the second observation is exactly this difference in Individualism and, to a smaller extent, Uncertainty Avoidance between IRT and PC ILSAs. It may be the case that the IRT methodology of scoring, being weighted, highlights the differences between individuals, or there may have been a shift in these countries that participated in IRT in reality.

In any case, a major contribution of this section has been the introduction of the ILSA PC data in the correlation with cultural factors. Their inclusion has shown that while the conventional PISA-Individualism relation is there, it is not there for PC. In contrast, **Long Term versus Short Term Orientation is** *a better predictor* **for achievement in assessment**, as it is consistently positively correlated with statistical significance in both IRT and PC ILSAs for the 81 countries studied here from 1963 to 2015. This is a very important finding.

In [13], Long Term versus Short Term orientation is defined as follows:

> Long-term orientation deals with change. In a long-time-oriented culture, the basic notion about the world is that it is in flux, and preparing for the future is always needed. In a short-time-oriented culture, the world is essentially as it was created, so that the past provides a moral compass, and adhering to it is morally good.

That is, countries that have a culture that prepares for the future in anticipation of inevitable change are more likely to perform better in assessment and have a high achievement level in reading, mathematics, and science.

Bivariate plots of long-term orientation and individualism with IRT and PC scores can be found at Fig. 4.3; the coefficient α that represents the slope of each regression

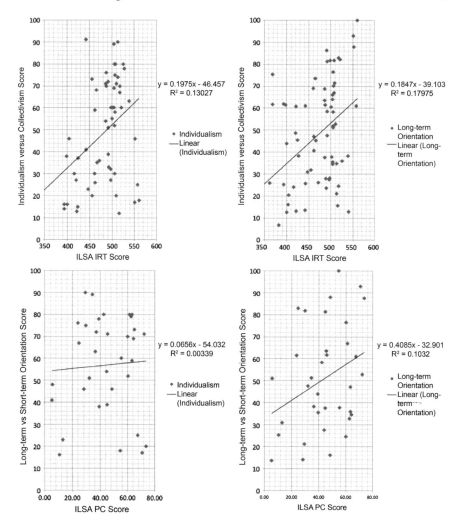

Fig. 4.3 Bivariate plots of long-term orientation and individualism with IRT and PC scores. Right hand side of the image: long-term oriented culture score; Left: Individualism, Top: ILSA IRT achievement; and Bottom: ILSA PC achievement. Long-term orientation is a better overall predictor for achievement than individualism

line, and R^2 values show that long-term orientation is a better overall predictor for achievement, as it is comparable to individualism for IRT assessments and very significantly better for PC assessments, thus better for ILSA in total.

For maps that can be used to compare Individualism, Long-term orientation and achievement in ILSAs, in order to visually glance which cultural dimension is a better achievement predictor, see Fig. 4.4. It can be seen that long-term orientation explains the results of, say, the U.S.A. or Russia better than individualism.

Fig. 4.4 Map of the countries of the study showing **a** long-term oriented score, **b** individualist score, **c** ILSA IRT, and **d** ILSA PC achievement. Darker is higher value. It can be seen that there is a smaller correlation between individualism and achievement than with long-term orientation

4.4 Discussion

The literary author Amos Oz has famously written that "facts have a tendency to obscure the truth" and proceeds to explain by the example of his grandmother's health, which was factually affected by a heart attack, but truthfully by her reacting extremely after migrating to another country with a disturbing, to her, climate [24].

Have governments and policy makers a similar issue with ILSA data? It can be argued that yes, for at least a few reasons. Firstly, there exists an emphasis on PISA alone, simply by the fact that the OECD reports contain policy recommendations; however, it has been reported that these are typically misinterpreted [8] and do not consider cultural factors [8, 28], thus creating a "PISA shock" in several countries [28]. Moreover, the same sources and an internet search will demonstrate how the media treat the results with little nuance, often reporting more on the political reactions (which appear to often have little nuance themselves) rather than the meaning behind the results—e.g., [15]. However, differences in reporting and policy-making do exist amongst countries, ranging from including Teacher Unions in PISA-related reforms in Ireland, to adopting PISA-based governmental reports without referring to researchers of Educational Assessment in some Canadian provinces [8].

The variety of responses in different countries shows that culture is an important factor in both ILSA results, and the responses to them. In addition, the insistence to focus on a single ILSA, PISA, at the expense of others makes for a biased dataset, as demonstrated at the above section by the predictors for success in assessment. How are policy makers to go about such a fuzzy topic as culture without getting lost in a convoluted philosophical argument about its definition and role? Is it possible for ILSA facts to not obscure the 'truth'? This section aims to limit the universe of discourse around culture and result to practical guidelines for designing systems that will be culturally appropriate.

4.4.1 Culture, Education, and Assessment

Culture affects Educational Assessment Technology (EdAT) in two main ways. Firstly, it affects the educational assessment, in the sense that education is affected by culture in the content that is being taught, how it is being assessed, and the processes followed during the educational lifecycle. Secondly, culture affects how technology is being designed and deployed, including technology about assessment. This section explains these two main points.

4.4.1.1 Culture and Education

Concerning the relation between culture and education, the following have been said, excluding specific instructional practices, as the scope of this section is the bigger ideas that are the basis of instruction. Therefore, the discussion around the terms 'culturally appropriate' or 'culturally relevant' or 'culturally responsive' are going to be considered equivalent here, and two broad schools of thought are going to be presented: the descriptive and the exploratory.

The descriptive school consists of those who think that the role of education is to explain to new members of a society the state of things [2]. This school does not deny that there can be an additional stage where exploration is permitted, but a learner is to be taught the rules of a certain culture first even when they are expected to break them later [2]. In addition, the relation of education to democracy has been highlighted: "Education makes the difference between mob rule and legitimate majority rule" [27]. Especially in thinkers where democracy is conceived as a way of life and a regime, instead of a decision-making process, education is thought of as of paramount importance to the collective and not only to individual learners [5].

Exploratory theories believe that humans of any age are autonomous and capable of exploring the world in order to learn. Some have seen how negative cultural trends can sometimes exploit exploratory theories, as in an application of exploratory theories to the extreme relativism can—and has—rendered any attempts to explain the state of things as an attempt to impose power to the learners [3, 25]. Thus, despite their theoretical intentions, in practice exploratory theories can think of all instruction and assessment as 'reactionary' and part of a power struggle between instructors and learners. As a result it is difficult to evaluate the impact of said theories concerning assessment.

It can be said that despite the popularity of exploratory theories among individual teachers most official school systems are based on a descriptive model.

4.4.1.2   Culture, Assessment, Technology

With regard to culture and assessment, the following points are of particular interest in this thesis. Kim [19], has linked cultural aspects with assessment and performance. In particular, she has pointed out that some cultures *cultivate attitudes*, which in turn improves *thinking skills*: she has studied a set of East Asian countries that have often been called collectively *Confucian countries*, namely China (PRC), Taiwan (RoC), South Korea, Japan, Hong Kong, and Singapore (see Fig. 4.5) and their performance; also, Jewish education and the performance of Jewish people, who are an extensively international people and not concentrated in specific countries[3] (see Fig. 4.6 for a map showing the distribution of Jewish population).

Specifically, she has found both these cultures to be positively related to performance, as indicated for example by the high amount of Jewish winners of the Nobel Prize; about $\approx 23\%$, while being less than 0.2% of the world population [19]. Similarly, Confucian countries perform consistently better than other countries in International Large-Scale Assessments (ILSAs), such as PISA.

---

[3] Israel is the single Jewish-majority country, but it represents only $\approx 25\%$ of the Jewish population; Jewish people are a minority everywhere else, hence the difficulty to measure the impact of Jewish Education.

Fig. 4.5 Countries described in bibliography as Confucian Countries are coloured: China (PRC), Taiwan (RoC), South Korea, Japan. Hong Kong and Singapore are circled with a dark red stroke

While one cannot be so hasty as to attribute technological progress in these societies at exactly the same cultural dimensions of 6CD, there is agreement that values do play their role, such as Confucianism for the technological advancement of East Asia [6] or for Israel and the Jewish Diaspora being the technological 'startup nation' [26].

4.4.2 Castoriadean Magmas

Culture was defined by Castoriadis 'all that, in the public domain of a society, goes beyond that which is simply functional and instrumental in the operation of that society' [5].

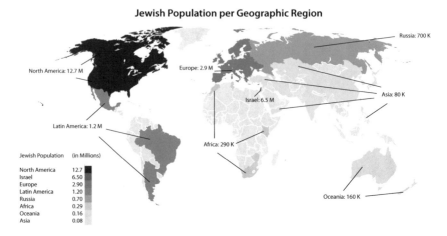

Fig. 4.6 Countries/regions and their Jewish population. Israel is the single Jewish-majority country, but it represents only ≈25% of the Jewish population; Jewish people are a minority everywhere else, hence the difficulty to measure the impact of Jewish Education

In order to understand better what can be done with this definition, it is necessary how an entity such as culture can be classified. Also in [5] exist and will be examined here definitions for useful classifications for entities such as culture. Two types of entities are described, *sets* and *magmas*, and one operator, *marking*.

Set: Or ensemble, is 'a gathering together into a whole of definite, distinct objects of our perception [Anschauung] or of our thought, which are called elements of the set' [4].

Magma: That from which one can extract (or in which one can construct) an indefinite number of ensemblist organisations but which can never be reconstituted (ideally) by a (finite or infinite) ensemblist composition of these organisations.

Marking: [in French original: *repérer*] A 'term/relation, whose valence is both unary and binary', for example: "to mark X'; 'X marks Y'; 'to mark X in Y' (to mark a dog; the collar marks the dog; to mark or locate the dog in the field)' [5].

A magma has the following properties:

M1: If M is a magma, one can mark, in M, an indefinite number of ensembles.
M2: If M is a magma, one can mark, in M, magmas other than M.
M3: If M is a magma, M cannot be partitioned into magmas.[4]
M4: If M is a magma, every decomposition of M into ensembles leaves a magma as residue.
M5: What is not a magma is an ensemble or is nothing.

---

[4]The ensemblistic 'part' is 'everywhere dense' in natural language.

A note on notation: hereinafter the symbol ⊛ will be introduced for the 'marking' operator[5]; Castoriadis himself does not use the above symbol for the operator but uses the word *mark* spelt our instead.

Since it can be said that a culture C is a magma of societal values (derived from property $M5$, as culture is not nothing, and is not a definite set of objects), it can be shown that subtracting measurable aspects S of a society such as 6CD [13] or other similar aspects, still leave a magma M as residue:

$$
\begin{aligned}
&(1)\ C \circledast M && M2 \\
&(2)\ C \circledast S + M && M1, M4 \\
&(3)\ C \circledast \sum_{n=1}^{\infty} dim + M && M4 \\
&(4)\ C \circledast \sum_{n=1}^{6} dim + \sum_{n=7}^{\infty} dim + M\ \square &&
\end{aligned}
\tag{4.1}
$$

where $S = \sum_{n=1}^{\infty} dim$ are measurable cultural dimensions.

Hofstede has implied that there are indefinite measurable cultural dimensions [13], thus the ones his team has measured are noted as $n = 1 \to 6$ while the other possible dimensions as $n = 7 \to \infty$. It should be noted that this doesn't imply the existence of infinitely many cultural dimensions at the moment, but rather the potential to measure an indefinite number of dimensions.[6]

In Eq. 4.1 above, it is important to note that while there is magma as residue and unmeasured dimensions $n = 7 \to \infty$, it is still possible to shed light to some dimensions of culture, dimensions $n = 1 \to 6$. These dimensions may be indicative and help us elucidate the nature of the residual magma and the other cultural values.

4.4.3 Principles for Designing Culturally-Appropriate EdAT

In our previous research, various Educational Assessment Technologies (EdAT) were described. One was a self-assessment app for K-12 student skills [17], and one was a peer-feedback app for workplace competencies [18]. These apps were trialled in diverse environments and evaluated.

In both cases it was shown that EdAT alone is not sufficient if not actively supported by the institutions that use them. This was the case in both schools and workplaces, and support needs to come from teachers and team leads/managers, respectively. This support is not only about explaining how assessment is done best,

---

[5]The symbol ⊛ has been used before for convolution in functional analysis; however, since ∗ is used more commonly for it, there should be no conflict in its use.

[6]To give an example of the difference between infinitely many and indefinite number of, let us consider the surface of a sphere: it is not infinite, as it is limited by the physical dimensions of the sphere, but one can be indefinitely traversing its surface for ever.

but also about nurturing a culture of assessment where it is understood that assessment can be a vehicle for educational improvement and development and is not merely a tool for measurement.

Nurturing a culture of assessment is not an easy task, but is feasible. At least, some aspects of culture such as Hofstede's 6 cultural dimensions can be correlated with achievement in international large-scale assessment, such as PISA and others, by correlating publicly available datasets. These correlations can indicate which cultural dimensions are good predictors for achievement in these assessments.

System designers, educators, and policy makers, can take advantage of the outcomes of these correlations as follows. It is tempting to think that, if Long-term Orientation or Individualism are good predictors for high achievement, then all countries should look out to become long-term oriented or individualist; however, this is not always achievable or even desirable. Rather, countries can be inspired by these cultural dimensions and interpret them and apply them in their own way. Therefore, instead of presenting a definitive list of prescriptive guidelines that would attempt to "solve" culture, the following list offers principles that are specific enough to serve as a blueprint for designing EdAT, but are also open to interpretation depending on each specific culture:

4.4.3.1 Design for the Educative Relationships Triangle

The Educative Relationships Triangle informs us that the Learner-Instructor-Knowledge relationships are of paramount importance in their entirety for learning to flourish. Learner-centric design should facilitate all three points of the triangle, rather than be learner-only design; the latter would be at the expense of the Learner-Instructor, Learner-Knowledge, and Instructor-Knowledge[7] relationships.

Specifically, a design that would focus on the Learner-Teacher-Content/Knowledge relationships, would facilitate long-term orientated thinking, where the Teacher does not only help Learners consume Content, but to own it as part of their culture.

Thus, a learner-centric design for EdAT should have a scope that includes the teacher and content, so as to facilitate meaningful assessment (see Fig. 4.7). In this case the EdAT stakeholders would know that the assessment is done for content that is appropriate, by teachers who have organised the delivery of it, and by students that are prepared for it.

In terms of the algebra of magmas, it can be said that the aim would be to design EdAT where $\{Learner \circledast Instructor\}$ && $\{Instructor \circledast Knowledge\}$ && $\{Learner \circledast Knowledge\}$.

---

[7]Knowledge and Content are used interchangeably in most depictions of the Triangle, and also here. The aim is to point that, in this context, Content should not be understood as a static piece of text in a book as in previous times, but could be an interactive activity on a tablet, a project in a laboratory, or an assessment. It is in this way that the more generic term Knowledge can be used instead.

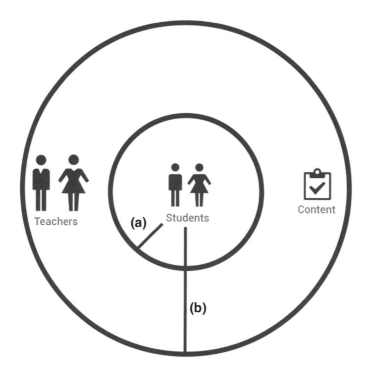

Fig. 4.7 **a** Inner circle: a student-centred system design whose scope excludes Teachers and content. **b** Outer circle: a student-centred design including the Educative Relationships entities

4.4.3.2   Focus on Assessment, Not Technology: Ed*A*T, not EdA*T*

Education is, ultimately, a human enterprise: technology should facilitate it, but it should not be an end in itself. Thus, the focus of designing EdAT must always be on education. We must design services to help people facilitate and enable learning. In magmatic terms we would say that *Education* ⊛ *Assessment* always, while *Education* ⊛ *Technology* only sometimes, and not necessarily.

Moreover, technology may sometimes be the "easy" piece of the puzzle in EdAT, since a single System Design can potentially solve multiple problems. This makes it possible, in addition to desirable, to actually focus on assessment instead of technology.

Example of focus points on assessment are: formative versus summative assessment, continuous assessment, gamification of assessment (not its technological implementation).

4.4.3.3　Allow for Institutional Support

Institutional Support is of paramount importance for EdAT. That is, EdAT needs to be treated as part of a system, which is integral to the whole system of an institution.[8] It cannot be an afterthought, and it cannot be applied as a last layer on top of existing activities.

In both [17, 18], it was shown that EdAT does not function well unless there is strong support from the institution or organisation performing the assessment. This institutional support needs to exist in terms of cultivating an appropriate culture of assessment, and not limit itself in support by providing expenditure. That is, institutional support needs to be constant and iterative rather than initiated once and then leaving assessment to its own devices.

In addition, the correlation of ILSA and 6CD data here, shows that the evidence from [17, 18] is not incidental. Indeed, the results agree with previous results in aspects of ILSA that are culture dependent; and, given that expenditure in education is also culture dependent [7], there is a strong need to cultivate appropriate assessment cultures. This is no small task, and institutional support should be a given in discussing such opportunities for culture cultivation. Moreover, it follows that this cultivation cannot remain only in the remit of assessment proper, but should also be extended in all things included in the assessment lifecycle, including EdAT.

This principle is listed last, but not because of lack of importance. Rather, it is listed last since many of the above are a prerequisite for it. For instance, institutional support will have little to no effect if a system design has not taken under consideration the Educative Relationships Triangle.

4.5　Conclusion

4.5.1　ILSA and Culture

This chapter presented firstly a data analysis on ILSA and Cultural Dimensions at a global scale. The analysis took an approach of including previously under-represented historical ILSA data, thus reaching beyond the usual PISA remit. While the methodology was the minimum required to produce useful results [12], the findings are interesting nonetheless.

Firstly, it was shown that countries that do well at PC ILSAs are likely to do well at IRT ILSAs as well. ILSAs that use the PC and the IRT scoring methods show correlations in achievement results at a country level. IRT is weighted scoring, and may thus emphasise individual differences, but at the country level the correlation holds.

---

[8]Here we take institutions to be magmas in their own right, as they cannot be reduced to their parts without loss of meaning.

Secondly, the scoring method does have some impact on the predictors of achievement indeed. This is demonstrated by the fact that Individualism is a better predictor for IRT, but has no effect on PC; in contrast, Uncertainty Tolerance is a better predictor for PC, but has only a moderate effect on IRT.

However, there is a good predictor for both IRT and PC: long term-versus short-term orientation. This cultural dimension shows a statistically significant strong correlation with both IRT and PC assessments. Therefore, one can conclude that countries with a long-term orientation are more likely to perform well at an ILSA, regardless if its scoring style is IRT or PC.

4.5.2   Design for (Cultural) Magmas

The chapter then proceeded to present a discussion on the meaning of these findings with regard to culture. Apart from using extensive datasets that are inclusive of different methodologies, an interpretation of ILSA data requires a rethinking of the theoretical framework applied, and practical guidelines in designing EdAT systems.

Castoriadean magmas were presented as a suitable framework for representing culture, as they excel at separating the quantitative and qualitative properties of such mixed concepts. In combination with the ILSA and culture data analysis, and with previous work on assessment, it was suggested that EdAT System Designers follow a set of guiding principles that facilitate designing for specific magmas, i.e., culture and institutions. In particular, institutional support is a decisive factor. Institutional support should reach beyond funding, as shown by previous research [17, 18] and by long-term orientation being an ILSA success predictor.

The aim is to minimise the reliance of policy makers on media reports, and to shift the responsibility to EdAT System Designers, who can improve their systems based on best practice and guidelines.

4.5.3   Future Work

Advancements in technology make possible even an Artificial Intelligence agent to successfully pass some University entrance examinations [1]. What would happen if culturally appropriate EdAT was applied? Would the same A.I. still pass? If yes, what does it mean for cultural assumptions that exist in the programming of the A.I.? If not, what would a culturally appropriate A.I. look like?

These are only some questions that arise. Of course, a different line of future research can investigate the necessary institutional support by surveying existing practices, identifying gaps, and suggesting improvements. Institutional support can come in many forms, and it may be itself culture dependent. This is a big topic to be explored.

Whichever shape or form the future work takes, it can only be anticipated that it is a topic that will attract increasing interest and exploring it further will be an exciting opportunity.

References

1. Arai, N.H., Matsuzaki, T.: The impact of A.I. on education—can a robot get into the University of Tokyo? In: The Proceedings of The 22nd International Conference on Computers in Education, pp. 1034–1042 (2014)
2. Arendt, H.: The crisis in education. In: Between Past and Future, pp. 181–182 (1961)
3. Bawer, B.: The Victims' Revolution: The Rise of Identity Studies and the Closing of the Liberal Mind. HarperCollins (2012)
4. Cantor, G.: Beiträge zur Begründung der transfiniten Mengenlehre. Mathematische Annalen **46**, 481–512 (1895)
5. Castoriadis, C.: The Castoriadis Reader. Trans. and Ed. David Ames Curtis. Blackwell Publishing (1997)
6. Chan, M.: Economic Growth in Japan Cultural (Neo-Confucianism) Analysis. Honors Theses. Paper 109. Southern Illinois University Carbondale (1991)
7. French, J.J., French, A., Li, W.X.: The relationship among cultural dimensions, education expenditure, and PISA performance. Int. J. Educ. Dev. **42**, 25–34 (2015)
8. Froese-Germain, B.: The OECD, PISA, and the Impacts on Educational Policy, Virtual Research Centre, Canadian Teachers' Federation (2010)
9. Ganimian, A.J.: Why do some school-based management reforms survive while others are reversed? The cases of Honduras and Guatemala. Int. J. Educ. Dev. **47**, 33–46 (2016)
10. Ganimian, A.J., Koretz, D.M.: Dataset of International Large-Scale Assessments. Harvard Graduate School of Education, Cambridge, MA, Last updated, 8 Feb 2017
11. Gruber, N.: (2017) [in Hebrew:] נעם גרובר: גורמים להישגים הנמכים של תלמידי ישראל: מחקרי שורש
12. Hagedorn, L.S., Purnamasari, A.V.: The Role of culture in predicting students' achievement: international perspectives. Int. J. Glob. Educ. **1**(1) (2012)
13. Hofstede, G.: Culture's Consequences: Comparing Values, Behaviors, Institutions and Organizations Across Nations, 2nd edn. Sage Publications, Thousand Oaks CA (2001)
14. Hu, Y.: Cultural Perspectives on Learning Environments. MA thesis on Educational Theory and Policy, Penn State University (2013)
15. Hungary Today, Hungarian Students Lagging Behind International Average, PISA 2015 Study Reveals 7 Mar 2016. http://hungarytoday.hu/news/hungarian-students-lag-bihend-oecd-averge-pisa-2015-study-reveals-39831
16. Inclusive Design Research Centre: What it Inclusive Design. OCAD University (2016)
17. Kapros, E., Kipp, K.: Microinteractions and a Gamification framework as a mechanism for capturing 21st Century Skills. In: Proceedings of HCI International Conference Learning and Collaboration Technologies (2016)
18. Kapros, E., Neelen, M., Walsh, E.: Designing a peer feedback mobile application as a professional development tool. In: Proceedings of HCI International Conference Learning and Collaboration Technologies (2016)
19. Kim, K.H.: The Creativity Challenge: How We Can Recapture American Innovation. Prometheus Books, New York (2016)
20. Morony, S., Kleitman, S., Lee, Y.P., Stankov, L.: Predicting achievement: confidence vs self-efficacy, anxiety, and self-concept in Confucian and European countries. Int. J. Educ. Res. **58**, 79–96 (2013)
21. Murphy, S.: The pull of PISA: uncertainty, influence, and ignorance. Inter-Am. J. Educ. Democr. **3**(1), 27–44 (2010)

22. National Council for Curriculum and Assessment: Key Skills framework. NCCA (2009)
23. Organization for Economic Cooperation and Development (OECD): Pisa 2015 Draft Collaborative Problem Solving Framework. OECD (2013)
24. Oz, A.: A Tale of Love and Darkness. Mariner Books (2005)
25. Pluckrose, H.: How French "Intellectuals" Ruined the West: Postmodernism and Its Impact, Explained. Areo Magazine. Mar 2017. https://areomagazine.com/2017/03/27/how-french-intellectuals-ruined-the-west-postmodernism-and-its-impact-explained/
26. Senor, D., Singer, S.: Start-up Nation: The Story of Israel's Economic Miracle. Twelve, New York, NY and Boston, MA (2009)
27. Straume, I.: Castoriadis, Education and Democracy. Creation, Rationality and Autonomy: Essays on Cornelius Castoriadis. NSU Press, Denmark (2013)
28. The Guardian, OECD and Pisa tests are damaging education worldwide—academics, 6 May 2014. https://www.theguardian.com/education/2014/may/06/oecd-pisa-tests-damaging-education-academics
29. Weirich, S., Haag, N., Hecht, M., Böhme, K., Siegle, T., Lüdtke, O.: Nested multiple imputation in large-scale assessments. Larg. Scale Assess. Educ. **2**(1), 9 (2014)

Chapter 5
What Do Prospective Students Want? An Observational Study of Preferences About Subject of Study in Higher Education

Alessandro Vinciarelli, Walter Riviera, Francesca Dalmasso, Stefan Raue and Chamila Abeyratna

Abstract This article presents an observational study about the choice of Higher Education of 4,885 prospective students attending an Open Day organised by the University of Glasgow. In particular, this work analyses the distribution of the preferences across the 109 subjects (e.g., Mechanical Engineering, English Literature, etc.) that have been presented during the Open Day. The results suggest that factors like the graduate prospects (percentage of students that have a job or continue education within one year after graduation in a given area) or the first expected salary do not play a major role in the distribution of the preferences. When it comes to the difference between female and male prospective students, the results show that the science-humanities divide is less significant than the care-technical one. Finally, the data shows that the level of education in the environment where someone lives is the dimension along which the prospective students are most unequal.

5.1 Introduction

Higher Education is important because it leads to "*a wide range of personal, financial, and other lifelong benefits*" [8]. Thus, it is not surprising to observe that there

A. Vinciarelli (✉) · W. Riviera · F. Dalmasso
University of Glasgow, Glasgow, UK
e-mail: Alessandro.Vinciarelli@glasgow.ac.uk

W. Riviera
e-mail: walteriviera@gmail.com

F. Dalmasso
e-mail: fdalmasso1991@libero.it

S. Raue · C. Abeyratna
Bizvento, Glasgow, UK
e-mail: stefan@bizvento.com

C. Abeyratna
e-mail: cham@bizvento.com

© Springer Nature Switzerland AG 2019
A. Esposito et al. (eds.), *Innovations in Big Data Mining and Embedded Knowledge*,
Intelligent Systems Reference Library 159,
https://doi.org/10.1007/978-3-030-15939-9_5

is a *"trend toward regarding higher education as a private good, which benefits individuals, and which individuals should therefore pay"* [37]. As a consequence, Higher Education is becoming a significant, often underestimated cost [10] and its choice is a delicate moment not only for its impact on life, but also for the major financial burden it represents. In such a context, it has been observed that prospective students and their families, before they make any commitment, try to acquire as much information as possible [11] and, in particular, they attribute major value to the possibility of interacting with faculties and students [28]. This is the reason why Higher Education institutions regularly organise events—typically called *Open Day* or *Open House*—during which it is possible to visit campuses and their facilities while interacting with teachers and students.

The observations above suggest that Open Days provide an opportunity to investigate the factors underlying the choice of a particular subject of study. For this reason, this article focuses on the preferences of the people that have participated in an Open Day of the University of Glasgow. Unlike in similar previous works, the data has not been collected with questionnaires, but with a mobile application (an *app* in the everyday language). This latter was aimed at a different purpose, but has generated the data analysed in this work as a side-product. Such an approach resulted into two main advantages, namely the possibility to involve a large number of individuals (4, 885 in total) and the possibility to consider 109 individual subjects of study rather than a few, broad areas like, e.g., *Technology* and *Humanities* (like it happens in most previous works).

The analysis—based on data analytics methodologies—addresses three main issues, namely distribution of the preferences across different subjects, gender segregation and relationship between social class and access to Higher Education. In the first case, the questions addressed are whether there are subjects that tend to attract more interest than the others and, if yes, what are the possible explanations behind the observed differences. For what concerns gender segregation, the analysis aims at showing whether female and male Open Day participants tend to express preferences for different subjects and, if yes, what are the directions along which the divide is most evident. Finally, the investigation of factors related to social class compares the contribution of multiple dimensions (education level, income, occupation, etc.) to the overall inequality observed across the Open Day participants.

The results confirm several phenomena observed earlier in the literature—thus showing that the data collection approach is reliable—while providing new insights about issues that, to the best of our knowledge, were still subject of debate in the literature. In particular, the analysis suggests that graduate prospects and first expected salary do not play a major role in the choice of a subject and that the main divide between female and male students is the care-technical one. These results are important because understanding the preferences of Open Day participants and their interplay with factors like gender or social class allows Higher Education institutions (and other bodies concerned with education) to outline better strategies, whether the goal is to attract more applicants, to ensure the widest and fairest possible access to Higher Education, or to align the educative offer with the needs of prospective students.

The rest of this paper is organised as follows: Sect. 5.2 describes the data collection approach, Sect. 5.3 analyses the preferences and their distribution across subjects, Sect. 5.4 describes the observed gender segregation effects, Sect. 5.5 reports on factors related to social class and the final Sect. 5.6 draws some conclusions.

5.2   Data Collection

To the best of our knowledge, the works presented so far in the literature collect information through questionnaires and interviews (see, e.g., [10, 22, 28]) that include explicit questions about preferences, personal characteristics (age, gender, status, etc.) and any other information relevant to the problem under exam. This work adopts a different approach and the data is the side-product of a mobile application—an *app* in the everyday language—allowing its users to set an agenda for the Open Day. In other words, while not being aimed at the collection of the data, the app still gathers information of interest through its working.

A user sets the agenda by selecting the events that she or he plans to attend during the Open Day (talks, demonstrations, visits, etc.). Every event is related to one of the 109 subjects that have been presented during the day (e.g., *Physics*, *Computer Science*, *English Literature*, etc.). Thus, the agenda provides information about the subjects that a user is interested to study or, at least, to know more about. Such an approach has the main advantage that it allows one to consider individually a large number of subjects—109 in this case—and not a few broad groups of domains [1, 14, 18]—e.g., *Technology* and *Humanities*—or a comparison between different universities [28, 35] like most previous studies. Furthermore, such an approach is likely to attenuate the effect of biases that typically affect the use of questionnaires [12, 21].

Before being allowed to set the agenda, the users are invited to *register*, i.e., to provide information that includes, in particular, first name and postcode. The first name is important because it allows one to infer whether the user is male or female. Thus, it is possible to investigate the interplay between gender and field of study, one of the most investigated aspects in choice of Higher Education (see, e.g., [1, 7, 13–16, 18, 27]). Similarly, the postcode is important because it provides information about the social class of the users. Thus, it is possible to analyse the relationship between social class and choice of Higher Education, another aspect that has been widely investigated in the literature [2, 3, 9, 15, 23, 24, 32–34]. In this case, the main advantage of using the app is that it should reduce the effect of phenomena such as stereotype threat [20] and priming [30]—the tendencies to act according to stereotypes and expectations about one's gender and class—that have been observed in the case of traditional questionnaires.

The main risk of the data collection approach is that the use of a mobile application can lead to a selection bias, meaning that the Open Day participants that use smartphones or other suitable mobile devices might not be representative of the population under exam. However, the statistics for 2015 (year of the Open Day under exam) of the *International Telecommunication Union* state that "*there are more than 7 billion*

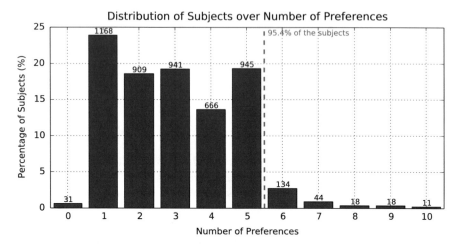

Fig. 5.1 The chart shows the percentage of users that have inserted a given number of events in their agenda (the numbers above the bars are the absolute numbers of users). Since the timetable of the Open Day does not allow one to attend more than 5 events, the distribution shows a drop in correspondence of 6. This seems to confirm that most of the users have tried to set a realistic agenda

mobile cellular subscriptions, corresponding to a penetration rate of 97%".[1] Furthermore, according to the 2015 data of *Ofcom*[2]—the communication regulator in UK—90% of the people of age between 16 and 24 years own a smartphone, 61% of them claim to be "*hooked*" to their phone, and 59% of them say that the smartphone is the device they miss most when they cannot have one at disposition [31]. According to these figures, the use of smartphones does not appear more likely to produce selection biases than more traditional data collection approaches.

The number of users that have registered to use the app in view of the Open Day analysed in this work is 4, 885. The percentage of registered users that have inserted at least one event in their agenda is 99.4% and, on average, every user has selected 3.11 events for a total of 15, 208 items. Figure 5.1 shows that the large majority of the users (4, 629 out of 4, 885) have inserted less than 6 events. The reason is that the timetable of the Open Day does not allow one to attend more than 5 events. Thus, the distribution of Fig. 5.1 seems to confirm that the users have tried to set a realistic agenda and, hence, the data can be considered a reliable source of information.

5.3 Preference Analysis

Section 5.2 shows that the mobile application allows its users to select the events they plan to attend during the Open Day. Since each event is related to one subject, it is possible to say that every time an event has been selected by one of the users, the

---

[1] http://www.itu.int/en/ITU-D/Statistics/Documents/facts/ICTFactsFigures2015.pdf.
[2] https://www.ofcom.org.uk/about-ofcom/what-is-ofcom.

corresponding subject has attracted a *preference*. The analysis of the 15, 208 collected preferences can provide information about the choice of Higher Education, i.e., the process leading Open Day participants to study one subject rather than another.

To the best of our knowledge, most of the previous works tend to focus on the choice between different universities or between broad areas of study (see, e.g., [1, 14, 18, 28, 35]). This article takes into account the 109 individual subjects of study that have been presented during the Open Day under exam. In this respect, the first important question is whether there are subjects that tend to attract more preferences than the others. Figure 5.2 shows the distribution of the preferences over the subjects. The deviation with respect to the uniform distribution is statistically significant ($p \ll 0.01$ according to a χ^2 test), hence, there are subjects that tend to be selected more frequently than the others to a statistically significant extent. In particular, while the 10 most selected subjects account for 30.0% of the preferences, the 10 least selected subjects account for only 0.6% of them, a 50 to 1 ratio.

The literature proposes two main alternative explanations for the distribution of Fig. 5.2. The first is that prospective students and their families "*compare the outcomes of the different options and choose the option with the highest return*" [36], i.e., they tend to select subjects expected to maximise the chances of achieving a career goal of interest, whether this means to be competitive on the job market or to have a certain salary. The second is that the Open Day is a "*preference stage, where [...] academic achievement has the strongest correlation with students' educational aspirations*" [35] and, hence, the focus is on intellectual interests rather than on career or occupational targets.

One possible way to show which of the two explanations above better accounts for the observations is to estimate the correlation between, on one side, the number of preferences that a given subject attracts and, on the other side, the *Graduate Prospects* (GP) and the first expected salary. The GPs correspond to the fraction of students that, one year after graduation in a certain subject, are studying at a higher level or have a job that actually requires a University degree. The GPs are provided by the *University Central Admission System* (UCAS), "*an independent charity providing information, advice, and admissions services to inspire and facilitate educational progression*" (quoted from the UCAS website).[3] The correlation—measured with the Spearman Coefficient—is 0.27 and it is statistically significant ($p < 0.01$ according to single-tailed t-test). The first expected salary for a given subject is available through the *The Complete University Guide*, a popular website that provides information to prospective students.[4] In this case, the correlation is 0.19 and it is not statistically significant.

The findings above suggest that, at the moment of the Open Day, prospective students and their families do not focus on financial and occupational goals, but take the opportunity to explore subjects that, while not necessarily maximising the chances of obtaining a highly remunerative job, might still be attractive from other points of

---

[3]https://www.ucas.com.

[4]http://www.thecompleteuniversityguide.co.uk/careers/what-do-graduates-do-and-earn/what-do-graduates-earn/.

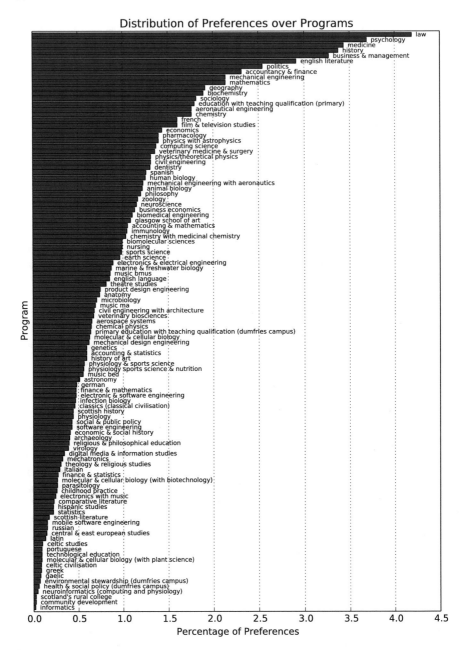

Fig. 5.2 The chart shows the percentage of preferences for every program presented during the Open Day under exam

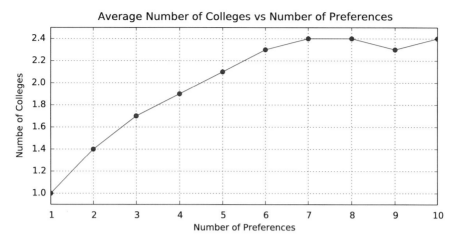

Fig. 5.3 The plot shows the average number of colleges as a function of the number of preferences. Whenever this latter is larger than 1, the app users take into account, on average, subjects that are taught in more than one college

view. Figure 5.3 appears to further confirm such a hypothesis. The plot shows the relationship between the number of events that a given app user has included in the agenda and the number of colleges that those events cover. Such a plot provides an indication of the diversity of the preferences because every subject belongs to one of the four Colleges of the University of Glasgow[5] and these are designed to cover areas as different as possible from one another. When the preferences are more than one, the average number of colleges is always larger than one, meaning that a significant fraction of Open Day participants take into account highly diverse subjects. In other words, prospective students and their families appear to be more worry about exploring multiple opportunities than to commit towards a particular subject or area.

5.4   Gender Segregation and Choice of Higher Education

A large number of observational studies show that female and male students tend to make different choices when it comes to Higher Education and that, in particular, "*while men prefer engineering or natural sciences, women more often opt for humanities or education*" [16]. More recent studies suggest that gender segregation has more facets than previously thought and "*the divide between humanistic and scientific fields [...] accounts for no more than half of the association between gender and college major [...] a second, equally important gender divide [...] can be*

---

[5]*College of Science and Engineering* (SE), *College of Medicine, Veterinary and Life Sciences* (MVLS), *College of Social Sciences* (SS) and *College of Arts* (A).

described as the care-technical divide" [7]. One limitation of the studies presented so far is that they tend to work at the level of macro-areas (e.g., *technology* and *humanities*) that often encompass widely diverse topics. This section tries to go beyond such a limitation and takes into account the 109 individual subjects presented during the Open Day under exam.

The gender of the users is inferred from the first name they provide at the moment of the registration. The first name is mapped into a gender (*male* or *female*) using a publicly available online service[6] that claims an accuracy of 99.9% in those cases where unambiguous name-gender mapping is possible, i.e., when a name or nickname is spelled correctly, it is not abbreviated and it is not used indifferently for both males and females. Unambiguous gender attribution is possible for 84.2% of the users (4, 112 out of 4, 885) and, among these, female and male users are 60.8% and 39.2%, respectively. These figures are in line with the latest data of the UK *Higher Education Statistics Authority*—according to which female students in higher education are 57.4% of the total in UK[7]—and of the Scottish Government—according to which female students are the majority both at entrance (54% of the total in Scotland) and at graduation (60.1% of the total in Scotland) [25]. In other words, while resulting into a loss of data (773 users out of the original 4, 885 cannot be considered), the inference of the gender from the first name appears to produce reliable results.

According to the figures above, the *females-to-males* ratio among the users for which gender inference is possible is 1.55. In absence of gender segregation effects, such a ratio should be the same—within statistical fluctuations—for every group of app users that has attributed a preference to a particular subject. When this is not the case and the ratio diverges to a statistically significant extent from the value above, it is possible to say that the gender distribution is not balanced for that particular subject. This latter will be said to be *female-dominated* if the ratio is larger than 1.55 and *male-dominated* otherwise. Figure 5.4 shows that the gender distribution is unbalanced for 35 subjects out of the total 109. In other words, the gender distribution deviates to a statistically significant extent ($p < 0.01$ according to a χ^2 test with Bonferroni correction) with respect to the population of the app users for roughly one third of the subjects, corresponding to 40.3% of the 15, 208 preferences analysed in Sect. 5.3.

Overall, the data of Fig. 5.4 confirms the "*care-technical divide*" [7] (none of the male-dominated subjects relate to care and, vice versa, none of the female-dominated ones relate to technology). In the case of "*the divide between humanistic and scientific fields*" [7], the evidence is more nuanced. If it is true that none of the male-dominated subjects belongs to the College of Art (where humanities are grouped at the University of Glasgow), it is true as well that several female-dominated subjects are taught in the *College of Medical, Veterinary and Life Sciences* (one of the two colleges where all scientific subjects are grouped at the University of Glasgow). Finally, it is important to consider that there are 74 subjects (59.7% of the total preferences) scattered across all colleges for which the gender distribution is balanced.

---

[6]http://www.gender-api.com.

[7]https://www.hesa.ac.uk/files/student_1516_table_B.xlsx.

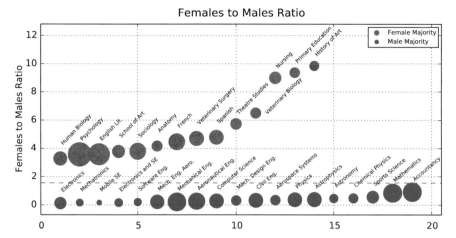

Fig. 5.4 Every bubble corresponds to a subject for which the gender distribution is unbalanced. The vertical axis is the Females to Males ratio and the size of the bubble is proportional to the number of people that have expressed a preference for the subject. The horizontal dashed line corresponds to 1.55, the Females to Males ratio observed in the population of the Open Day participants. For the sake of clarity, the plot does not include Education and Childhood Practice—both female dominated—with ratio 14.1 and 31.0, respectively

The literature proposes two main explanations for gender segregations effects like those of Fig. 5.4, one is the tendency of female students to avoid subjects that require deeper mathematical background [1, 13–15, 18, 27], and the other is the different distribution of Empathy and Systemising skills (see below) across women and men [4]. In, particular several studies have observed the tendency of female students to underestimate their mathematical skills irrespectively of their actual performance in school tests [1, 13]. Such an an effect is typically attributed to widely shared cultural beliefs—lacking any factual basis [16]—about male students performing better than female ones in mathematics [20, 27, 29] or about *"what is understood as gender appropriate behaviour by girls and boys at schools during adolescence"* [14] .

Besides the above, recent studies have shown that Empathy Quotient (EQ) [6]—a psychometric instrument aimed at measuring the ability to establish empathic relationships—and Systemising Quotient (SQ) [5]—a psychometric instrument aimed at measuring the ability to detect patterns and regularities—predict choice of Higher Education better than gender: *"[...] there is an association between degree subject and SQ and EQ scores [...] individuals' scores on EQ and SQ were better predictors [...] than gender"* [17]. In other words, people with higher EQ tend to study care oriented subjects while people with higher SQ tend to study technical subjects, irrespectively of their gender. However, women and men tend to score higher in EQ and SQ, respectively [4]. Thus, gender segregation might be an indirect effect of gender differences in terms of the two quotients.

Higher Education is, in most cases, the first step of one's professional life, meaning that different subject choices channel people to different careers. This means that the gender segregation effects above might be at the origin of gender-related

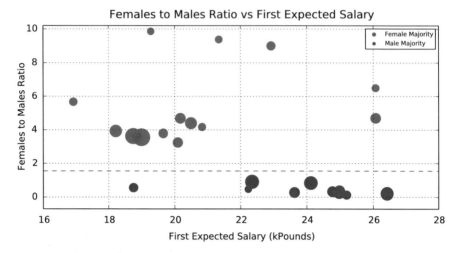

Fig. 5.5 The bubble plot shows the relationship between the Females to Males Ratio and the first expected salary for those subjects where gender imbalance has been observed. The horizontal dashed line corresponds to 1.55, the Females to Males ratio observed in the population of the Open Day participants. The size of the bubble is proportional to the number of students that express a preference for the subject. For the sake of clarity, the plot does not include Education and Childhood Practice—both female dominated—with ratio 14.1 and 31.0, respectively

differences observed in the job market. Figure 5.5 shows the relationship between females-to-males ratio and first expected salary for the 35 subjects characterised by gender imbalance in this study. The data shows that, on average, the expected first salary is £25, 000 and £20, 300 for male-dominated and female-dominated subjects, respectively (the difference is statistically significant with $p \ll 0.01$ according to a two-tailed t-test). This corresponds to a difference of 19.7%, not far from the gender pay gap of 19.3% that the *UK Government Department for Culture, Media and Sport* has measured in 2013 [26]. Overall, "*gender segregation in fields of study is not questionable per se [...] But as soon as gender segregation is connected to systematic disadvantages for one group (in this case women), it constitutes a social problem*" [16]. Whether the best way to address the issue is to eliminate the segregation patterns observed in this work or reducing the pay differences across jobs (possibly both) it is still an open issue.

5.5 Social Status and Access to Higher Education

A large number of studies have shown that access to Higher Education has never stopped widening in the last decades [19]. However, there is consensus in the literature that "*[...] educational inequalities tend to persist despite expansion because those from more advantaged social class backgrounds are better placed to take up new educational opportunities that expansion affords [...]*" [9].

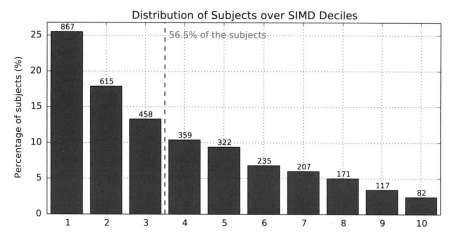

Fig. 5.6 The chart shows the distribution of the subjects over the SIMD deciles, from the least deprived to the most deprived. The number of Open Day participants falling in every decile is written above the bars

In general, the works presented so far in the literature measure the social class of a student in terms of three main parental characteristics, namely type of occupation [3, 24], income [2, 34] and, most frequently, attained education level [2, 15, 23, 32, 33]. This work adopts a different approach and relies on the *Scottish Index of Multiple Deprivation* (SIMD), a publicly available indicator provided by the Scottish Government.[8] The main advantage of the SIMD is that it is a weighted sum of multiple factors and, hence, it allows one to compare the relative importance of different aspects of social class. The main disadvantage is that the SIMD does not apply to individuals, but to *data zones*, i.e., small geographic areas of the Scottish territory (the average population is 802.8) where the standard of living has been observed to be relatively uniform. In this respect, the SIMD can be considered a good approximation of the social class of an individual living in a given data zone.

The total number of data zones is 6, 505 and they cover exhaustively the Scottish territory. Every data zone includes entirely the areas corresponding to one or more postcodes. Thus, if an app user provides the postcode during the registration step (see Sect. 5.2), it is possible to know the SIMD value for the data zone where she or he lives. Overall, the number of app users that have provided a Scottish postcode is 3, 433, corresponding to 70.3% of the 4, 885 registered users (the results of this section apply to these students). The SIMD allows one to rank the data zones, from the least to the most deprived, and count the number of app users that live in the data zones of the resulting deciles (from the 10% of the least deprived data zones to the 10% of the most deprived data zones). The resulting distribution is available in Fig. 5.6 and shows that 43.2% of the app users live in the 20% of the least deprived data zones while only 5.8% of them live in the 20% of the most deprived areas.

---

[8]http://www.gov.scot/Topics/Statistics/SIMD.

The SIMD is the weighted sum of 7 individual indicators that cover different aspects of social class, namely *Income* (In), *Education* (Ed), *Health* (He), *Employment* (Em), *Housing* (Ho), *Crime* (Cr) and *Access* (Ac). The data zones can be ranked according to each of these indicators and, for each of them, it is possible to obtain a distribution like the one of Fig. 5.6. The deviation with respect to the uniform distribution is statistically significant for all 7 distributions ($p \ll 0.01$ according to a χ^2 test with Bonferroni correction), but to a different extent. Figure 5.7 shows, for each individual indicator, the ratio between the number of app users that live in the 10% of the least deprived data zones and the number of app users that live in the 10% of the most deprived data zones. This provides an indication of how skewed the 7 distributions are (the larger the ratio, the more skewed the distribution).

The distribution for the Education indicator—that accounts for the average education level in a data zone—appears to be more skewed than the others, in line with the observation that, in general, the literature tends to measure the social class in terms of parental education [2, 15, 23, 32, 33]. Figure 5.7 shows that, at least in Scotland, such a factor contributes more than any other to the inequalities observable in Fig. 5.6. Employment, Housing and Income—three highly correlated indicators that account for the economic conditions of a data zone—show a similar ratio, but their distribution is significantly less skewed than the Education indicator, in line with the observation that income tends to be used less in the literature as a measure of social class. The remaining indicators appear to play a less important role and, not surprisingly, they have not been considered in the literature as a potential source of inequality when it comes to access to Higher Education.

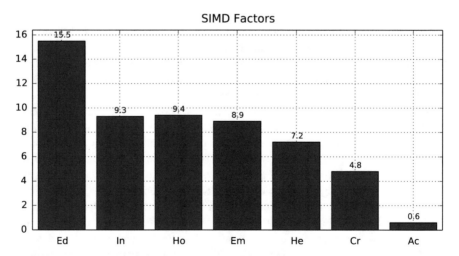

Fig. 5.7 For every dimension of the Scottish Index of Multiple Deprivation (SIMD), the plot shows the ratio of the number of Open Day participants falling in the least deprived decile to the number of students falling in the most deprived decile. The higher the bar, the larger the inequality. The value of the ratio is written above each bar of the chart

5.6   Conclusions

This work has presented an observational study about the choice of Higher Education, i.e., the process that leads a prospective student to select a particular subject of study (or a particular University). In particular, the analysis of this work revolves around the preferences expressed by the participants of an Open Day—an event organised by a Higher Education institution to let prospective students to visit the campus while interacting with faculties and students—about the subjects they intend to study. The work has adopted data analytics methodologies and it has investigated the distribution of the preferences across different subjects, gender segregation effects and interplay between social class and access to Higher Education.

The observations have been inferred from the items that the Open Day participants have inserted in their agenda, thanks to the use of a mobile application. In this respect, the data collection and analysis methodologies are alternative to questionnaires and surveys typically adopted in studies of the same type. The main advantage of the new approach is the possibility to collect information about a large number of persons (4, 885 in total) during the very brief duration of an Open Day. Furthermore, the use of an agenda rather than a questionnaire has made it possible to consider individually 109 subjects. The observation of several effects and phenomena described earlier in the literature is an indirect confirmation that the collected data is a reliable source of information.

The results show that prospective students and their families tend to use the Open Day to explore a wide and diverse spectrum of subjects without necessarily taking into account criteria like, e.g., the chances to obtain a job or the first salary that can be expected after graduation. Such an observation confirms previous observations about the tendency to gather as much information as possible before making a definitive choice [28, 35]. Furthermore, it suggests that the particular information of interest at an Open Day is the content of the educational offer, i.e., the type of subjects that can be studied, more than the outcomes following the graduation.

When it comes to gender segregation, the results show that there is no significant difference between female and male participants for most of the subjects (74 out of 109). For what concerns the other 35 subjects, the results suggest that there is a care-technical divide while the science-humanities divide is not as evident as previous works in the literature suggest it to be. Still, female participants are less likely to attribute their preferences to subjects that involve a significant mathematical component.

Finally, this work has shown that the dimension along which the Open Day participants are most unequal is the education level of the environment where someone lives. While being skewed, the distributions of the economic factors (income, housing and occupation) still account for less inequality. One possible reason is that access to Higher Education is free for Scottish residents and, hence, the only difference that can play a role is the education level. Moreover, parental education has been shown to be an important factor in other education systems as well [33] and this is probably the reason why widening the access to Higher Education has not been sufficient, in general, to eliminate inequalities [32].

Developing instruments for analysis and understanding of the choice of Higher Education is important to better address the needs of prospective students, to develop more effective marketing strategies, to better organise an Open Day or to track the progress of initiatives aimed, e.g., at attenuating gender segregation and at making the access to Higher Education more inclusive. In this respect, the findings of this work are interesting not only because they provide insights about a specific case (the participants of an Open Day organised by a particular university), but also because they provide data collection and analysis methodologies that can be applied, at least partially, to a large number of initiatives held by Higher Education institutions.

Acknowledgements This work was supported by The Data Lab (www.thedatalab.com) through the project "*Knowledge Extraction for Business Opportunities*" (agreement 70539). Two of the authors (Stefan Raue and Chamila Abeyratna) work for Bizvento (www.bizvento.com), the company that has developed the mobile application used to collect the data.

References

1. Ayalon, H.: Women and men go to university: mathematical background and gender differences in choice of field in higher education. Sex Roles **48**(5–6), 277–290 (2003)
2. Ayalon, H., Yogev, A.: Field of study and students' stratification in an expanded system of higher education: the case of Israel. Eur. Sociol. Rev. **21**(3), 227–241 (2005)
3. Ball, S., Davies, J., David, M., Reay, D.: "Classification" and "judgement": social class and the "cognitive structures" of choice of higher education. Br. J. Sociol. Educ. **23**(1), 51–72 (2002)
4. Baron-Cohen, S.: The Essential Difference. Penguin (2004)
5. Baron-Cohen, S., Richler, J., Bisarya, D., Gurunathan, N., Wheelwright, S.: The systemizing quotient: an investigation of adults with Asperger syndrome or high-functioning autism, and normal sex differences. Philos. Trans. R. Soc. Lond. B Biol. Sci. **358**(1430), 361–374 (2003)
6. Baron-Cohen, S., Wheelwright, S.: The Empathy Quotient: an investigation of adults with Asperger syndrome or high functioning autism, and normal sex differences. J. Autism Dev. Disord. **34**(2), 163–175 (2004)
7. Barone, C.: Some things never change: gender segregation in higher education across eight nations and three decades. Sociol. Educ. **84**(2), 157–176 (2011)
8. Baum, S., Ma, J., Payea, K.: Education pays, 2010: the benefits of higher education for individuals and society. Tech. rep., College Board Advocacy & Policy Center (2010)
9. Boliver, V.: Expansion, differentiation, and the persistence of social class inequalities in British higher education. High. Educ. **61**, 229–242 (2011)
10. Briggs, S.: An exploratory study of the factors influencing undergraduate student choice: the case of higher education in Scotland. Stud. High. Educ. **31**(6), 705–722 (2006)
11. Briggs, S., Wilson, A.: Which university? A study of the influence of cost and information factors on Scottish undergraduate choice. J. High. Educ. Policy Manage. **29**(1), 57–72 (2007)
12. Choi, B., Pak, A.: A catalog of biases in questionnaires. Prev. Chronic Dis. **2**(1), A13 (2005)
13. Correll, S.: Gender and the career choice process: the role of biased self-assessments 1. Am. J. Sociol. **106**(6), 1691–1730 (2001)
14. Durndell, A., Siann, G., Glissov, P.: Gender differences and computing in course choice at entry into higher education. Br. Educ. Res. J. **16**(2), 149–162 (1990)
15. Egerton, M., Halsey, A.: Trends by social class and gender in access to higher education in Britain. Oxf. Rev. Educ. **19**(2), 183–196 (1993)

16. Lörz, M., Schindler, S., Walter, J.: Gender inequalities in higher education: extent, development and mechanisms of gender differences in enrolment and field of study choice. Ir. Educ. Stud. **30**(2), 179–198 (2011)
17. Manson, C., Winterbottom, M.: Examining the association between empathising, systemising, degree subject and gender. Educ. Stud. **38**(1), 73–88 (2012)
18. Maple, S., Stage, F.: Influences on the choice of math/science major by gender and ethnicity. Am. Educ. Res. J. **28**(1), 37–60 (1991)
19. Marginson, S.: The worldwide trend to high participation higher education: dynamics of social stratification in inclusive systems. High. Educ. **72**(4), 413–434 (2016)
20. Nguyen, H.H., Ryan, A.: Does stereotype threat affect test performance of minorities and women? A meta-analysis of experimental evidence. J. Appl. Psychol. **93**(6), 1314–1334 (2008)
21. Podsakoff, P., MacKenzie, S., Lee, J.Y., Podsakoff, N.: Common method biases in behavioral research: a critical review of the literature and recommended remedies. J. Appl. Psychol. **88**(5), 879–903 (2003)
22. Poock, M., Love, P.: Factors influencing the program choice of doctoral students in higher education administration. J. Student Aff. Res. Pract. **38**(2), 203–223 (2001)
23. Reay, D.: "Always knowing" and "never being sure": familial and institutional habituses and higher education choice. J. Educ. Policy **13**(4), 519–529 (1998)
24. Reay, D., Davies, J., David, M., Ball, S.: Choices of degree or degrees of choice? class, 'race' and the higher education choice process. Sociology **35**(4), 855–874 (2001)
25. Scottish government equality outcomes: gender evidence review. Social research series, The Scottish Government (2013)
26. Secondary analysis of the gender pay gap. Tech. rep., UK Government, Department for Culture, Media and Sport (2014)
27. Siann, G., Callaghan, M.: Choices and barriers: factors influencing women's choice of higher education in science, engineering and technology. J. Furth. High. Educ. **25**(1), 85–95 (2001)
28. Simões, C., Soares, A.: Applying to higher education: information sources and choice factors. Stud. High. Educ. **35**(4), 371–389 (2010)
29. Spencer, S., Steele, C., Quinn, D.: Stereotype threat and women's math performance. J. Exp. Soc. Psychol. **35**(1), 4–28 (1999)
30. Steele, J., Ambady, N.: "Math is hard!" The effect of gender priming on women's attitudes. J. Exp. Soc. Psychol. **42**(4), 428–436 (2006)
31. The communications market report. Tech. rep., Ofcom (2015)
32. Triventi, M.: The role of higher education stratification in the reproduction of social inequality in the labor market. Res. Soc. Stratif. Mobil. **32**, 45–63 (2013)
33. Triventi, M.: Stratification in higher education and its relationship with social inequality: a comparative study of 11 European countries. Eur. Sociol. Rev. **29**(3), 489–502 (2013)
34. Van de Werfhorst, H., Kraaykamp, G.: Four field-related educational resources and their impact on labor, consumption, and sociopolitical orientation. Sociol. Educ. **74**(4), 296–317 (2001)
35. Vrontis, D., Thrassou, A., Melanthiou, Y.: A contemporary higher education student-choice model for developed countries. J. Bus. Res. **60**(9), 979–989 (2007)
36. Webbink, D., Hartog, J.: Can students predict starting salaries? Yes!. Econ. Educ. Rev. **23**(2), 103–113 (2004)
37. Wilkins, S., Shams, F., Huisman, J.: The decision-making and changing behavioural dynamics of potential higher education students: the impacts of increasing tuition fees in England. Educ. Stud. **39**(2), 125–141 (2013)

Chapter 6
Speech Pause Patterns in Collaborative Dialogs

Maria Koutsombogera and Carl Vogel

Abstract This chapter discusses the multimodal analysis of human behavioral data from the big data perspective. Though multimodal big data bring tremendous opportunities for related applications, we present current challenges in the domain of multimodal and behavioral analytics. We argue that in the case of analysing human behavior in interaction, we need to shift to the analysis of samples, smaller datasets, before scaling up to large data collections. We describe a dataset developed to study group collaborative interaction from measurable behavioral variables. As a case study, we investigate speech pauses and their patterns in the data, as well as their relationship to the topics of the dialog and the turn-taking mechanism, and we discuss their role in understanding the structure of collaborative interactions as well as in interpreting the behavior of the dialog participants.

6.1 Introduction

Human communication is rich and complex, in that it consists of an interplay between speech, body activity and cognition. It is not only the content of words, but also the way and the time in which they are uttered that contribute to the successful delivery of the message. Human behavior in interaction depends on an individual's communicative intent and is influenced by other people as well as by the context of the interaction. Therefore, interactions may largely vary depending on the context or the content of the interaction or the nature of the task that the interaction participants are involved in, resulting in heterogeneous multimodal signals. This heterogeneity and variability in human behavior signals make the understanding and automatic decoding of human behavior cues a challenging engineering problem [33].

M. Koutsombogera (✉) · C. Vogel
Trinity College Dublin, Dublin 2, Ireland
e-mail: koutsomm@cs.tcd.ie

C. Vogel
e-mail: vogel@cs.tcd.ie

© Springer Nature Switzerland AG 2019
A. Esposito et al. (eds.), *Innovations in Big Data Mining and Embedded Knowledge*,
Intelligent Systems Reference Library 159,
https://doi.org/10.1007/978-3-030-15939-9_6

Multimodal interaction analysis provides information about the way different expressive modalities shape the structure of the interaction (i.e. turn management) and convey the speakers' cognitive and affective state in any given moment (including feedback responses and emotions), thus demonstrating the speakers' interactional and social behavior [12, 47].

In recent years, new methodologies and tools have emerged to quantitatively understand and model human behavior in interaction, by exploiting the richness of multimodal signals that humans convey, and based mostly on affective and social behavior cues, as expressed through speech, acoustic and visual activity features. Aspects of behavior are highly represented in multimodal data, as aspects of distinct data sources that are nevertheless related to each other in a common context.

From this perspective, multimodality is important in the context of big data in the sense of developing models useful for analyzing unstructured behavioral data and responding to challenging optimization problems. Multimodal data analytics provides great opportunities to build better computational models to mine, learn, and analyze enormous amounts of social signals, i.e. data related to human behavior, because of the richness of the data in terms of content, context and speakers [35]. While multimodal analysis is usually focused on micro-interaction, the analysis may scale up when linked with broader application areas, such as learning analytics [2] and healthcare analytics [36]. For example, multimodal learning analytics is about using advanced sensing and artificial intelligence technologies to measure, collect, analyse and report data about learners and their contexts, with the purpose to improve pedagogical support for students' learning and the optimisation of learning and learning environments. The latter can benefit from multimodal analyses due to the heterogeneity of the data sources available [31], while new high-frequency data collection technologies and machine learning analysis techniques could offer new insights into learning, and students' learning trajectories [2].

6.1.1 Challenges in Multimodal Analytics

Advances in multimodal and multimedia big data are expected to provide more opportunities to build better computational models to mine, learn and analyse enormous volumes of data. Multimodal analytics is related to behavioral analytics and brings new insights to the analysis of human behavior, focusing on the understanding of how humans behave and why, and enabling accurate predictions of how they are going to act in the future. The rationale behind multimodal and behavioral analytics is to utilize the massive volumes of data collections where users interact in various social or professional contexts, and have the ability to query data in a number of ways and create predictive models of behavior.

The collection of behavioral data from either real-world or laboratory settings offers novel processing and modelling opportunities to extract measurable behavioral cues and develop predictive models of behavior on a large scale. This is a challenge per se, in the sense of developing tools to acquiring multimodal data, controlled or in-the-wild, together with the necessary contextual information that will allow for robust processing and prediction [33]. However, an ongoing challenge remains the representation and modelling of multimodal data, as well as the development of computing methods that effectively analyse data. This challenge in the representation is due to the unstructured and heterogeneous nature of data (i.e. data come from multiple sources with different representations), and mainly because of the complexity in understanding the semantic gap between the low-level behavioral features and the high-level semantics they bear [48]. Because of their nature, another challenge in multimodal big data are the real-time requirements in the related applications and services that demand more efficient processing and large-scale computation, but also in the optimization of storage and networking resources.

6.1.2   From Big Data to Sample Data

Because of the multiple levels of information multimodal data conveys, big data analytics need to focus not only on large volumes of multimodal data, but also on the high quality of this data [35]. Thus, a limitation that has been acknowledged in big data is that, although enormous data volumes are exciting, data quality is not always guaranteed and that data quality matters more than quantity [4]. An important methodological issue in the analysis of multimodal data is sampling. It is essential to investigate samples of multimodal interaction, i.e. datasets of human interactions in a specific context, to better understand what data is about and to be in a position to make valid claims about aspects of human behavior, e.g. examine the frequencies of certain representative features or account for outliers. Sample data therefore enable the extrapolation of arguments and claim representativeness in a specific context [4].

Also, big data introduces the possibility of analyzing whole datasets, but this is extremely complex to do in multimodal analytics, as it is impossible to have access to all possible behaviors in human interaction. Again, because of the variability in human behavior, an out-of-context investigation cannot guarantee whether some behaviors are over- or underrepresented or whether data collections have been created with the same methodology.

Thus, the advantages of working with samples is that we know where the data comes from, the conditions under which they were captured and their quality. We also understand the meaning of the representations they carry, because we are aware of the context in which they have been created. And most importantly, we know the purpose for which the data was collected and the questions we are aiming to answer, the parameters we want to measure and the methodology we follow.

Also important aspect in multimodal data analysis is understanding the context, i.e. the context of the interaction, the physical and discourse setting on which the interaction takes place. As Goffman notes, the awareness of social framings is critical, since speakers adapt their speech depending on who they are speaking with and the expectation that the context raises [17]; therefore, framing the research of interaction behavior in a specific environment enables its interpretation.

Thus, when it comes to multimodal human behavior, there are social and cognitive parameters complex to model, at least on a big scale. The investigation of samples of behavior, of smaller datasets, is important in understanding the underlying structure and intentionality, in developing and testing hypotheses and then considering the possibilities of scaling up the research questions to big data.

6.1.3 Multimodal Group Dialog: Turn-Taking and Pauses

While the most common setup of human multimodal interaction, the two-party dialog, is already a rich and informative setup, multiparty interaction is even more challenging because of the dynamics developed among group members [16]. The importance of analyzing collaborative dialog datasets lies in decoding communicative patterns involving verbal and non-verbal modalities. By definition, group dialogue is a canvas where different communicative intentions, personalities, lexical choices that may affect the outcome and the effectiveness of the interaction are manifested by the participants. In terms of behavior modelling, efforts focus on automatically analyzing various facets of group interactions and collecting this knowledge to improve the quality of the interaction either in human-human or in human-machine settings. Related work that exploits group dialogue and multiparty corpora studies prediction of the next speaker in multiparty meetings [20], dominance and leadership [21, 32], and personality traits [30], among others.

The speech signal is a modality that offers important cues for the investigation of turn-taking, the process where there is a speaker change in the conversation. Understanding the turn-taking mechanism is especially important in group dialogue. To achieve smooth turn-taking, dialog participants need to allocate a turn or predict the person who will speak next, and need to consider a strategy for themselves to achieve accurate timing for their own turn. In this respect, the information coming from the speech signal may clarify the behavior that contributes to smooth turn-taking, but also inform about the degree of participation of the speakers in the discussion, revealing possibly unbalanced participation or signs of dominance from certain speakers. This information consists of the speakers' words, the number and duration of their turns, but also of the silence intervals, the pauses that occur during the dialog.

The investigation of pauses and their functions in dialog helps achieving better understanding of human communication and their role in turn organisation, including the examination of who takes a turn and when, how turn allocation is performed and the association of pauses to overlapping speech. Pauses are frequent in spoken language, and in addition to their use and functions, the local context where they occur, but also the linguistic and cultural contexts are also important. As mentioned in detail in the next section, there are numerous studies that address the importance of speech pauses in delivering the discourse message in a successful way, as well as aspects of language specificity. Furthermore, understanding speech pauses contributes to the design and implementation of dialog systems that can perceive and generate natural turn exchanges to interact with humans in a successful and cognitively natural way.

In the remainder of this chapter we will present a study that was carried out in what we called in the previous section *sample* human behavioral data, i.e. a corpus of multimodal group dialogs. In this study, we are looking into silent and breath pauses, i.e. non-filled pauses. The main goals of this study are (a) to discover possible relations among the pauses and the topic of the dialog, (b) to investigate the extent to which the context of the pauses contributes to the identification of next speakers, (c) quantitative aspects in the use and frequency of pauses and (d) the effects of the English native and non-native linguistic background of speakers on the duration of their pauses. In the next sections we discuss related literature, the dataset used for this study, the analysis of the speech pauses and the results of the study.

6.2 Speech Pauses: Background

Speech pauses can be categorized in general in two groups: silent pauses that may include breath, and filled pauses, which usually include disfluencies, i.e. non-lexicalised words such as *ahm*, *ehm*, etc. Speech pauses are considered as a mechanism related to internal cognitive processes that speakers employ in their messages. They have been described as signals of discourse planning, used as markers of discourse structure [7, 19], and indicating the way speakers are planning their message and speech [18, 28]. Most importantly, pauses have been considered as interaction management markers through which speakers regulate the interaction [8, 9]. In this respect, pauses have also been associated to the communicative functions of feedback and turn management [1]. Other functions related to cognitive processing are those associated to lexical retrieval [25] or difficult concepts that speakers need to think about [39]. The effects of filled pauses to memory for discourse have been also investigated and it has been shown that they facilitate recall [14].

From the multimodal perspective, research literature has reported on the temporal, semantic and functional relations in the co-occurrence of pauses and gestures [3, 11, 23, 29]. Pauses have also been considered an important functional cue to be taken into account in the design and development of conversational agents, in terms of both perception of human speech pauses as well as in the generation of pauses that contribute to the naturalness of the agents' output [6, 27, 37].

As far as automatic prediction is concerned, it has been reported that words preceding pauses are reliable predictors of their function as clause boundary markers [13, 34] and that discourse structure can to some extent be predicted from characteristics of filled pauses [45]. In tasks related to automatic detection of conversational dominance, the duration of silence intervals (pauses) is important when looking for instances of floor grabbing that occur right after these intervals [38]. Similarly, the events of taking a turn during silence or breaking a silence are variables that are associated with dominant speakers [32]. Also, approaches on the semi-automatic recognition of filled pauses have been proven to reduce the effort of manual transcription of filled pauses [10].

Silent and filled pauses have also been explored in relation to personality aspects, especially regarding the fluency in speech and the quantity of pauses produced with regard to personality traits such as extraversion or neuroticism, but it has been also stressed that these relations may be affected by other factors such as social skills [40, 43].

Different cultural and language backgrounds have different effects on speech production. Since speech pauses are part of a speaker's linguistic production, these effects are also evident in pause patterns. Language-specificity and speaker-specificity in pause production have been investigated in the literature. Speakers adopt their own variants of filled pauses [26], while cross-linguistic analyses have provided evidence about the language-specific patterns in the vocalic quality of filled pauses [5]. A study about the impact of different factors on pause length (region, gender, ethnicity, age) has shown that region and ethnicity have significant influences on pause duration [22]. Also, a comparative study of speech rate in nine languages has shown that the probability of pauses occurring before nouns is about twice as high than before verbs [42]. Different pause tolerance (i.e. perception of what is considered e.g. a long pause) in the conversations among speakers of different languages may result in difficulties in communication [41] and, as pause tolerance can vary distinctly between different cultures, different pause patterns may cause problems in intercultural communication [15]. An investigation of turn transitions with regards to pauses and overlaps in ten different languages has given evidence about differences across the languages in the average gap between turns [44].

6.3   Data Description

In this work we used the MULTISIMO corpus, a multimodal corpus consisting of 23 sessions of collaborative group interactions where two players need to provide answers to a quiz and are guided by a human facilitator. Players work together while the facilitator monitors their progress and provides feedback and hints when needed. In this setup, collaboration refers to the process where the two players coordinate their actions to achieve their shared goal, i.e. find the appropriate answers and rank them.

The scenario was designed in a way that would elicit the desired behavior from the participants, that is, encourage their collaboration towards a goal. We thus designed sessions, in which 3 members of a group, 2 players and 1 facilitator, collaborate with each other to solve a quiz. The sessions were carried out in English and the task of the players was to discuss with each other, provide the 3 most popular answers to each of 3 questions (based on survey questions posed to a sample of 100 people), and rank their answers from the most to the least popular. The players expressed and exchanged their personal opinions when discussing the answers, and they announced the facilitator the ranking once they reached a mutual decision. They were also assisted by the facilitator who coordinated this discussion, i.e. provided the instructions of the game and confirmed participants' answers, but also helped participants throughout the session and encouraged them to collaborate.

The corpus consists of a set of audio and video recordings that are fully synchronised. During the recording of the sessions the participants were seated around a table and were captured with three HD cameras, one 360 camera, three head-mounted microphones, one omnidirectional microphone and one Kinect 2 sensor. The head-mounted microphones were recording the individual audio signals (SR 44.1 kHz), while the omnidirectional microphone was used as a backup audio source (SR 44.1 kHz). Thus, the audio files that were used for the present study come from the individual head-mounted microphones.

The overall corpus duration is approximately 4 h and the average session duration is 10 min. Overall, 49 participants were recruited and the pairing of players was randomly scheduled. 46 were assigned the role of players and were paired in 23 groups. The remaining 3 participants shared the role of the facilitator throughout the 23 sessions. In most of the sessions the participants don't know each other, although there are a few cases (i.e. in four groups) where the players are either friends or colleagues. The average age of the participants is 30 years old. Furthermore, gender is balanced, i.e. with 25 female and 24 male participants. Nevertheless, the gender distribution varies, depending on the pairing of the players. For example, there are groups where both of the players are female, or groups with male players, and groups with both genders. The participants come from different countries and span eighteen nationalities, one third of them being native English speakers. More information about the corpus is provided in [24] and at a dedicated webpage.[1]

This dataset addresses multiparty collaborative interactions and aims at providing tools for measuring collaboration and task success based on the integration of the related multimodal information, including collaborative turn organization, i.e. the multimodal turn managing strategies that members of a group employ to discuss and collaborate with each other. The corpus will serve as the knowledge base for identifying measurable behavioral variables of group members with the goal of creating behavioral models. These models may be exploited in human-computer interfaces, and specifically in the design of embodied conversational agents, i.e. agents that need to be able to extract information about their interlocutors to increase the intuitiveness and naturalness of the interaction.

6.4 Data Analysis

The pauses were manually annotated during the transcription process. The audio signal of the files was then transcribed by 2 annotators using the `Transcriber` tool.[2] The annotators listen to the audio files, segment the speech in turns and transcribe the speakers' speech. Apart from the speaking activity, pause intervals are

---

[1] https://www.scss.tcd.ie/clg/MULTISIMO/.
[2] http://trans.sourceforge.net/ last accessed 28.02.2018.

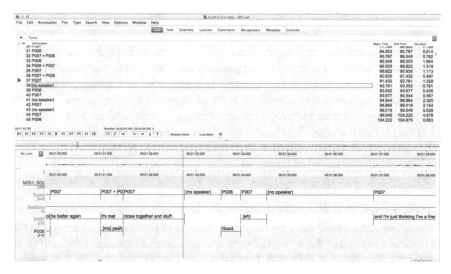

Fig. 6.1 Screenshot of a sample file in ELAN. At the top part of the editor the sequence of speaker turns may be viewed, including pause intervals (no speaker). The bottom part includes the timeline with the transcript annotations for words and turns for each speaker, as imported from Transcriber

also annotated by using a 'no speaker' value. The annotators also segmented each group dialogue in 5 topics, i.e. introduction, question 1, question 2, question 3 and closing. Transcripts were then imported into the ELAN annotation editor,[3] so that all the information coming from the transcript was visible and further editable (cf. Fig. 6.1).

6.4.1 Manual and Automatic Extraction of Pauses

Using the ELAN search functionality, the transcripts were exploited to create an index where pauses are searchable within their context. We thus exported the list of pauses (i.e. the segments that were annotated with the 'no speaker' value) and their context as a set of concordance lists. Each concordance line includes (a) the filename where the pause occurs, (b) the speaker turn id that precedes the pause, (c) the speaker turn id that follows the pause, (d) the topic or substructure in which the pause occurs, i.e. whether it is in the introduction, the closing or one of the 3 questions part, (e) the start time of the pause, (f) the end time of the pause and (g) its duration. Table 6.1 presents a sample.

Since the speech pauses were extracted from the manual transcriptions, the transcript annotation serves as the gold standard. However, to test whether pauses can be also automatically detected in a way that is comparable to human annotations

---

[3] https://tla.mpi.nl/tools/tla-tools/elan/ last accessed 28.02.2018.

Table 6.1 A sample of exported information about pauses, that includes the filename, the speaker turns before and after the pause, the topic in the discussion where the pause occurs, and its begin and time and duration in milliseconds

Filename	Turn before	Pause	Turn after	Topic	Begin time	End time	Duration
S10	P20	no speaker	P22	Introduction	6433	6895	462
S10	P21	no speaker	P22	Question 1	93991	94299	308
S10	P21	no speaker	P20	Question 2	214733	215300	567
S10	P22	no speaker	P20	Question 3	236384	237770	1386

in terms of accuracy, we automatically extracted silent pauses using the `auditok` tool.[4] Voice activity detection was then extracted, with the energy threshold for the perceived voice loudness set to 55, where values below this threshold are considered silence, and values above are considered speech. This resulted in a list of time intervals where the voice activity occurs, and this list was used to measure the pauses by calculating the silence intervals that are located between two successive voice activity segments. The results were very similar to the manual annotations, indicating that the set threshold gives reliable results for the specific audio quality of the dataset, by generating silence intervals above 0.2 s. However, since the threshold depends on the audio quality and the sensitivity of the microphone used for the recordings, the threshold tuning proves to be an important factor for automatic silence detection.

6.4.2 Speech Pause Frequency

Overall, 1719 silent and breath pauses were extracted from the 23 dialogs of the corpus in a duration of approximately 4 h. Because the duration of dialogs varies, the frequency of pauses was calculated with respect to each individual dialog duration. Specifically, we calculated the number of pauses occurring in one minute of a dialog, and the percentage of silence in a dialog, by exploiting the number of pauses and their duration respectively, cf. Table 6.2. The results confirm previous findings in that speech pauses are frequent in dialog, with a median value of 7 speech pause occurrences per minute, and pause intervals occupy (on median basis) a 14% of the overall dialog duration.

The duration of a silence interval in the data may vary from 0.2 to 10 s. However, the median values for the whole corpus fall within the range of 0.5–1.7 s. A detailed presentation of the silence interval durations is depicted in Fig. 6.2, which includes the distribution of Min, Max and Median values of the duration of pause segments in the 23 files. Therefore, speakers in the corpus have the tendency to produce short pauses, of less than 2 s.

---

[4]https://github.com/amsehili/auditok last accessed 28.02.2018.

Table 6.2 Frequency information for speech pauses, i.e. MIN, MAX and MEDIAN values for (a) the number of pauses occurring in 1 min, and (b) percentage of the duration of pause intervals in the dialogs

Speech pauses	MIN	MAX	MEDIAN
Pauses per minute	2	13	7
Duration in the file (%)	7%	24%	14%

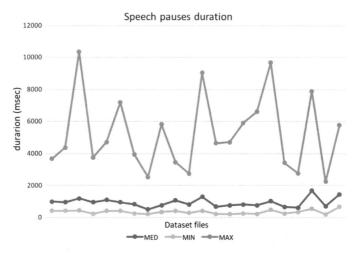

Fig. 6.2 Distribution of MIN, MAX and MEDIAN values (in msec) for pause duration in the 23 corpus dialogs

6.4.3 Silence Intervals and Dialog Topics

All corpus dialogs have a uniform structure, that is, they consist of the introduction to the game, the discussion of the 3 questions and the closing. We then examined the distribution of pauses in the dialog topics, to investigate whether there is an association of the speech pauses with a particular topic in the dialog. Figure 6.3 shows the distribution of pauses in the four of the five topics for each of the dialogs, i.e. the percentage of pause interval occurrences in each topic with regard to the total number of pauses in each of the 23 dialogs. No pause segments were identified in the closing section, therefore this topic was left out of the plot. A possible reason for the absence of pauses in the closing section is that it is about a very brief section where the facilitator thanks the players for their participation, therefore there are almost no turn exchanges.

Table 6.3 lists the Min, Max and Median values of (a) the percentage of the duration of the pause intervals with regard to the duration of each section, and (b) the percentage of pause interval occurrences in each topic with regard to the total number of pauses in each dialog (i.e. a summary of Fig. 6.3). The shortest (in duration)

Table 6.3 Distribution of MIN, MAX and MEDIAN values for (a) the percentage of duration of pauses occurring in each of the dialog topics, and (b) the percentage of the quantity of pauses in each of the dialog topics

Topic	Pause duration			Pause number		
	MIN	MAX	MEDIAN	MIN	MAX	MEDIAN
Introduction	1%	11%	5%	1%	11%	3%
Question 1	2%	28%	14%	19%	56%	31%
Question 2	4%	28%	15%	8%	47%	25%
Question 3	11%	34%	17%	16%	64%	39%
Closing	0	0	0	0	0	0

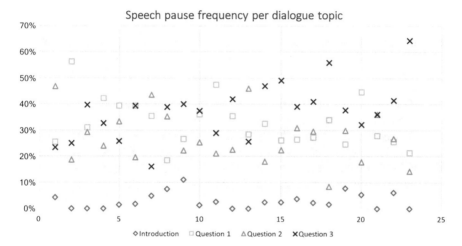

Fig. 6.3 Distribution of pause occurrences (percentage) in 4 dialog topics (Introduction, Questions 1, 2 and 3) in the 23 corpus dialogs

and fewest (in number) pauses are included in the introduction section. This may be due to the fact that this section is very predictable in terms of turn-taking and has few turn exchanges: it is the section where the facilitator introduces the game and the instructions to the players. The players often acknowledge what the facilitator is explaining with brief verbal feedback, or they may ask brief clarification questions. Nevertheless, the majority of pauses in this topic are performed by the facilitators themselves, who pause briefly to elicit feedback from the players that they follow them, to mark the sequence of instructions, or to take a break as they holds the floor for a long time.

The three questions are the core topics of the dialog, where the participants need to discuss and agree upon the appropriate answers. As expected, pauses are frequent during the question topics; what is interesting though is that Question 2 presents the lowest number of pauses among all 3 questions, and that Question 3 includes both the highest number of pauses, as well as the longest pauses. This observation is interesting in that it is an indicator that Question 3 is more complex than the other

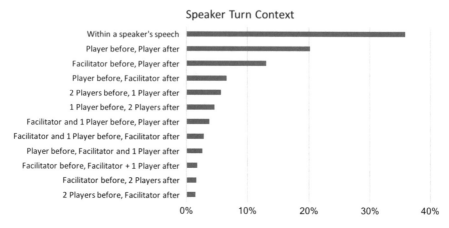

Fig. 6.4 Distribution of the speaker turn context (i.e. who speaks) that precedes and follows a speech pause. All possible combinations and the related percentage are listed

two, either because it is more difficult to address or it requires an elaborate discussion. Both in terms of quantity and duration, the high percentages of pauses in Question 3 may indicate what the literature has often claimed, that pauses are related to cognitive discourse planning; in our case, the speakers may for example need more time to think about potential answers, or may have difficulty in identifying answers to the questions, and speech pauses are a mechanism they employ for this.

6.4.4 Local Context

As mentioned in Sect. 6.4.1, the list of pauses was extracted together with the speaker id of the turns that are located before and after the pause interval. This was done to examine the position of pauses with regard to speaker change, and to identify whether we can derive any conclusions about the speaker who takes a turn after a speech pause. Pauses may occur within the same speaker's turns, may be located between the turns of two different speakers, but may also occur before or after simultaneous talk. Figure 6.4 presents the percentages of occurrences of the various combinations of the speaker ids that were found before and after a pause interval in the dataset, including cases of overlapping talk.

The majority of pauses (46%) are located among the turns of the same speaker. This case is most probably related to the fact that pauses are a mechanism that speakers employ to plan their discourse and mark its structure, but also to the cognitive processing aspects, i.e. the time speakers need to find appropriate words or think about difficult concepts. Pauses that occur between the two players of the game are also frequent (20%) and indicate that speakers with this role (i.e. player) exchange turns more frequently with each other than with the facilitator. An interesting aspect that cannot be clarified from the pause measurements alone, is that of the way the turn change occurs. It would be important to be able to infer whether the transition from one player to the other is done in a smooth way, i.e. in the case where a player offers the turn to his/her co-player, or whether a speaker takes advantage of a pause to grab the floor.

The next most frequent pattern is that of the pause happening after a facilitator's turn, followed by a turn from one of the players (13%). This is the most frequent case with the facilitator in the left context, and it highlights the role of the facilitator, i.e. to give the floor to the participants after addressing a question or providing helping cues. A 7% of the occurrences refers to cases where the player pauses and the facilitator takes the floor. These cases usually indicate either feedback elicitation, i.e. the players have provided the right answer and await for confirmation, or that the players need help to address the questions posed as they cannot keep on further guessing or discussing potential answers.

An interesting case for further investigation is that of overlapping talk, either before or after the pause. For example, when two players talk simultaneously and after they pause, one of them takes the floor (6% of the cases), it is very possible that an interruption has taken place and it is resolved after the pause. In this respect, pauses may be associated with conversational dominance detection, in the sense that they help identify the person that takes the floor after overlapping talk. However, the frequency of the rest of the patterns where overlapping talk is involved is similar, hence the context is not very informative in providing helpful cues, and information from language or other features is needed to interpret behaviors before or after the overlaps.

6.4.5 Language Effects on Pause Duration

Silence in conversations is a critical communication device and the beliefs expressed in talk and silence are culture dependent [46]. Language and cultural differences are important factors in the analysis and interpretation of pauses. Although the conversations in our dataset were carried out in English, the dataset consists of speakers that are both native and non-native English speakers. At the same time, since most of the cross-lingual and cross-cultural studies on pauses are focused on the pause length aspect [22, 42, 44], we specifically tested the effects of the native and non-native linguistic background of the corpus speakers on the duration of their pauses.

The dataset includes 16 native English speakers and 33 non-native English speakers. The non-native speakers have a fluent command of English and they span 15 nationalities from Europe, Asia and South America. The majority of the native speakers are Irish[5] (13), while there are 2 speakers from the UK and one speaker from the US.

---

[5] All Irish speakers in the corpus are English native speakers and use English in their conversations and elsewhere, and not Irish Gaelic, the first official language in the Republic of Ireland.

We took into account two features related to the speaker who holds the floor before and after a pause: the nationality of the speaker and whether that speaker is a native speaker of English or not. In the case of overlapping talk occurring before or after a pause, we considered the aforementioned features for both of the speakers who talk simultaneously.

The results show that there is a significant difference ($p = 0.01$) in pause durations for pauses that are preceded and followed by speech from native speakers of the same language, whatever that language is, and those that are not. The cases where the speech before and after the pause comes from speakers of the same language may refer to pauses that occur within a speaker's speech, or in cases where there is a pause between two turns of speakers of the same linguistic background. In those cases, pauses turned out to be shorter than the pauses that are preceded and followed by speech from native speakers of different languages, with a difference of 267 ms in the mean duration of pauses.

A more detailed analysis of the pause production of the English native speakers shows that there are differences in pause duration patterns among Irish, British and American speakers. Irish and British speakers have longer pause durations for their own pauses than the pauses that they yield, the latter e.g. in cases when the speaker changes after they pause (a difference of 177 and 199 ms respectively in the pauses mean duration). The American speaker has shorter pause durations for pauses he owns than those he yields (a difference of 244 ms in the pauses mean duration). Those differences suggest that any generalizations about pause durations appear to require dialect-level articulation rather than larger-language level articulation.

6.5 Conclusion

In this chapter, we presented a study about silent and breath pauses that occur in a multimodal dataset of collaborative group dialogs. We presented data related to the frequency and the duration of the speech pauses in the corpus as a whole and in the distinct sections of each dialogue where different topics are discussed. Also, we provided measurements related to the context that precedes and follows a speech pause in terms of speaker turns. Finally, we investigated the effects of the linguistic background of the speakers on the duration of pauses they produce. Our observations confirm that speech pauses are frequent in human dialogs, that they serve several functions and that they are employed by the speakers as a means to structure and emphasise their discourse and give time to reflect the conversation messages. Furthermore, we argue that pauses may provide important cues about the complexity of a given topic, in that this topic elicits more discussion time from the participants or requires more reflexion. We also suggest that, when investigating language and culture specificity in relation to duration aspects of pauses, one should consider both the language and the dialectal variation of the speakers. While exploring the context of the pauses, we focused on the structure of the turns (the sequence of speakers) and not their content. Although in some patterns the sequences are informative per se, in the majority of the cases additional information is needed to draw conclusions about who could be the speaker after a pause. Such information could be drawn from linguistic cues, i.e. the words or the syntactic boundaries of the utterances, but also from information related to other modalities, such as prosodic features or gestures co-occurring with speech. Additional information is therefore needed in cases where pause context patterns, pause frequency and duration do not provide sufficient information. Furthermore, psychological aspects are equally important to co-investigate, as personality and social skills features contribute to the interpretation of speakers' behavior.

Although more information apart from the pause intervals is needed to further investigate the speakers' actions and intentions, we believe that the investigation of pauses is important to achieve better understanding of the human communication and exploit this knowledge in the development of models used in dialog systems and conversational agents. We consider this as a challenging topic and addressing this scientific question in sample data will provide significant input to multimodal big data computing, especially in terms of dealing with cognition and understanding complexity. Since multimodal data analysis is focused on how to fuse the information from the different modalities and different features within modalities, to form a coherent decision, speech pause cues should be further explored and included in future investigations on human behavior model development.

Acknowledgements The research leading to these results has received funding from the European Union's Horizon 2020 research and innovation programme under the Marie Sklodowska-Curie grant agreement No 701621 (MULTISIMO).

References

1. Allwood, J.: The structure of dialog. In: Taylor, M.M., Neel, F., Bouwhuis, D. (eds.) The Structure of Multimodal Dialogue II, pp. 3–24. John Benjamins (1999)
2. Blikstein, P.: Multimodal learning analytics. In: Proceedings of the Third International Conference on Learning Analytics and Knowledge, pp. 102–106. ACM, New York, NY, USA (2013)
3. Boomer, D., Dittmann, A.: Hesitation pauses and juncture pauses in speech. Lang. Speech **5**, 215–220 (1962)
4. Boyd, D., Crawford, K.: Critical questions for big data. Inf. Commun. Soc. **15**(5), 662–679 (2012)
5. Candea, M., Vasilescu, I., Adda-Decker, M.: Inter- and intra-language acoustic analysis of autonomous fillers. In: DISS 05, Disfluency in Spontaneous Speech Workshop, pp. 47–52. Aix-en-Provence, France (2005)
6. Cassell, J., Pelachaud, C., Badler, N., Steedman, M., Achorn, B., Becket, T., Douville, B., Prevost, S., Stone, M.: Animated conversation: rule-based generation of facial expression, gesture and spoken intonation for multiple conversational agents, pp. 413–420 (1994)
7. Chafe, W.: Cognitive constraint on information flow. In: Tomlin, R.S. (ed.) Coherence and Grounding in Discourse, pp. 21–51. John Benjamins (1987)
8. Clark, H., Fox Tree, J.E.: Using uh and um in spontaneous speaking. Cognition **84**(1), 73–111 (2002)

9. Duncan, S.J., Fiske, D.W.: Face-to-Face Interaction: Research, Methods, and Theory. Lawrence Erlbaum Associates (1977)
10. Egorow, O., Lotz, A., Siegert, I., Bock, R., Krüger, J., Wendemuth, A.: Accelerating manual annotation of filled pauses by automatic pre-selection. In: 2017 International Conference on Companion Technology (ICCT), pp. 1–6 (2017)
11. Esposito, A., Esposito, A.M.: On speech and gestures synchrony. In: Esposito, A., Vinciarelli, A., Vicsi, K., Pelachaud, C., Nijholt, A. (eds.) Analysis of Verbal and Nonverbal Communication and Enactment. The Processing Issues, pp. 252–272. Springer, Berlin, Heidelberg (2011)
12. Esposito, A., Esposito, A.M., Vogel, C.: Needs and challenges in human computer interaction for processing social emotional information. Pattern Recognit. Lett. **66**, 41–51 (2015)
13. Esposito, A., Stejskal, V., Smékal, Z., Bourbakis, N.: The significance of empty speech pauses: cognitive and algorithmic issues. In: Mele, F., Ramella, G., Santillo, S., Ventriglia, F. (eds.) Advances in Brain, Vision, and Artificial Intelligence, pp. 542–554. Springer, Berlin, Heidelberg (2007)
14. Fraundorf, S.H., Watson, D.G.: The disfluent discourse: effects of filled pauses on recall. J. Mem. Lang. **65**(2), 161–175 (2011)
15. Fujio, M.: Silence during intercultural communication: a case study. Corp. Commun. Int. J. **9**(4), 331–339 (2004)
16. Gatica-Perez, D., Aran, O., Jayagopi, D.: Analysis of small groups. In: Burgoon, J.K., Magnenat-Thalmann, N., Pantic, M., Vinciarelli, A. (eds.) Social Signal Processing, pp. 349–367. Cambridge University Press (2017)
17. Goffman, E.: Frame Analysis: An Essay on the Organization of Experience. Harper and Row (1974)
18. Goldman-Eisler, F.: Pauses, clauses, sentences. Lang. Speech **15**(2), 103–113 (1972)
19. Hirschberg, J., Nakatani, C.: Acoustic indicators of topic segmentation. In: Proceedings of the International Conference on Speech and Language Processing (1998)
20. Ishii, R., Otsuka, K., Kumano, S., Yamato, J.: Prediction of who will be the next speaker and when using gaze behavior in multiparty meetings. ACM Trans. Interact. Intell. Syst. **6**(1), 4:1–4:31 (2016)
21. Jayagopi, D., Hung, H., Yeo, C., Gatica-Perez, D.: Modeling dominance in group conversations from non-verbal activity cues. IEEE Trans. Audio Speech Lang. Process. **17**(3), 501–513 (2009)
22. Kendall, T.: Speech Rate, Pause, and Sociolinguistic Variation: Studies in Corpus Sociophonetics. Palgrave Macmillan (2013)
23. Kendon, A.: Gesture: Visible Action as Utterance. Cambridge University Press (2004)
24. Koutsombogera, M., Vogel, C.: Modeling collaborative multimodal behavior in group dialogues: the MULTISIMO corpus. In: Proceedings of the Eleventh International Conference on Language Resources and Evaluation (LREC 2018). European Language Resources Association (ELRA), Paris, France (in press)
25. Krauss, R.M., Chen, Y., Gottesman, R.F., Krauss, R.M., Chen, Y., Gottesman, R.F.: Lexical gestures and lexical access: a process model. In: Mcneill, D. (ed.) Language and Gesture, pp. 261–283. University Press (2000)
26. Künzel, H.: Some general phonetic and forensic aspects of speaking tempo. Int. J. Speech Lang. Law **4**(1) (2013)
27. Maatman, R.M., Gratch, J., Marsella, S.: Natural behavior of a listening agent. In: Panayiotopoulos, T., Gratch, J., Aylett, R., Ballin, D., Olivier, P., Rist, T. (eds.) Intelligent Virtual Agents, pp. 25–36. Springer, Berlin, Heidelberg (2005)
28. Maclay, H., Osgood, C.: Hesitation phenomena in spontaneous english speech. Word **15**, 19–44 (1959)
29. McNeill, D.: Hand and Mind: What Gestures Reveal About Thought/David Mcneill. University of Chicago Press, Chicago (1992)
30. Mohammadi, G., Vinciarelli, A.: Automatic personality perception: prediction of trait attribution based on prosodic features. IEEE Trans. Affect. Comput. **3**(3), 273–284 (2012)

31. Muñoz-Cristóbal, J.A., Rodríguez-Triana, M.J., Bote-Lorenzo, M.L., Villagrá-Sobrino, S., Asensio-Pérez, J.I., Martínez-Monés, A.: Toward multimodal analytics in ubiquitous learning environments. In: Mmla-crosslak@lak, vol. 1828, pp. 60–67. CEUR-WS.org (2017)
32. Nakano, Y., Fukuhara, Y.: Estimating conversational dominance in multiparty interaction. In: Proceedings of the 14th ACM International Conference on Multimodal Interaction, pp. 77–84. ACM, New York, NY, USA (2012)
33. Narayanan, S., Georgiou, P.G.: Behavioral signal processing: deriving human behavioral informatics from speech and language. Proc. IEEE **101**(5), 1203–1233 (2013)
34. Navarretta, C.: Pauses delimiting semantic boundaries. In: Proceedings of the 6th IEEE International Conference on Cognitive Infocommunications (CogInfoCom2015), pp. 533–538. IEEE Signal Processing Society (2015)
35. Oviatt, S., Cohen, P.R.: The Paradigm Shift to Multimodality in Contemporary Computer Interfaces. Morgan & Claypool Publishers (2015)
36. Raghupathi, W., Raghupathi, V.: Big data analytics in healthcare: promise and potential. Health Inf. Sci. Syst. **2**(1), 3 (2014)
37. Rehm, M., Nakano, Y., André, E., Nishida, T.: Culture-specific first meeting encounters between virtual agents. In: Prendinger, H., Lester, J., Ishizuka, M. (eds.) Intelligent Virtual Agents, pp. 223–236. Springer, Berlin, Heidelberg (2008)
38. Rienks, R., Heylen, D.: Dominance detection in meetings using easily obtainable features. In: Renals, S., Bengio, S. (eds.) Machine Learning for Multimodal Interaction, pp. 76–86. Springer, Berlin, Heidelberg (2006)
39. Rochester, S.: The significance of pauses in spontaneous speech. J. Psycholinguist. Res. **2**(1), 51–81 (1973)
40. Scherer, K.R.: Personality markers in speech. In: Scherer, K.R., Giles, H. (eds.) Social Markers in Speech, pp. 147–209. Cambridge University Press (1979)
41. Scollon, R., Scollon, S.B.K.: Narrative, Literacy, and Face in Interethnic Communication. Ablex Pub., Corp Norwood, N.J. (1981)
42. Seifart, F., Strunk, J., Danielsen, S., Hartmann, I., Pakendorf, B., Wichmann, S., Witzlack-Makarevich, A., de Jong, N.H., Bickel, B.: Nouns slow down speech across structurally and culturally diverse languages. In: Proceedings of the National Academy of Sciences (2018)
43. Siegman, A.W., Pope, B.: Effects of question specificity and anxiety-producing messages on verbal fluency in the initial interview. J. Pers. Soc. Psychol. **2** (1965)
44. Stivers, T., Enfield, N.J., Brown, P., Englert, C., Hayashi, M., Heinemann, T., Hoymann, G., Rossano, F., de Ruiter, J.P., Yoon, K.-E., Levinson, S.C.: Universals and cultural variation in turn-taking in conversation. Proc. Nat. Acad. Sci. **106**(26), 10587–10592 (2009)
45. Swerts, M.: Filled pauses as markers of discourse structure. J. Pragmat. **30**(4), 485–496 (1998)
46. Ting-Toomey, S.: Communicating Across Cultures. The Guilford Press, New York, London (1999)
47. Vinciarelli, A., Esposito, A., André, E., Bonin, F., Chetouani, M., Cohn, J.F., Cristani, M., Fuhrmann, F., Gilmartin, E., Hammal, Z.: Open challenges in modelling, analysis and synthesis of human behaviour in human–human and human–machine interactions. Cogn. Comput. **7**(4), 397–413 (2015)
48. Zhu, W., Cui, P., Wang, Z., Hua, G.: Multimedia big data computing. IEEE MultiMedia **22**(3), 96–c3 (2015)

Chapter 7
Discovering Knowledge Embedded in Bio-medical Databases: Experiences in Food Characterization and in Medical Process Mining

Giorgio Leonardi, Stefania Montani, Luigi Portinale, Silvana Quaglini and Manuel Striani

Abstract In this paper, we will explore the potential of knowledge discovery from bio-medical databases in health safeguard, by illustrating two specific case studies, where different knowledge extraction techniques have been exploited. Specifically, we will first report on how machine learning and data mining algorithms can address the problem of food adulteration. Then, we will show how process mining techniques can be adopted to analyze the quality of patient care provided at a specific health care organization. Working in the bio-medical application domain has not only led to consistent and concretely useful experimental outcomes, but has also required some significant methodological advances with respect to the existing literature.

7.1 Introduction

Knowledge extraction from biological or medical databases can provide crucial support in health safeguard, and in patient care improvement. Health safeguard, in particular, is strongly related to nutrition and food quality. Knowledge discovery techniques

G. Leonardi (✉) · S. Montani · L. Portinale
DISIT, Computer Science Institute, University of Piemonte Orientale, Alessandria, Italy
e-mail: giorgio.leonardi@uniupo.it

S. Montani
e-mail: stefania.montani@uniupo.it

L. Portinale
e-mail: luigi.portinale@uniupo.it

S. Quaglini
Department of Electrical, Computer and Biomedical Engineering,
University of Pavia, Pavia, Italy
e-mail: silvana.quaglini@unipv.it

M. Striani
Department of Computer Science, University of Torino, Turin, Italy
e-mail: manuel.striani@unito.it

© Springer Nature Switzerland AG 2019   117
A. Esposito et al. (eds.), *Innovations in Big Data Mining and Embedded Knowledge*,
Intelligent Systems Reference Library 159,
https://doi.org/10.1007/978-3-030-15939-9_7

can be applied to bio-chemical food property data, in order to identify food adulteration. As an example, mashed fruit may contain a higher quantity of sugar, or a different type of fruit may be used in (partial) substitution of the nominal one (e.g., because it has a lower cost). These adulterations may threaten consumers' health, especially if they need to follow a diet regimen with a reduced sugar consumption, or if they suffer from allergies—and the potentially dangerous ingredients are not declared.

In the context of medical data, on the other hand, knowledge mining techniques can support various tasks, ranging from a better patient characterization, to inter-patient comparison, to the overall evaluation of the service offered by a health care organization. This last task, in particular, can strongly benefit from mining algorithms able to discover the medical process model that is actually being implemented at the hospital, when managing a given disease or patient type. In fact, even though medical personnel is typically supposed to implement reference clinical guidelines, the presence of local constraints (like, e.g., human or instrumental resource limitations) or local best practices may lead to differences between the actual process model and the guideline itself. Knowledge mining, and more specifically *process mining* techniques [28], can extract the actual model, and help the user to check guideline conformance, or to understand the level of local adaptation. A quantification of the existing differences or non-compliances (and maybe a ranking of a set of hospitals derived from it) can be exploited for several purposes, like, e.g., auditing purposes, performance evaluation and funding distribution.

In this paper, we will explore the potential of knowledge extraction from bio-medical databases, by illustrating two specific case studies, where different data mining methodologies have been exploited to fulfil the health safeguard goal.

Specifically, the first case study will show how to exploit standard chemical lab analysis, in particular Fourier-Transform Infrared (FTIR) spectroscopy, to identify and detect alimentary frauds. We will consider the case of potential adulteration of pure raspberry purees previously studied in [13]; we will report on how machine learning and data mining techniques can suitably address the problem, and we will discuss how such approaches can provide advantages with respect to standard chemometric methodologies in terms of performance, model construction and interpretation of the results.

In the second case study we will show how process mining techniques can be adopted to analyze the quality of patient care provided at a specific health care organization, by illustrating differences with respect to standard prescriptions, and by allowing for inter-hospital comparison. Interestingly, this medical application domain has required significant advances with respect to the state of the art in process mining. In particular, we have introduced a *semantic approach*, where ground data logged in the database are *abstracted*, on the basis of medical knowledge, in order to ignore details, while keeping all the relevant information. Experimental results in field of stroke management will be presented as well.

The paper is organized as follows: in Sect. 7.2 we present the first case study, on adulteration detection in agrifood products; in Sect. 7.3 we describe the second case study, on medical process mining. Section 7.4 summarise some related work, while Sect. 7.5 is devoted to conclusions and future research directions.

7.2 Counterfeit and Adulteration Detection in Agrifood Products

As mentioned in the introduction, food integrity is an important issue related to nutrition and health. The detection of food adulteration becomes important when such adulterations potentially threaten consumers' health, for example if a dietary regimen has to be followed or if specific allergies provide particular limitations, in terms of food, to people.

Data mining techniques may prove very relevant in discovering alimentary frauds or food counterfeits. As a specific case study we consider the detection of adulteration in pure raspberry purees, previously studied, with a standard chemometrical methodology, in [13]. Raspberry purees can be adulterated through the addition of a relevant amount of sucrose, or fake versions can be produced by using other kinds of fruits; as observed above, this may cause troubles in people having to follow a dietary regimen with a reduced sugar consumption, or if they suffer from allergies, and the potentially dangerous ingredients are not declared.

The study presented in [13] exploited FTIR spectroscopy to generate a dataset of instances containing both pure raspberry samples and adulterated ones. FTIR spectroscopy identifies chemical bonds in a molecule by producing an infrared absorption spectrum. An example is shown in Fig. 7.1. The dataset is produced by constructing the features as follows: given an FTIR spectrum, the range of frequencies (wave number) is divided into specific bins, each one representing a feature; the value of the feature is the area under the spectrum curve between the extrema of the bin. As an example, Fig. 7.1 shows the construction of the feature denoted $F_{[3700,3800]}$ representing the signal between the two given frequencies.

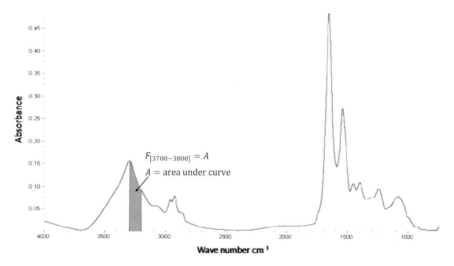

Fig. 7.1 FTIR absorbance spectrum of a Gram-negative bacterium, Salmonella enterica (adapted from [3])

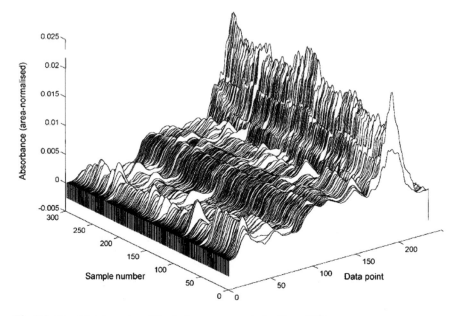

Fig. 7.2 The dataset spectra of the fruit purees case study (from [13])

In the considered case study, in preparing the dataset, different types of pure fruits (raspberry, plum, apple, strawberry, blackberry, cherry, apricot, and black currant) were considered, as well as different adulterants (sucrose, apple puree and plum puree) that have been used for spiking raspberry at different percentage levels. The dataset includes 1023 fruit puree samples, each one characterized by 234 features, corresponding to baseline-corrected and area-normalized fingerprint regions, in the range of 899.327–1802.564 wave numbers (cm^{-1}) with a bin width of 3.86 cm^{-1} from infrared spectra (see Fig. 7.2). In addition, each sample is labeled with the class attribute with values R (raspberry), NR (non-raspberry). The problem of adulteration detection can then be modeled as a binary classification problem, with NR being the positive class and R the negative one (i.e. the null hypothesis). Among the 1023 instances, 200 are in the negative and 823 in the positive class. The study has been organized as follows:

- **Feature reduction**: since the feature construction process has been performed at a very fine granularity (thus producing a large number of features), we decided to select the more relevant features, in order to reduce the number of attributes to deal with. We performed a filter approach [24], based on a correlation-based feature selection [9], resulting in the selection of 21 attributes (the majority of which concentrated into the 1100 and 1650–1780 frequency regions).
- **Unsupervised analysis**: given the feature-reduced dataset, we investigated whether adulterated or pure samples were naturally grouped into separate clusters. As dis-

cussed in the following, the obtained results suggested the need for supervised data analysis.

- **Classification (supervised analysis)**: we finally tested different types of classifiers on the feature-reduced dataset, by considering different aspects, such as the relative importance of Type I and Type II errors.

7.2.1 Unsupervised Analysis: Clustering

As mentioned above, we first tested whether the two different classes of the samples were able to separate on the basis of the measured features. We tested two different clustering approaches: *k-means* and *Expectation Minimization (EM)* with a target number of clusters $k = 2$ in both cases. The idea was to verify if two clusters corresponding to the target classes could be identified. Results of cluster assignments with respect to classes are shown in Table 7.1; in both cases class R (the negative class) is assigned to cluster 1 and class NR (the positive class corresponding to adulterations) to cluster 0. By looking at the clustering results we can notice a very large false negative rate, meaning that several adulterated products can be misclassified as good ones. One reason for that can be the unbalanced number of instances in the classes (the ratio between NR and R instances is 4:1), making the total accuracy not very significant. Moreover, in the current problem, type II error (related to FN) is much

Table 7.1 k-means and EM: cluster assignments (clusters = {0, 1}, classes = {NR, R}); FP = false positive rate, FN = false negative rate

	k-means	
	0	1
R	11	189
NR	688	135
Tot.	699	324
FP = 1.6%		
FN = 41.7%		
Accuracy: 85.73%		
	EM	
	0	1
R	11	189
NR	674	149
Tot.	685	338
FP = 1.6%		
FN = 44.1%		
Accuracy: 84.36%		

more important than type I error, since the main goal is to avoid the assignment to the pure raspberry class of an adulterated sample. Results from unsupervised analysis seem then to suggest to better exploit class information and not only regularities in the data. For this reason we resorted to supervised data mining methods as reported in the next subsection.

7.2.2 Supervised Analysis: Classification

Concerning supervised analysis, we considered the following classifiers:

- a kernel-based classifier (a Support Vector machine or SVM) based on Sequential Minimal Optimization (SMO) [23] and the related AdaBoost version (AB_SMO);
- a random forest classifier (RF) [2].

The reason for this choice is the following: SVMs are one of the most successful categories of classifiers, showing a general high degree of robustness in several problems; however, the unbalanced nature of the considered dataset suggests to consider the adoption of an ensemble learning approach as well. In addition, SMO (both in the base and in the AdaBoost version) is not an "interpretable model". To check the suitability of a more interpretable model, we then decided to test RF as well, where an ensemble of decision trees (that are graphical and more interpretable models) is built for classification.

In addition, to address the problem of unbalanced data from another perspective, we also considered a cost-based version of both SMO and RF algorithm that we denoted as C_SMO and C_RF respectively. In particular, the cost of misclassifying an adulterated sample has been considered twice the cost of misclassifying a pure sample, by taking into account the relative importance of Type II errors with respect to Type I errors. The approach taken in this case is *cost-sensitive learning*, meaning to learn a classifier with a datasets where the instances of the positive class (i.e., adulterations) are counted twice [6, 34] to reflect the cost. In the following, we will report the experimental results. The experimental setting considered the feature-reduced dataset consisting in 1023 instances (200 negatives and 823 positives) with 21 attributes obtained from FTIR spectra; each classifier has been learned using the above dataset, and evaluated by means of a 10-fold cross validation. Accuracy and Cohen's Kappa statistics are reported in Table 7.2; all the considered classifiers provide a very good accuracy, even if the Kappa not very close to 1 indicates a limited reliability of the parameter. This is as expected, because of the unbalanced nature of the problem. For this reason, precision and recall (and the relative F-scores) are considered more adequate to evaluate the classifiers performance. In particular, since the emphasis is to identify suitable data mining techniques for food integrity and protection, we concentrate on the behavior of the classifiers in recognizing adulaterations (the class NR). Figures 7.3 and 7.4 show the Precision-Recall (PR) curves, and the area under the PR curve for the three tested kernel-based and the two random forest classifiers, respectively. Table 7.3 reports values of precision and recall, together

Table 7.2 Classifiers general accuracy and Kappa statistics

	Accuracy (%)	Kappa
SMO	94.6	0.83
AB_SMO	96.2	0.88
C_SMO	93.5	0.78
RF	94.4	0.82
C_RF	92.9	0.76

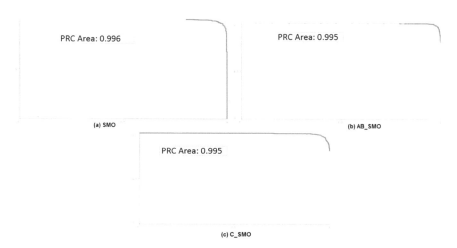

Fig. 7.3 Precision-Recall curves for kernel-based classifiers

Fig. 7.4 Precision-Recall curves for random forest classifiers

with F1-score, F2-score and AUC (Area under ROC) for class NR. In addition to the standard F1-score (the harmonic mean of precision and recall), we consider the F2-score as well; it weighs recall higher than precision, by placing more emphasis on false negatives which is exactly what we need in the described application.

Table 7.3 Precision, Recall and F-scores for class NR

	Precision (%)	Recall (%)	F1-score	F2-score	AUC
SMO	96.6	96.7	0.966	0.967	0.985
AB_SMO	97.8	97.4	0.976	0.975	0.986
C_SMO	94.2	97.9	0.960	0.971	0.979
RF	96.4	96.7	0.965	0.966	0.981
C_RF	93.6	97.8	0.957	0.969	0.979

As we can see from Table 7.3, both precision and recall, and the F-scores are very large, showing a satisfactory behavior for the considered classifiers. Finally, to further confirm this, also the AUC parameters show values very close to 1 in all the tested classifiers.

7.3 Medical Process Mining: A Semantic Approach

Process mining [10] describes a family of a-posteriori analysis techniques that exploit the so-called *event log*, a database which records information about the sequences (*traces* henceforth) of actions executed at a given organization. The most relevant and widely used process mining technique is *discovery*; process discovery takes as an input the event log and produces in output a process model.

Medical process mining, in particular, is a research field which is gaining attention in recent years (see, e.g., [16, 25]). Indeed, the complexity of healthcare [17] demands for proper techniques to analyze the *actual* patient management process applied at a given hospital setting, to identify bottlenecks and to assess the correct application of clinical guideline directives in practice.

Classical process mining algorithms, however, provide a purely syntactical analysis, where actions in the traces are processed only referring to their names. Action names are strings without any semantics, so that identical actions, labeled by synonyms, will be considered as different, or actions that are special cases of other actions will be processed as unrelated.

Leveraging process mining to the conceptual layer can enhance existing algorithms towards more advanced and reliable approaches. As a matter of fact, **semantic process mining**, defined as the integration of semantic processing capabilities into classical process mining techniques, has been recently proposed in the literature (see Sect. 7.4). However, while more work has been done in the field of semantic *conformance checking* (another branch of process mining) [5, 8], to the best of our knowledge semantic *process model discovery* needs to be further investigated.

In this section of the paper, we present a **semantic process mining** approach, tailored to medical processes. The approach exploits a **knowledge-based abstraction mechanism** (see Sect. 7.3.1), able to operate on event log traces. In our approach:

- actions in the log are mapped to the ground terms of an **ontology**;
- a **rule base** is exploited, in order to identify which of the multiple ancestors of an action should be considered for abstracting the action itself. **Medical knowledge and contextual information** are resorted to in this step;
- when a set of consecutive actions on the trace abstract as the same ancestor, they are merged into the same abstracted *macro-action*, labeled as the common ancestor at hand. This step requires a proper treatment of delays and/or actions in-between that descend from a different ancestor.

Our abstraction mechanism is then provided as an input to the *process model discovery* algorithms embedded in the open source framework ProM [30] (see Sect. 7.3.2), made semantic by the exploitation of domain knowledge in the abstraction phase.

We will describe our experimental work (see Sect. 7.3.3) in the field of stroke care, where the application of the abstraction mechanism on log traces has allowed us to mine simpler and more understandable process models, allowing for easier comparisons.

7.3.1 Knowledge-Based Trace Abstraction

In our framework, trace abstraction has been realized as a multi-step mechanism. The following subsections describe the various steps.

7.3.1.1 Ontology Mapping

As a first step, every action in the trace to be abstracted is mapped to a ground term of an **ontology**, formalized resorting to domain knowledge.

In our current implementation, we have defined an ontology related to the field of stroke management, where ground terms are patient management actions, while abstracted terms represent medical goals. Figure 7.5 shows an excerpt of the stroke domain ontology, formalized resorting to the Protègè editor.

In particular, a set of classes, representing the main goals in stroke management, have been identified, namely: "Administrative Actions", "Brain Damage Reduction", "Causes Identification", "Pathogenetic Mechanism Identification", "Prevention", and "Other". These main goals can be further specialized into subclasses, according to more specific goals (e.g., "Parenchima Examination" is a subgoal of "Pathogenetic Mechanism Identification", while "Early Relapse Prevention" is a subgoal of "Prevention"), down to the ground actions, that will implement the goal itself.

Some actions in the ontology can be performed to implement different goals. For instance, a Computer Assisted Tomography (CAT) can be used to check therapy

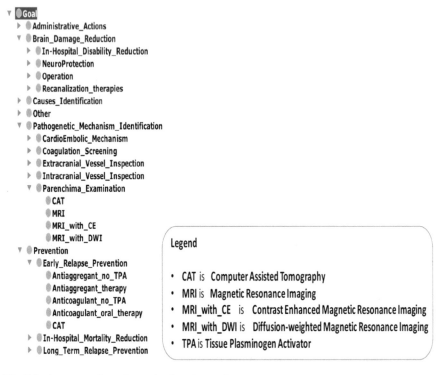

Fig. 7.5 An excerpt from the stroke domain ontology

efficacy in "Early Relapse Prevention", or to perform "Parenchima Examination" (see Fig. 7.5).

The proper goal to be used in the abstraction phase will be selected on the basis of the context of execution, as formalized in the rule base, described in the following subsection.

7.3.1.2 Rule-Based Reasoning for Ancestor Selection

As a second step in the trace abstraction mechanism, a **rule base** is exploited to identify which of the multiple ancestors of an action in the ontology should be considered for abstracting the action itself. The rule base encodes medical knowledge. Contextual information (i.e., the actions that have been already executed on the patient at hand, and/or her/his specific clinical conditions) is used to activate the correct rules. The rule base has been formalized in Drools [15].

As an example, referring to the CAT action mentioned in the previous subsection, the following rule states that, if intra-venous (ev_tPA) or intra-arterial (ia_tPA) anti-thrombotic therapies have been administered, then CAT implements the "Early Relapse Prevention" goal.

```
rule "CAT"
    when
        (groundActionIsBefore("ev_tPA") ||
        groundActionIsBefore("ia_tPA"))
    then
        macroAction.setAncestorName("Early_Relapse_Prevention");
end
```

where "groundActionIsBefore" is a function that, given the name of a ground action, returns true if this action precedes CAT in the trace, false otherwise.

On the contrary, if the context is different (i.e., anti-thrombotic therapy was not administered), CAT has to be intended as a means for "Parenchima Examination" (see Fig. 7.5).

More complex situations, where it is necessary to activate a chain of multiple rules, can also be managed by our system.

7.3.1.3 Trace Abstraction

Once the correct ancestor of every action has been identified, trace abstraction can be completed.

In this last step, when a set of consecutive actions on the trace abstract as the same ancestor, they have to be merged into the same abstracted ***macro-action***, labeled as the common ancestor at hand. This procedure requires a proper treatment of *delays*, and of actions in-between that descend from a different ancestor (*interleaved actions* henceforth).

Specifically, the procedure to abstract a trace operates as follows:

- for every action i in the trace:

  - i is abstracted as one of its ancestors (the one identified by the rule based reasoning procedure), at the ontology level chosen by the user; the macro-action m_i, labeled as the identified ancestor, is created;
  - for every element j following i in the trace:
    if j is a delay (or an interleaved action), its length is added to a variable, that stores the total delay duration (or the total interleaved action duration, respectively) accumulated so far during the abstraction of i;
    if j is an action that, according to domain knowledge, abstracts as the same ancestor as i, m_i is extended to include j, provided that the total durations mentioned above do not exceed domain-defined thresholds. j is then removed from the actions in the trace that could start a new macro-action, since it has already been incorporated into an existing one.
  - the macro-action m_i is appended to the output abstracted trace which, in the end, will contain the list of all the macro-actions that have been created by the procedure.

7.3.2   Interface to the ProM Tool

In our approach, process mining, made semantic by the exploitation of the abstraction mechanism illustrated above, is implemented resorting to the well-known process mining tool ProM, extensively described in [30]. ProM (and specifically its newest version ProM 6) is a platform-independent open source framework that supports a wide variety of process mining and data mining techniques, and can be extended by adding new functionalities in the form of plug-ins.

For the work described in this paper, we have exploited ProM's Heuristic Miner [31]. Heuristic Miner is a plug-in for process model discovery, able to mine process models from event logs. It receives in input the log, and considers the order of the actions within every single trace. It can mine the presence of short-distance and long-distance dependencies (i.e., direct or indirect sequence of actions), and information about parallelism, with a certain degree of reliability. The output of the mining process is provided as a graph, known as the "dependency graph", where nodes represent actions, and edges represent control flow information. The output can be converted into other formalisms as well.

Currently, we have chosen to rely on Heuristics Miner, because it is known to be tolerant to noise, a problem that may affect medical event logs (e.g., sometimes the logging may be incomplete). Anyway, testing of other mining algorithms available in ProM 6 is foreseen in our future work.

7.3.3   Experimental Results

In this section, we describe the experimental work we have conducted for semantic process mining, in the application domain of stroke care.

The available event log is composed of more than 15000 traces, collected at 40 stroke units (SUs) located in northern Italy. Traces are composed of 13 actions on average.

We have tested whether our capability to abstract the event log traces on the basis of their semantic goals allowed to obtain process models where unnecessary details are hidden, but key behaviors are clear. Indeed, if this hypothesis holds, in our application domain it becomes easier to compare process models of different SUs, highlighting the presence/absence of common paths, regardless of minor action changes (e.g., different ground actions that share the same goal) or irrelevant different action ordering or interleaving (e.g., sets of ground actions, all sharing a common goal, that could be executed in any order).

Figure 7.6 compares the process models of two different SUs (SU-A and SU-B), mined by resorting to Heuristic Miner, operating on ground traces. Figure 7.7, on the other hand, compares the process models of the same SUs as Fig. 7.6, again mined by resorting to Heuristic Miner, but operating on traces abstracted according to the

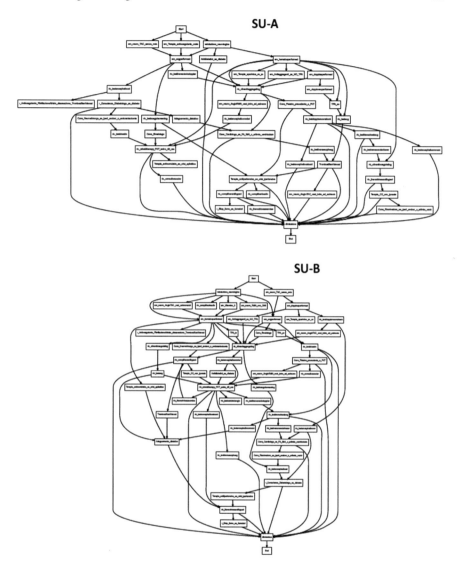

Fig. 7.6 Comparison between two process models, mined by resorting to Heuristic Miner, operating on ground traces. The figure is not intended to be readable, but only to give an idea of how complex the models can be

goals of the ontology in Fig. 7.5. In particular, abstraction was conducted up to level 2 in the ontology (where level 0 is the root, i.e.. "Goal").

Generally speaking, a visual inspection of the two graphs in Fig. 7.6 is very difficult. Indeed, these two ground processes are "spaghetti-like" [28], and the extremely large number of nodes and edges makes it hard to identify commonalities in the two models.

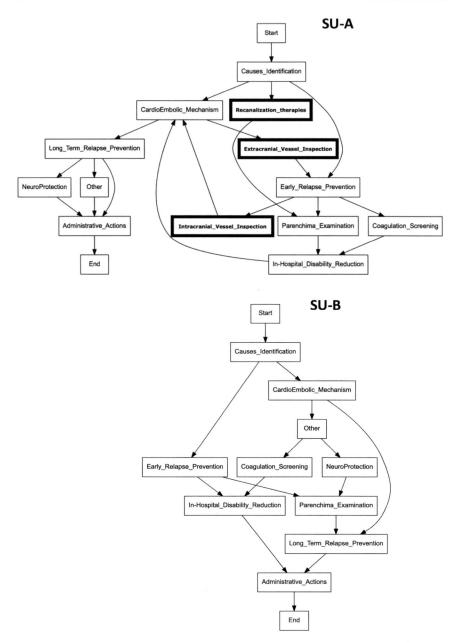

Fig. 7.7 Comparison between the two process models of the same SUs as Fig. 7.6, mined on abstracted traces. Additional macro-actions executed at SU-A are highlighted in bold

Table 7.4 Average fitness values calculated on the mined process models, when operating at different levels of abstraction

Ground	Abs. level 2	Abs. level 1
0.54	0.78	0.89

The abstract models in Fig. 7.7, on the other hand, are much more compact, and it is possible for a medical expert to analyze them.

In particular, the two graphs in Fig. 7.7 are not identical, but in both of them it is easy to a identify the macro-actions which corresponds to the treatment of a typical stroke patient, as precribed by the guideline.

However, the model for SU-A at the top of Fig. 7.7 exhibits a more complex control flow (with the presence of loops), and shows three additional macro-actions with respect to the model of SU-B, namely "Extracranial Vessel Inspection", "Intracranial Vessel Inspection" and "Recanalization". This finding can be explained, since SU-A is a better equipped SU, where different kinds of patients, including some atypical/more critical ones, can be managed, thanks to the availability of different skills and instrumental resources. These patients may require the additional macro-actions reported in the model, and/or the repetition of some procedures, in order to better characterize and manage the patient's situation.

On the other hand, SU-B is a more generalist SU, where very specific human knowledge or technical resources are missing. As a consequence, the overall model control flow is simpler, and some actions are not executed at all.

Finally, Table 7.1 reports our results on the calculation of *fitness* [28] on the process models mined for our 40 SUs, at different levels of abstraction. Fitness evaluates whether a process model is able to reproduce all execution sequences that are in the event log. If the log can be replayed correctly, fitness evaluates to 1. In the worst case, it evaluates to 0. Fitness calculation is available in ProM, and acts as an objective and quantitative metric for process mining evaluation (Table 7.4).

As it can be observed from the table, the more the traces are abstracted, the more the average fitness value increases in the corresponding models obtained by the mining algorithm.

In conclusion, our abstraction mechanism, while hiding irrelevant details, allows to still appreciate relevant differences between models, such as, e.g., the presence/absence of relevant actions. Moreover, very interestingly, abstraction proves to be a means to significantly increase the quality of the mined models, measured in terms of fitness, a well known and largely adopted quantitative indicator.

7.4 Related Works

The discovery of alimentary frauds through data mining techniques is a very young area of research, since the detection of frauds and adulteration in food models has been traditionally performed analyzing the chemical or imaging profiles using Principal

Component Analysis (PCA) or regression models. In the very recent years, the switch from these analysis techniques to data mining classification models proved that this direction can be considered as a promising advance of the state of the art.[1]

PCA and regression models have been applied in many studies. In [14], the authentication of extra virgin olive oil and its adulteration with lower-priced oils is investigated using partial least squares and genetic programming on the results of Raman spectroscopy. Partial Least Squares Regression (PLSR) is vastly applied for the identification of meat adulterations: [7, 26] detect adulterations of beefs, starting from Fourier transform infrared (FTIR) spectroscopy and Multispectral image analysis data, respectively, while [12] applies PLSR on visible near-infrared hyperspectral imaging data to identify adulterations in chicken minced meat.

Machine learning-based models have been used to improve the outcome of the fraud detection frameworks. Here, the classification models are used as a nonlinear multivariate calibration procedure, in order to assess the parameter settings allowing for better recognition of frauds or adulterations. Examples of this procedure can be found in [1], where a least-squares support vector machine (LS-SVM) has been used to improve the recognition rate of common adulterants in powdered milk, and in [27], where SVM is used to elaborate PCA scores with the aim of optimizing the detection of meat and fat adulteration.

The application of machine learning classification models to separate and identify adulterated food samples on the basis of their chemical and/or imaging profiles has been proposed, e.g., in [35], where Near-Infrared (NIR) spectroscopy combined with chemometrics methods has been used to detect adulteration of honey samples and to compare five classification modeling methods: SVM, LS-SVM, back propagation artificial neural network (BP-ANN), linear discriminant analysis (LDA), and K-nearest neighbors (KNN). An improved SVM (I-SVM) and an improved and simplified KNN (IS-KNN) have been introduced in [33], to detect adulteration of milk samples containing different pseudo proteins and thickeners, based on near infrared (NIR) spectra data.

Despite the use of machine learning classifiers is not yet widely applied in food fraud and adulteration detection, the recent results showed a very good potential and form a good basis for further investigations if this area.

On the other side, the use of semantics in business process management, with the aim of operating at different levels of abstractions in process discovery and/or analysis, is a relatively young area of research as well, where much is still unexplored.

The topic was studied in the SUPER project [22], within which several ontologies were created, such as the process mining ontology and the event ontology [21]; these ontologies define core terminologies of business process management, usable by machines for task automation. However, the authors did not present any concrete implementations of semantic process mining or analysis.

---

[1] This does not mean that techniques like PCA are not used in machine learning approaches, however they are often adopted as preliminary step (e.g. for feature reduction) rather than as main techniques to identify classes as in SIMCA analysis [32].

Ontologies, references from elements in logs to concepts in ontologies, and ontology reasoners (able to derive, e.g., concept equivalence), are described as the three essential building blocks for semantic process mining in [5]. This paper also shows how to use these building blocks to extend ProM's LTL Checker [29] to perform semantic auditing of logs.

In [4] an example of process discovery at different levels of abstractions is presented. It is however a very simple example, where a couple of ground actions are abstracted according to their common ancestor. The management of interleaved actions or delays is not addressed, and multiple inheritance is not considered. A more recent work [11] introduces a methodology that combines domain and company-specific ontologies and databases to obtain multiple levels of abstraction for process mining. In this paper data in databases become instances of concepts at the bottom level of a taxonomy tree structure. If consecutive tasks in the discovered model abstract as the same concepts, those tasks are aggregated. However, also in this work we could find neither a clear description of the abstraction algorithm, nor the management of interleaved actions or delays.

Referring to medical applications, the work in [8] proposes an approach, based on semantic process mining, to verify the compliance of a computer interpretable guideline with medical recommendations. In this case, semantic process mining refers to conformance checking rather than to process discovery (as it is also the case in [5]). These two works are thus only loosely related to our contribution, which grounds on our previous papers [19, 20] and extends them. In particular, in [19] we proposed a simpler version of the framework, where rule based reasoning was not adopted. With respect to [19], on the other hand, we provide here some new experimental results.

7.5 Conclusions

In this paper, we have illustrated how knowledge extraction from bio-medical databases can provide crucial support in different applications, sharing the goal of guaranteeing health safeguard.

In particular, as a first case study, we have reported our experiments on the use of machine learning and data mining algorithms to detect food adulteration. Experiments have been conducted on chemometrical analyses of raspberry purees, in order to discriminate between adulterated and non-adulterated fruit samples. Supervised classification strategies, in particular, performed very well in recognizing the adulterated samples, even in presence of an unbalanced data-set. In the future, we will consider the adoption and validation of a larger number of classification strategies, for example model stacking or specific models for one-class problems. Furthermore, experiments will be conducted on other food models, whose adulteration is related not only to health problems, but also to economical damages (as it is the case of fish, honey and wine, which are among the mostly adulterated food models in Europe).

As a second case study, we have presented a framework for semantic medical process mining, resorting to knowledge-based abstraction of event log traces. Experimental results in the field of stroke management have proved that the capability of abstracting the event log traces on the basis of their semantic goal allows to mine clearer process models, where unnecessary details are hidden, but key behaviors are clear. In the future, we plan to conduct further experiments, by quantitatively comparing different process models (of different SUs) obtained from abstracted traces. Comparison will resort to knowledge-intensive process similarity metrics, such as the one we described in [18]. We will also extensively test the approach in different application domains.

Bio-medical databases span in a very large area and present peculiar challenges, to be tackled from different points of view. Interestingly, our experience in this field has not only led to consistent and concretely useful experimental outcomes, but also to significant methodological advances with respect to the existing research.

References

1. Borin, A., Ferro, M.F., Mello, C., Maretto, D.A., Poppi, R.J.: Least-squares support vector machines and near infrared spectroscopy for quantification of common adulterants in powdered milk. Anal. Chimica Acta **579**(1):25 – 32, 2006
2. Breiman, L.: Random forests. Mach. Learn. **45**(1), 5–32 (2001)
3. Davis, R., Mauer, L.J.: Fourier transform infrared (ftir) spectroscopy: a rapid tool for detection and analysis of foodborne pathogenic bacteria. Curr. Res. Technol. Educ. Top. Appl. Microbiol. Microb. Biotechnol. **2**, 1582–1594 (2010)
4. de Medeiros, A.K.A., van der Aalst, W.M.P.: Process mining towards semantics. In: Dillon, T.S. Chang, E., Meersman, R., Sycara, K.P. (eds.) Advances in Web Semantics I—Ontologies, Web Services and Applied Semantic Web. Lecture Notes in Computer Science, vol. 4891, pp. 35–80. Springer (2009)
5. de Medeiros, A.K.A., van der Aalst, W.M.P., Pedrinaci, C.: Semantic process mining tools: Core building blocks. In: Golden, W., Acton, T., Conboy, K., van der Heijden, H., Tuunainen, V.K. (eds.) 2008 16th European Conference on Information Systems, ECIS, pp. 1953–1964. Galway, Ireland (2008)
6. Elkan, C.: The foundations of cost-sensitive learning. In: Proceedings of the 17th International Joint Conference on Artificial Intelligence, IJCAI 01, pp. 973–978. Seattle, WA (2001)
7. Ellis, D.I., Broadhurst, D., Goodacre, R.: Rapid and quantitative detection of the microbial spoilage of beef by fourier transform infrared spectroscopy and machine learning. Anal. Chimica Acta **514**(2):193–201 (2004)
8. Grando, M.A., Schonenberg, M.H., van der Aalst, W.M.P.: Semantic process mining for the verification of medical recommendations. In: Traver, V., Fred, A.L.N., Filipe, J., Gamboa, H. (eds.) HEALTHINF 2011–Proceedings of the International Conference on Health Informatics, pp. 5–16. SciTePress Rome, Italy, 26–29 Jan 2011
9. Hall, M.A.: Correlation-based feature subset selection for discrete and numeric class machine learning. In: Proceedings of the 17th International Conference on Machine Learning, ICML, 2000, pp. 359–366. Stanford, CA (2000)
10. IEEE Taskforce on Process Mining: Process Mining Manifesto. http://www.win.tue.nl/ieeetfpm/doku.php?id=shared:process_mining_manifesto. Accessed 22 Nov 2017
11. Jareevongpiboon, W., Janecek, P.: Ontological approach to enhance results of business process mining and analysis. Bus. Proc. Manag. J. **19**(3), 459–476 (2013)

12. Kamruzzaman, M., Makino, Y., Oshita, S.: Rapid and non-destructive detection of chicken adulteration in minced beef using visible near-infrared hyperspectral imaging and machine learning. J. Food Eng. **170**, 8–15 (2016)
13. Kemsley, E.K., Holland, J.K., Defernez, M., Wilson, R.H.: Detection of adulteration of raspberry purees using infrared spectroscopy and chemometrics. J. Agric. Food Chem. pp. 3864–3870 (1996)
14. Lpez-Dez, E.C., Bianchi, G., Goodacre, R.: Rapid quantitative assessment of the adulteration of virgin olive oils with hazelnut oils using raman spectroscopy and chemometrics. J. Agric. Food Chem. PMID: 14518936, **51**(21):6145–6150 (2003)
15. De Maio, M.N., Salatino, M., Aliverti, E.: Mastering JBoss Drools 6 for Developers. Packt Publishing (2016)
16. Mans, R., Schonenberg, H., Leonardi, G., Panzarasa, S., Cavallini, A., Quaglini, S., et al.: Process mining techniques: an application to stroke care. In: Andersen, S., Klein, G.O., Schulz, S., Aarts, J. (eds.) Proceedings of the MIE. Studies in Health Technology and Informatics, vol. 136, pp. 573–578. IOS Press, Amsterdam (2008)
17. Mans, R., van der Aalst, W., Vanwersch, R., Moleman, A.: Process mining in healthcare: Data challenges when answering frequently posed questions. In: Lenz, R., Miksch, S., Peleg, M., Reichert, M., Riaño, D., ten Teije, A. (eds.) ProHealth/KR4HC. Lecture Notes in Computer Science, vol. 7738, pp. 140–153. Springer, Berlin (2013)
18. Montani, S., Leonardi, G., Quaglini, S., Cavallini, A., Micieli, G.: A knowledge-intensive approach to process similarity calculation. Expert Syst. Appl. **42**(9), 4207–4215 (2015)
19. Montani, S., Leonardi, G., Striani, M., Quaglini, S., Cavallini, A.: Multi-level abstraction for trace comparison and process discovery. Expert Syst. Appl. **81**, 398–409 (2017)
20. Montani, S., Striani, M., Quaglini, S., Cavallini, A., Leonardi, G.: Towards semantic process mining through knowledge-based trace abstraction. In: Ceravolo, P., van Keulen, M., Stoffel, K. (eds.) Proceedings of the 7th International Symposium on Data-driven Process Discovery and Analysis (SIMPDA 2017). CEUR Workshop Proceedings, vol 2016, pp. 98–112. CEUR-WS.org, Neuchâtel, Switzerland, 6-8 Dec 2017
21. Pedrinaci, C., Domingue, J.: Towards an ontology for process monitoring and mining. In: Hepp, M., Hinkelmann, K., Karagiannis, D., Klein, R., Stojanovic, N. (eds.) Proceedings of the Workshop on Semantic Business Process and Product Lifecycle Management SBPM 2007, held in conjunction with the 3rd European Semantic Web Conference (ESWC 2007). CEUR Workshop Proceedings, vol. 251, Innsbruck, Austria, 7 June 2007
22. Pedrinaci, C., Domingue, J., Brelage, C., van Lessen, T., Karastoyanova, D., Leymann, F.: Semantic business process management: Scaling up the management of business processes. In: Proceedings of the 2th IEEE International Conference on Semantic Computing (ICSC 2008),pp. 546–553. IEEE Computer Society, Santa Clara, California, USA, 4–7 Aug 2008
23. Platt, J.: Fast training of support vector machines using sequential minimal optimization. In: Schölkopf, B., Burges, C.J.C., Smola, A.J. (eds.) Advances in Kernel Methods, pp. 185–208. MIT Press (1999)
24. Portinale, L., Saitta, L.: Feature selection. Technical Report D.14.1, Mining Mart Project (2002). http://mmart.cs.uni-dortmund.de/content/publications.html
25. Rojas, E., Munoz-Gama, J., Sepulveda, M., Capurro, D.: Process mining in healthcare: a literature review. J. Biomed. Inform. **61**, 224–236 (2016)
26. Ropodi, A.I., Pavlidis, D.E., Mohareb, F., Panagou, E.Z., Nychas, G.-J.E.: Multispectral image analysis approach to detect adulteration of beef and pork in raw meats. Food Res. Int. **67**, 12–18 (2015)
27. Schmutzler, M., Beganovic, A., Bhler, G., Huck, C.W.: Methods for detection of pork adulteration in veal product based on FT-NIR spectroscopy for laboratory, industrial and on-site analysis. Food Control **57**, 258–267 (2015)
28. van der Aalstm, W.: Process Mining. Data Science in Action. Springer (2016)
29. van der Aalst, W.M.P., de Beer, H.T., van Dongen, B.F.: Process mining and verification of properties: an approach based on temporal logic. In: Meersman, R., Tari, Z., Hacid, M., Mylopoulos, J., Pernici, B., Babaoglu, Ö., Jacobsen, H., Loyall, J.P., Kifer, M., Spaccapietra, S. (eds.) On the

Move to Meaningful Internet Systems 2005: CoopIS, DOA, and ODBASE, OTM Confederated International Conferences CoopIS, DOA, and ODBASE 2005. Lecture Notes in Computer Science, Proceedings, Part I, vol. 3760, pp. 130–147. Agia Napa, Cyprus, Oct 31 – 4 Nov 2005. Springer 2005

30. van Dongen, B., Alves De Medeiros, A., Verbeek, H., Weijters, A., van der Aalst, W.: The proM framework: a new era in process mining tool support. In: Ciardo, G., Darondeau, P. (eds.) Knowledge Mangement and its Integrative Elements, pp. 444–454. Springer, Berlin (2005)

31. Weijters, A., van der Aalst, W., Alves de Medeiros, A.: Process Mining with the Heuristic Miner Algorithm, WP 166. Eindhoven University of Technology, Eindhoven (2006)

32. Wold, S., SJSTRM, M.: SIMCA: A Method for Analyzing Chemical Data in Terms of Similarity and Analogy, Chap. 12, pp. 243–282

33. Zhang, L.-G., Zhang, X., Ni, L.-J., Xue, Z.-B., Xin, G., Huang, S.-X.: Rapid identification of adulterated cow milk by non-linear pattern recognition methods based on near infrared spectroscopy. Food Chem. **145**, 342–348 (2014)

34. Zhou, Z-H., Liu, X-Y.: On multi-class cost-sensitive learning **26**(3):232–257 (2010)

35. Zhu, X., Li, S., Shan, Y., Zhang, Z., Li, G., Donglin, S., Liu, F.: Detection of adulterants such as sweeteners materials in honey using near-infrared spectroscopy and chemometrics. J. Food Eng. **101**(1), 92–97 (2010)

Chapter 8
A Paradigm for Democratizing Artificial Intelligence Research

Erwan Moreau, Carl Vogel and Marguerite Barry

Abstract This proposal outlines a plan for bridging the gap between technology experts and society in the domain of Artificial Intelligence (AI). The proposal focuses primarily on Natural Language Processing (NLP) technology, which is a major part of AI and offers the advantage of addressing problems that non-experts can understand. More precisely, the goal is to advance knowledge at the same time as opening new communication channels between experts and society, in a way which promotes non-expert participation in the conception of NLP technology. Such interactions can happen in the context of open-source development of languages resources, i.e. software tools and datasets; existing usages in various communities show how projects which are open to everyone can greatly benefit from the free participation of enthusiastic contributors (participation is not at all limited to software development). Because NLP research is mostly experimental and relies heavily on software tools and language datasets, this project proposes to interconnect the societal issues related to AI with the NLP research resources issue.

8.1 Introduction

We propose a means of bridging the gap between technology experts and society in the domain of Artificial Intelligence (AI). Here we focus primarily on Natural Language Processing (NLP) science and technology, which is a major part of AI, for

E. Moreau (✉) · C. Vogel
Computational Linguistics Group, Trinity College Dublin & the SFI ADAPT Centre,
Dublin, Ireland
e-mail: moreaue@tcd.ie

C. Vogel
e-mail: vogel@tcd.ie

M. Barry
School of Information and Communication Studies, University College Dublin,
Dublin, Ireland
e-mail: marguerite.barry@ucd.ie

© Springer Nature Switzerland AG 2019
A. Esposito et al. (eds.), *Innovations in Big Data Mining and Embedded Knowledge*,
Intelligent Systems Reference Library 159,
https://doi.org/10.1007/978-3-030-15939-9_8

137

practical reasons explained in Sect. 8.2.1.2. However, the proposal we articulate takes the form of a paradigm[1] which may be adapted to other areas of artificial intelligence, retaining the key organizational principles. "Big data" creates the promise of gaining information not available through more sampling, but at the expense of extreme processing challenges. Under this perspective, our proposal focuses on extending the possibilities for scientific exploration and technology development outside the confines of academic and commercial labs to include the general public in a kind of principled "crowd-sourcing".

One goal that we take to be shared generally with the research community, concomitant with advancing knowledge, is to have a dialogue with the society about new findings and the direction of scientific progress. We propose to open new communication channels between experts and society, in a way which promotes non-expert participation in the conception of NLP technology. Such interactions can happen in the context of open-source development of languages resources, i.e. software tools and datasets; existing uses in various communities show how projects which are open to everyone can greatly benefit from the free participation of enthusiastic contributors (participation is not at all limited to software development). Because NLP research is mostly experimental and relies heavily on software tools and language datasets, this project proposes to interconnect the societal issues related to AI with the NLP research resources issue. Thus the key objectives of the proposal are:

1. Improve the reproducibility of NLP research, especially via more systematic publication of software and data resources;
2. Improve the reusability and dissemination of language resources coming from NLP research, in a way which improves openness and transparency in NLP research;
3. Reach out to a larger public, and encourage interactions between experts and the general public using open-source research resources as a support for discussion;
4. Integrate contributions from the general public into the research life cycle, i.e. allow the general public to become an actor of innovation[2] rather than a passive user only;
5. Credit contributions from individuals from all participating sectors in a manner that respects each individual's right to privacy as well as fair recompense.

In Sect. 8.2 we go over the reasons why there is a large consensus about the interest, if not the need, to include the general public in the "AI revolution" as much as possible; we also look more closely at some specific challenges related to resources in the field of Natural Language Processing (NLP). Then in Sect. 8.3 we explain how these two challenges can be addressed together; we propose an original and pragmatic approach based on involving the general public very early in the making of the resources used in NLP research. Finally in Sect. 8.4 we examine the potentially

---

[1]This is one of a number of senses of "paradigm" as used by Kuhn [25, p. 23]; see also, [31].

[2]The word *innovation* is used here in a general sense, that is, including but not restricted to commercial applications.

far-reaching benefits that can arise from the proposal on many levels: ethics, society, research and industry.

8.2 Motivations

8.2.1 Familiarizing the General Public with AI

8.2.1.1 Society and AI

AI technology is ubiquitous. The development of AI carries a number of societal and ethical questions (e.g. [32, 56]). AI is already introducing major disruptions in many areas of life (e.g. [7]), and is likely to cause more in the future. There are legitimate concerns about how the general public might react to the potential problems caused by AI [47], as well as about potential negative biases or flaws in AI systems (e.g. [2, 43]) which might prove hard to detect.

We think that lack of information about AI technology is a major source of confusion, misunderstanding and even potentially fear among the general public. As a matter of fact, lack of transparency itself in a legitimate cause for concern from a democratic perspective. Currently, the main providers of AI that the public is exposed to are large private companies, typically the "GAFA"[3]; these companies lead the field in terms of breaking the barriers of AI (e.g. [11]), but the AI technology used in their everyday products is not open to scrutiny; this can lead to various kinds of suspicions, based on reality or not [53]. Most of the information that mainstream media give to the general public is understandably short and focuses on major achievements or societal questions. Overall, there are very few ways for the general public to familiarize themselves with AI technology, unless they are willing to enroll for a university degree on the topic.

Clearly, the question of *how* AI works requires a level of expertise which is worth acquiring almost only for professionals. Nevertheless, the *what* question can address many concerns: by "familiarizing" the general public with AI, we mean making it possible and accessible for non-experts to get a general sense of *what* AI can do (and what it cannot), using *what* kind of information.[4] The accessibility of knowledge about AI is currently limited mostly to experts. We advocate for opening access to non-experts, through realistic feasible means.

---

[3] Acronym for the four major technology companies: Google, Apple, Facebook and Amazon.

[4] As a simple analogy, how many people know how modern car engines work? Certainly very few, but the vast majority of people are comfortable with using cars nonetheless.

8.2.1.2 Why Natural Language Processing (NLP) Technology Is a Good Starting Point

The challenge of popularizing AI is obvious: it is a very advanced technology which requires some scientific background. Even in the perspective that we propose of "familiarization" to AI (defined in Sect. 8.2.1.1), some applications include domain-specific complexities which require specialized expert background (e.g. in physics or biology). However some applicative domains are more intuitive; compared to other fields, NLP has several advantages with respect to popularization of scientific knowledge:

- NLP focuses on automatizing language-related tasks, and nearly every adult human knows how to speak, and possibly read and write at least one language. In other words, the object of study is something that everyone is familiar with. Thus, understanding the main NLP problems and applications does not require any advanced knowledge. For instance, everyone does not know how machine translation works, but they understand what machine translation is supposed to do. This is a great advantage compared to domains in which understanding the problems themselves requires advanced knowledge.
- NLP contributes to applied science—its outcomes have concrete applications in the form of software libraries or products. Nowadays a large proportion of the population is computer-savvy (at least in developed countries). Additionally the immaterial nature of software technology makes it an ideal product in terms of logistical costs, since it can be reproduced and distributed cost-free (as opposed to any physical-device dependent technology).
- NLP is largely an experimental science. The concept of experiment is intuitive and can be appealing to many people. Thanks to the previously mentioned advantages (little domain-specific knowledge required and software tools), there is no major obstacle to non-experts being able to understand, reproduce and do their own NLP experiments (see also Sect. 8.2.2.1 below).
- NLP is large. Developing and testing theories of the world's languages—past, present and future—involves an expanse of data that no one individual can ever hope to comprehend in any literal sense. This invites the benefit of contributions from many contributors across many categories of expertise, experts and citizen scientists alike.
- NLP is AI-complete, in the sense that it involves problems (such as enumeration of ambiguities inherent in natural language expressions) that are in the worst-case intractable, and which therefore require approximations in order to achieve human-level performance, in equivalence with other domains of AI research. Therefore, gaining advantage from a "big team" approach in NLP may be expected to yield progress in other domains of AI as well.

The scope of NLP is quite flexible. It intersects with various other domains or sub-domains, e.g. information retrieval, knowledge representation, data mining, multimodality and machine learning (ML). We propose to focus on NLP as a starting point for the reasons explained above, and then to progressively broaden the range of domains/applications, for instance to other kinds of concrete documents (images, audio, video). In the remainder of this document we will use NLP as the target field of interest.

8.2.2   Boosting NLP Research by Enhancing NLP Tools

8.2.2.1   NLP: Experimental Research

Much of the research carried out in NLP involves a corpus-driven view of language (i.e. in which language is defined empirically) conveniently coupled with Machine Learning (ML) methods. Naturally, this empirical approach relies heavily on experimental results: by definition, corpus-driven NLP aims at automating the extraction of some form of "language knowledge" from some "language data"; an experimental process is also used to validate, compare and sometimes combine ML methods. Experiments usually involve some kind of software prototype, or at least some form of algorithm describing the steps of the experiment (i.e. potentially convertible to a software prototype), as well as some data resources used to train and/or validate the system.

As a means of verifying or extending the findings, research papers are supposed to make it possible to replicate and/or reproduce the experiments. In particular, they usually give some details about the data and software used, sometimes by referencing previously published material for some part. Recently, major NLP conferences (e.g. ACL,[5] COLING,[6] EMNLP[7] or LREC[8]) have increasingly encouraged authors to favor reproducibility in different ways, especially by making their software and their data resources publicly available. Additionally, in many research projects, resources

---

[5]*"Papers that are submitted with accompanying software/data may receive additional credit toward the overall evaluation score, and the potential impact of the software and data will be taken into account when making the acceptance/rejection decisions."* ACL 2018 Call For Papers, http://www. acl2018.org/call-for-papers/—last verified: February 2018.

[6]COLING invites papers in 6 categories including *"Reproduction papers"* and *"Resource papers"*. COLING Call For Papers, http://www.coling2018.org/final-call-for-papers/—last verified: February 2018.

[7]*"Each EMNLP 2018 submission can be accompanied by a single PDF appendix, one .tgz or .zip archive containing software, and one .tgz or .zip archive containing data. EMNLP 2018 encourages the submission of these supplementary materials to improve the reproducibility of results, and to enable authors to provide additional information that does not fit in the paper."* EMNLP 2018 Call For Papers, http://www.emnlp2018.org/calls/papers/—last verified: February 2018.

[8]*"Describing your LRs [Language Resources] in the LRE Map is now a normal practice in the submission procedure of LREC (introduced in 2010 and adopted by other conferences). To continue the efforts initiated at LREC 2014 about "Sharing LRs" (data, tools, web-services, etc.), authors will have the possibility, when submitting a paper, to upload LRs in a special LREC repository. This effort of sharing LRs, linked to the LRE Map for their description, may become a new "regular" feature for conferences in our field, thus contributing to creating a common repository where everyone can deposit and share data."* LREC 2018 web page, http://www.lrec2018.lrec-conf.org/en/—last verified: February 2018.

are part of the outcomes, typically meant to demonstrate the validity and feasibility of a method. Thus, there seems to be a positive trend towards publishing the corresponding software and data resources together with experimental research papers.

Data resources are often shared because they are costly to produce. However, probably most software prototypes are still either not published or not publicized, and consequently never re-used; once they have served their purpose as support for an experiment and the results have been published, unless their authors plan to re-use them in subsequent work, they are simply abandoned.[9] Of course, not every software prototype is worth being re-used, but out of the thousands of experimental papers published every year, very few prototypes get enough scientific and/or technological interest to be deemed worthy of devoting resources to their maintenance, development and documentation.

Hence the vast majority of software prototypes are either lost or re-used only by a small group of people, typically the team in which it originated. At the other end of the spectrum, only a handful of tools are widely used by the NLP community. We explain below in Sects. 8.2.2.2 and 8.2.2.3 why this is problematic from a scientific perspective.

8.2.2.2 Reproducibility

The scientific value of NLP research relies on a rigorous evaluation of experimental results. The experimental setup in which the results are obtained is a major part of this evaluation, as well as the choices related to the data used in the experiment (origin, type, size, annotations, etc.). Results can vary widely depending on the data used as input, since textual data is extremely diverse. In the typical context of supervised ML this has to be expected; but NLP methods also aim at a certain level of generality: a method which does not scale to various kinds of text data is of little use. This is a distinctive feature of NLP which often surprises ML experts who are more familiar with different types of data: the richness and diversity of language makes the task of finding generalizable patterns much harder.[10]

With ML applied to text data, it is common to reach a satisfying level of performance with a specific dataset by "cherry-picking" the features and meta-parameters of the model, intentionally or by mistake.[11] The reviewing process of experimental papers includes checking for this kind of scientific error; this is why authors have to describe the experimental setup, including the steps they take to avoid overfitting

---

[9]For example, it is very common that Ph.D. students or postdocs implement some software tool as part of their Ph.D. project, but it is rarely picked up when the author of the software leaves at the end of their contract.

[10]One might reflect on the observations of those who try to identify linguistic universals. One might imagine it possible to generalize that nouns and verbs are categories of words attested in every human language. However, even this candidate universal is controversial [14].

[11]This is indeed a common beginner's mistake: it is easy to overlook the risk of overfitting, and to mistakenly interpret good performance results on a test set drawn from the same dataset as the training set as a sign that the approach is valid.

for instance (if such details are missing in a paper, experienced reviewers take it into account in their final evaluation). However, because the space is limited in most papers, and because the paper is supposed to focus on the main contribution rather than give every possible technical detail, the reviewing process cannot guarantee that all the conclusions made by the authors are correct, let alone generalizable. In other words, the paper itself is not always a sufficient proof of the validity of a method.

Replicability and/or reproducibility[12] are major evaluation tools in the methodology of experimental sciences such as NLP. A researcher or a reviewer, even doing their job diligently and honestly, is subject to many potential biases [41]. Ultimately, the scientific value of a finding lies in the extent to which it is reproducible.[13] The field of NLP has seen significant progress in this areas, for example with the well established organization of competitions[14] as a means to compare different approaches on the same grounds; participants in these competitions are encouraged to publish their software tools, so that everyone can reproduce their results and build upon them. In general, an increasing amount of research software and data is published together with traditional papers in many conferences. Despite the existence of many such initiatives, a majority of the papers published in NLP describe research without providing the means to reproduce it; it might be reproducible, but the cost involved in developing a system to actually reproduce it is prohibitive. Besides, since NLP does not deal directly with critical issues such as health or aircraft safety, there is no major need for a strict and thorough validation of its findings. As a consequence reproducibility studies are very rare in NLP, and the state of the art can be fuzzy in certain areas about which method works best for a particular application.

8.2.2.3 Today's Software Prototypes Are Tomorrow's Research Instruments

The relatively low level of software outcomes from NLP research weakens reproducibility, but also impacts the pace of research itself, especially academic research. Near-future challenges in the domain will precisely rely more and more on robust, scalable and complex software components, at least for two reasons: the amount of data to process and the growing complexity of the tasks that we try to achieve (hence a growing need for a variety of modular components).

---

[12]The exact definition of replicability and reproducibility can differ by scientific field. Nevertheless the two always differ in a way which makes replicability a more general condition than reproducibility: according to Wikipedia, reproducibility is the ability to get the same research results using the exact same raw data and computer programs as the original researchers, whereas replicability is the ability to independently achieve similar conclusions when differences in sampling, research procedures and data analysis methods may exist (https://www.en.wikipedia.org/wiki/Reproducibility—last verified: November 2017).

[13]That is, it is easy to reproduce a demonstration that water flows downhill, but this is not a "finding". "Cold fusion" is a finding, but it is difficult to reproduce.

[14]Also called "challenges" or "shared tasks".

"Normal" scientific research[15] progresses iteratively by incremental improvements, a fact commonly summarized by the phrase *"standing on the shoulders of giants"*.[16] The re-use of previous findings is the bread and butter of NLP research, either to improve over existing research or to combine them in order to solve new problems. This is true about theoretical findings as well as empirical results, the latter being ubiquitous in NLP research (as explained in Sect. 8.2.2.1). As a consequence, research prototypes originally developed to demonstrate the validity of a method (or clones based on them) are often re-used to achieve new goals. Moreover, text processing traditionally involves a processing chain: different levels of linguistic information are extracted incrementally in order to obtain a rich linguistic representation of the structure of a sentence or document. Every level is analyzed in turn using a tool specialized for this task. The processing chain typically starts with word and sentence segmentation, followed by Part-Of-Speech (POS) tagging and lemmatization; depending on the task, other steps might consist in named entity recognition, dependency parsing or other more advanced tasks. This hierarchical process implies that the accuracy at a given stage depends on the quality of the results obtained at earlier stages. Thus, error propagation, i.e. the fact that a minor error at an early stage might cause more serious errors which accumulate down the stream, is an important issue: the more advanced the task is, the more the latest levels of the chain rely on the quality of the earlier levels. Arguably, the state of the art in NLP keeps progressing towards more and more complex tasks. As a consequence, not only this issue is becoming more and more crucial, but also the research tends to be more and more specialized: for instance, experts working at the semantic level cannot afford to spend time on the morphosyntactic analysis part, which is not their focus. Hence they rely on software tools made by others for the first components of their processing chain. The impact of their research depends on the quality of these tools and also their diversity, because a certain implementation might be more appropriate for certain usages than others.

In other words, the software prototypes become the instruments used for further research. In many experimental sciences, the importance of scientific instruments is acknowledged. Research resources are devoted to improve them, and making progress in the quality of the instruments is rewarded as a real contribution to the research: the Nobel prize awarded to Georges Charpak in 1992 for his invention and development of particle detectors is a striking example of such recognition. There is arguably not enough effort in NLP towards developing and using quality tools, despite their importance in the research process. In the next section we illustrate this observation with an example.

---

[15] According to [25], periods of "normal scientific progress" are interrupted by periods of "revolutionary science".

[16] According to Wikipedia, *the metaphor of dwarfs standing on the shoulders of giants expresses the meaning of "discovering truth by building on previous discoveries". This concept has been traced to the 12th century, attributed to Bernard of Chartres. Its most familiar expression in English is by Isaac Newton in 1675: "If I have seen further it is by standing on the shoulders of Giants."* https://www.en.wikipedia.org/wiki/Standing_on_the_shoulders_of_giants—last verified: February 2018.

8.2.2.4   An Example: TreeTagger

Part of speech (POS) tagging is a task which is included in the standard text processing chain, and it is an essential step in many tasks in NLP. It consists in labeling every word in a sentence with its morpho-syntactic category, called the POS tag. TreeTagger is the name of one of the first supervised Part-Of-Speech taggers: it was introduced in 1994 by Schmid [48]. The software tool is very popular among the NLP community: Google Semantic Scholar estimates that the paper was cited 2,735 times[17] (1,008 citations according to CiteSeer's database[18]); the original paper reached its peak of citations in 2014, i.e. 20 years after publication,[19] and it still gathers a healthy hundred of citations every year. While some of these citations might not correspond to experiments which actually use the software (e.g. simple mentions in literature reviews, etc.), most of them probably do.

TreeTagger was an excellent tool for POS tagging, but after more than 20 years of success the technology on which this venerable tagger is based is outdated[20]: there are more recent approaches in ML and especially in sequence labeling which perform better than the probabilistic decision trees method used by TreeTagger.[21] Thus it is likely that the popularity of TreeTagger nowadays has more to do with: on the one hand, its availability and the awareness thereof in the NLP community; on the other hand, its simplicity and usability as a software tool, including the fact that it is available for a fairly large set of languages (see also Sect. 8.4.1.3). In other words, a lot of NLP research is still based on a suboptimal POS tagger because it is popular and easy to use, potentially causing suboptimal results in the applications in which it is used. Incidentally, this implies that more recent POS taggers lack popularity (awareness) and/or usability.

This example demonstrates that there are gaps in the NLP research life cycle: the most adequate tool is not always immediately picked up by the community. Of course, such gaps would be filled eventually: new tools gain popularity progressively, and eventually the best tools are adopted, even if the qualities that determine what is "best" evolve. However, the process of adoption of the best tools can be accelerated, with potentially important scientific benefits (see Sect. 8.4.2).

---

[17]https://www.semanticscholar.org/paper/Probabilistic-Part-of-Speech-Tagging-Using-Decisio-Schmid/bd0bab6fc8cd43c0ce170ad2f4cb34181b31277d—last verified: February 2018.

[18]http://www.citeseerx.ist.psu.edu/viewdoc/summary?doi=10.1.1.28.1139—last verified: February 2018.

[19]https://www.semanticscholar.org/paper/Probabilistic-Part-of-Speech-Tagging-Using-Decisio-Schmid/bd0bab6fc8cd43c0ce170ad2f4cb34181b31277d—last verified: February 2018.

[20]Additionally the datasets on which it was trained are quite old.

[21]https://www.aclweb.org/aclwiki/POS_Tagging_(State_of_the_art)—last verified: February 2018.

8.3 Approach

8.3.1 Key Ideas

8.3.1.1 Key Objectives

The problems presented in Sect. 8.2 cover a broad spectrum ranging from ethical and societal questions (Sect. 8.2.1.1) to epistemological questions in experimental NLP like reproducibility (Sect. 8.2.2.2) and research instruments (Sect. 8.2.2.3). These issues might look like abstract problems barely related to each other, and each of them hard to address in a methodical way. This section will hopefully demonstrate the contrary: we propose a strategy intended to tackle these issues, through a framework in which they are deeply interconnected. The key objectives summarized below sketch the main ideas of this strategy:

1. Improve the reproducibility of NLP research, especially via more systematic publication of software and data resources;
2. Improve the reusability and dissemination of language resources coming from NLP research, in a way which improves openness and transparency in NLP research;
3. Reach out to a larger public, and encourage interactions between experts and the general public using open-source research resources as a support for discussion;
4. Integrate contributions from the general public into the research life cycle, i.e. allow the general public to become an actor of innovation rather than a passive user only;
5. Credit contributions from individuals from all participating sectors in a manner that respects each individuals' right to privacy as well as fair recompense.

The fact that these issues involve heterogeneous categories of people is the main challenge to be addressed. Naturally, the cornerstone to establishing a dialogue between experts and non-experts lies in making every stakeholder see the benefit they can gain from contributing. This creates an interest in identifying a shared language through which non-trivial dialogue may emerge.

8.3.1.2 Philosophy and Scope

Before going into further details, in such a wide-ranging proposal two main pitfalls must be avoided: on the one hand, an "idealistic" approach in which success depends solely on the individual willingness of the actors to contribute; and, on the other hand, a "bureaucratic" approach which discourages individual initiatives and small contributions. This is why this plan focuses on implementing effective structures intended to make every actor benefit from their participation, by leveraging these contributions into a global cooperative ecosystem. Such structures must also integrate seamlessly in their respective environments, in order to allow a progressive cultural shift towards the goal. Thus, the core of the proposal consists in progressively building a backbone structure which provides support and organizes the contributions of different actors, but is also flexible enough to evolve and scale up. While this plan requires resources for its deployment and sustainability, ultimately its aim is to help growing a broad self-reliant ecosystem which evolves with its own dynamics.[22] This is why the proposal focuses on building from existing resources and structures and encouraging and coordinating third-party efforts which are compatible with its goals.

We propose to follow the philosophy of agile development,[23] in this case applied not only to software development but to the whole proposal, in particular in its social aspects. The main implication is that the proposal is meant to be developed iteratively over life cycles, with continuous re-evaluation of the best way to progress towards the goals taking into account any relevant circumstances: trends in research and/or society, but also available resources and partners on the proposal. Hence, the exact scope of the proposal itself across time is meant to be flexible, in line with the guiding principles explained above.

8.3.1.3 Trigger a "Contagion" Effect

The philosophy of the proposal entails the involvement of heterogeneous categories of people in different sectors. This relies on the assumption that the proposal can trigger a "contagion" effect[24] across these population sectors, which we explain in this section. First, a brief definition of what "these different levels of population" represent is in order. Below we present a schematic view of what is meant by that in two perspectives. In one view, from the perspective of depth of awareness and expertise with AI/NLP technology, one may see a pyramid-shaped hierarchy (in terms of the number of individuals within a sector, relative to other sectors). In the other view, one may see something like a chain-link mesh, in that with respect to NLP science and technology, one relatively small group counts as experts, but these same individuals participate among the general public for other domains of inquiry. Thus, the sectors described in some way participate in a chain, but the links are ultimately connected in more than one dimension. Understanding that the chain-link mesh view ultimately provides part of the explanatory force of our proposal, we focus initially on a single chain of linked sectors within the mesh.[25] We start at one end (in the

---

[22] The basis for our model is based on the insight that the economic viability of local communities is linked to diversity. Communities that are tied to the fortunes of single industries or industry players face greater risks than those that are home to a diverse employment base, when industrial "disruption" takes root. In the more specific perspective of collaborative projects, this philosophy is also inspired by other successful projects, for instance Stack Exchange or Wikipedia: the infrastructure is provided, but the evolution of the content completely depends on the community.

[23] https://www.en.wikipedia.org/wiki/Agile_software_development—last verified: February 2018.

[24] Remark: this is not meant to have pejorative associations with disease, but rather the replication/repetition behaviors associated with mirror neurons and "memes".

[25] The explanatory force mentioned is this: because experts in one field are laypeople in another field it may be anticipated that at least some of the laypeople share all of (i) curiosity about domains

pyramid view, this is the top with the smallest number of NLP experts (people and organizations)) proceed through different population sectors, which get larger but less aware as one proceeds, ending with the general public, the largest population with the least depth knowledge about AI/NLP science and technology.

- **Research Experts**. Research experts create, build upon and improve the methods used in NLP technology. New methods are tested and compared in scientific experiments (this is generally when the first software prototypes are built, using new or existing data resources). These experts have the required resources (time and skills) to stay up to date with the state of the art; they are also the first users of languages resources (software and data), that they use to compare, extend or improve the state of the art. This category includes experts from organizations in academia, public institutions, large companies with research activities as well as smaller but very specialized innovation companies.
- **Professionals**. Professional experts, in particular in the industry, are familiar with the state of the art in NLP and its potential applications. They are mostly converting research methods into end-user products, this is why they need a good understanding of the underlying concepts, without usually having the resources to do the core research themselves.
- **Direct Users**. Professional users who have some connection to NLP innovation but within a specific topic or application. They are more on the user side of NLP technology, and their vision of the field and its evolution is limited. This level includes a wide range of SMEs (Small and Medium Enterprises) with some connection to the field, but which do not have the need or resources to keep up to date with the state of the art.
- **Potential Users**. This level includes heterogeneous categories of people or organizations: savvy IT professionals, technology enthusiasts, students in computer science or related topics, professionals in companies which are not specialized in NLP. This layer corresponds to people and organizations who are potential NLP tech users: they have some resources but might simply not be aware of the existing technology. In particular, this category includes SMEs which might not be aware that their business could benefit from NLP technology.
- **General Public**. Finally the largest group, which forms the foundation of the pyramid consists in the general population, i.e. people who are not educated about NLP

---

where they lack expertise; (ii) interest in availing of the best of technologies that emerge from those domains; (iii) empathy for the desire of domain experts to achieve even greater expert knowledge in the domain. In turn, these laypeople are visible in their lay interactions with other laypeople. Laypeople who are not socially defined as experts using traditional schemes (such as university degree qualifications) may be anticipated to involve themselves even without contact with experts acting outside their domains. Nonetheless, we do not anticipate a universal take-up of a call for participation, no more than there is universal participation in recycling initiatives. However, when people perceive a duty to participate and have information that others participate, then they are more likely to than if they lack information about others' participation [9]. A question, then, is whether the general population is most appropriately approached with a lesser duty to *contemplate* participation as citizen scientists or with a non-duty to engage with citizen science for its entertainment value, alone.

technology, although they occasionally use its applications directly or indirectly. There are societal issues regarding the lack of awareness in the population; for example, people usually ignore what can be done (or not done) with their data, thus being potentially abused or on the contrary reluctant to use some helpful technology (see Sect. 8.2.1.1).

The general idea is that these layers are intertwined in such a way that changes at one level potentially affect other levels. For example, the wider availability of language resources from experts is likely to impact professionals and to some extent direct users. The approach that we detail below aims to maximize the impact, so as to extend it to potential users as well as the general public (see Sect. 8.4). This approach relies on leveraging the diversity of the open-source communities, as a means to drive changes across various layers of society.

8.3.1.4   Ethical Design

Recent discussions around the ethical ramifications of AI research point to two distinct directions in research either a) educating scientists, engineers and designers in ethics or b) incorporating social scientists and ethicists in the design process. The first approach incorporates several important techniques for integrating ethical approaches to design that have been developed in Human-Computer Interaction (HCI) studies, such as reflective design [50], participatory design, value-sensitive design [17] and so on. These approaches often focus on how to identify bias or define guiding values at the outset, although bias and values can also arise as local phenomena that are "discovered" throughout design and development [27]. The second approach suggests that social scientists must be incorporated into the research process, which may produce more disagreement but arguably more discussion and deeper reflection on where ethics happens and where there are perceived to be "ethics free zones" that require attention [54]. Both approaches suit certain design settings better than others, with some studies noting "disjunctions" between AI and big data research methods and existing research ethics paradigms [33]. Others have found a tendency for such approaches to produce different levels of ethics expertise rather than establishing ethical consideration as the responsibility of all involved in a continuing process of reflection design [51]. Meanwhile user studies attempt to include user requirements and values in design, but studies suggest they are too often constructed for abstract ideal or "intensional" users [3] rather than actual users whose responses to digital technologies and interactivity are more often individualised and inherently strategic. Further, there is an increasing attention paid to the numbers of indirect users or stakeholders in the design of technologies, particularly those using AI/NLP, whose inputs are rarely sought in the design process. The challenge is how to better educate and incorporate all stakeholders in the consideration of ethics and values in design. Recent studies on AI in society recommend that policy makers and leaders in enterprise seek ways to encourage experimentation around ethics review inside and outside of university settings and recognising that the function of pausing for independent review and deliberation is indispensable [33].

This proposal follows a complementary approach to continuous ethical reflection in research, by incorporating the general public or "community" into ethics analysis during the entire research cycle from research design to data collection, analysis, implementation and reflection on impact. According to [54], ethics must be part of the whole technological design process from the very beginning. A community participation approach to ethics empowers the general public with the tools to participate in discussions and debates about AI/NLP advances and their uses throughout the research lifecycle. Integrating the public into the ethical reflection process from the outset avoids problems with unforeseen or "emerging bias" while complementing citizen science efforts in collaboration around both the process and the product of technological innovation. It also shifts ethical reflection from being a post hoc theoretical exercise to being an integrated aspect of the research process. This will produce new knowledge around ethical issues and how they arise in an NLP project, and also fresh inputs on the essential pragmatic issue of how to identify and integrate applied ethics directly into the development process.

8.3.1.5 Rethinking the Innovation Chain

Traditionally, the innovation chain is thought as a linear process which goes from the initial scientific concept to its applications into end-user products, progressing through various stages from scientific development to industrial technology and finally commercial products. This linear process is often represented in the form of a scale of "Technology Readiness Levels" (TRL). For instance, the EU Commission [13] defines the following TRL scale[26]:

1. Basic principles observed
2. Technology concept formulated
3. Experimental proof of concept
4. Technology validated in lab
5. Technology validated in relevant environment (industrially relevant environment in the case of key enabling technologies)
6. Technology demonstrated in relevant environment (industrially relevant environment in the case of key enabling technologies)
7. System prototype demonstration in operational environment
8. System complete and qualified
9. Actual system proven in operational environment (competitive manufacturing in the case of key enabling technologies; or in space).

---

[26]Remark: *key enabling technologies (KET)* (mentioned in TRLs 5, 6 and 9) are a group of 6 technologies that the EU wants to focus on [12]: micro and nanoelectronics, nanotechnology, industrial biotechnology, advanced materials, photonics, and advanced manufacturing technologies. These areas are not at all or only very distantly related to software technology, in particular AI or NLP.

According to Wikipedia,[27] the EU initially adopted the TRL levels defined by NASA for the European Space Agency. It was later extended to every kind of technology developed within the H2020 framework program. To the authors' knowledge, the EU-defined TRL levels do not include any additional information or description, but one may find an additional short description of every level in the original NASA definition.[28] Of course, these descriptions insist on testing the technology in space; this shows that the scale was designed with a very specific purpose in mind. Thus TRL levels are supposed to apply for a very broad range of technology including software systems, whereas it was initially designed for space technology only. In particular, it understandably focuses on hardware components and follows a very thorough testing process. The life cycle of a software component differs significantly from this,[29] since:

- Software is very flexible: it can be modified, reused for a different purpose, its reproduction and dissemination is practically cost-free, and bugs can be fixed even after a product has been released.
- Software components are very rarely error-free. In the case of AI software, the risk of error is inherent to the concept of AI itself. Thus, it is not expected (nor is it possible) to make AI software infallible.

These differences make the standard TRL scale poorly fit for reliably representing software development across the innovation chain: a software system is seldom fully "proven", and its life cycle does not have to be linear; in fact it rarely is. The same point can be made for data resources, which are often developed iteratively, with versioning used to identify a particular release in time.

This point matters because language resources do not require the kind of heavily controlled top-bottom development necessary for some space shuttle component. On the contrary, it allows a lot of flexibility; for example, bottom-top feedback from users or developers can be integrated into language resources quite easily. Therefore, the representation of language resources in the form of a TRL scale tends to restrict the wide range of possibilities that are available, especially in the case of new technology for which usage and applications are still to be invented. Thus strictly following the TRL perspective introduces a bias towards centralized top-bottom technology development. For instance, this model does not offer room for interactions between the researchers at the source of a new technology and its end-users. It was designed under the assumption that only professional experts participate in the development of the technology. When applied to cutting-edge AI software, this view entails that professional experts are in charge of developing every aspect of the technology, including how to use it, and the general public is left out of the loop in a role of mere

---

[27]https://www.en.wikipedia.org/wiki/Technology_readiness_level—last verified: February 2018.

[28] https://www.en.wikipedia.org/wiki/Technology_readiness_level#NASA_definitions—last verified: February 2018.

[29]https://www.en.wikipedia.org/wiki/Systems_development_life_cycle,   https://www.en.wikipedia.org/wiki/Software_development_process,   https://www.en.wikipedia.org/wiki/Software_prototyping—last verified: February 2018.

passive customers. Moreover, a large space in the AI innovation chain is occupied by big technology companies which might be commercially interested in keeping customers in such a passive role[30]; to some extent, the control these companies have on the design process might sometimes undermine the general public's interest. In other words, behind the technical aspects of something like the TRL scale model, there are democratic and ethical issues at play. As a consequence, while TRL is a useful representation of technology development, integrating the general public at various levels of the innovation chain requires a more open and more flexible model.

8.3.2 Strategy

8.3.2.1 Open-Source Software as a Gate Between AI Research and the General Public

Open-source resources (software and data) play a crucial role in the plan that we propose: they can be thought of as a vessel through which "bits of AI knowledge" are conveyed. The resources must be open-source, so as to allow exploration and scrutiny by the general public. Language resources used in NLP fit in well with this plan, since the object of study is understandable without any particular background knowledge (see Sect. 8.2.2.1); moreover, NLP research could really benefit from improving the quality and diversity of its resources, as motivated in Sects. 8.2.2.2 and 8.2.2.3. The publication of open-source resources is already a well established practice in the NLP research community, albeit far from universal. However, its impact so far has been mostly limited to the research community circles.

The approach that we propose relies heavily on improving the dissemination and re-use of NLP resources very early in the innovation process, i.e. at the stage of research prototypes for software and at the stage of collection for datasets. This is possible only through a well-thought open-source strategy. A common misconception about open-source is that publishing the software code (or any kind of data) in some public repository suffices to ensure that the development will be picked up by "the open-source community", provided the software is worth it. This is definitely not the case, for the same reasons that in industry a good product cannot be relied upon to sell itself. There are many reasons why a project would not get traction and would simply be abandoned, no matter its quality. Among the most obvious ones one might mention:

- Clearly a project with no visibility cannot attract contributors: if not properly indexed, or hosted on a page which receives very few visits, then nobody can find it and it cannot receive any attention;
- Similarly, if it is unclear what the piece of software aims to achieve or what its main features are, then it is unlikely that anybody would bother spending time

---

[30]"It's really hard to design products by focus groups. A lot of times, people don't know what they want until you show it to them." Steve Jobs [45].

guessing what it is about; the same applies to data released with no indication of its origin, format or potential usage.

- If the original author does not answer messages or questions about their project,[31] this will discourage interested users who might have become contributors otherwise.

Unfortunately these issues are common in the case of software prototypes originating from the research community, because prototypes are often not thought as software projects but only as requirements for an experiment, and in turn for a publication (see Sect. 8.2.2.1). Thus the problem does not lie so much in the lack of open-source software prototypes (even if this is also a serious issue) rather than, in most cases, in the lack of effort to integrate the software in the open-source ecosystem. In the case of innovative software, providing good documentation is an even more crucial step in order to interest people in it, since the purpose of the software might be significantly more difficult to comprehend than for some regular software.

In general, the research community is positively inclined towards the principles of open-source: openness is a cornerstone of academic research, and research projects where people work together using collaborative tools are very common. In fact, in computer-related fields a number of academics are involved in various open-source projects, related to their professional work or not. However software development is rarely seen as an important goal by itself, because it does not constitute a scientific contribution to the field, i.e. a new finding worth publication. We discuss means to encourage software development in research in Sect. 8.3.2.2.

A closer look at the open-source ecosystem is required in order to improve the penetration of research software prototypes and data resources into it. First it is important to remind the reader that open-source does not preclude commercialization or any commercial use of the software. There are many examples of companies which participate to developing open-source code for profit, with various possible business models [4].[32] Nevertheless the open-source world spans over a wide range of cases, from casual hobby projects to ubiquitous security-sensitive software components (e.g. OpenSSL[33]). In general, there are established advantages of open-source software development: the inherent transparency improves the reliability of the code, in particular bugs are fixed faster, thanks to the community feedback and contributions. The existence of a motivated community also helps develop the product in new directions that might not have been explored otherwise, see e.g. [52]. Interface, documentation and features can be based more directly on users' experience. In open-source development, the distinction between developers and users is not as binary: users can get involved in the development in many ways, even without any knowledge of coding; contributions can be as small as fixing a typo in the documentation, reporting a bug or proposing a new feature. This clearly allows a more direct and

---

[31]For instance because they have moved to a new job and do not use the email address anymore, which is a frequent occurrence in academia.

[32]https://www.en.wikipedia.org/wiki/Business_models_for_open-source_software—last verified February 2018.

[33]https://www.openssl.org/—last verified February 2018.

flexible communication between users and developers. Naturally, in order to enjoy these benefits an open-source proposal must be managed in a way which encourages the integration and participation of people who are external to the project. This involves following various kinds of good practice intended to make the proposal welcoming and motivating for its participants on the one hand, and maintain a efficient organization of work on the other hand [15].

Thus the plan that we propose aims to establish bridges between heterogeneous communities (as described in Sect. 8.3.1.3); summarily, the two main target populations are the research community and the general public. This will be implemented through different kinds of incentives which can be roughly classified into two groups, based on their target population.

8.3.2.2 Involving the Research Community

The first and perhaps most important incentive for experts to engage with non-experts on a project is the satisfaction of seeing their resources used and appreciated. While this rewarding feeling is important, concrete professional advantages are likely to play a bigger role to sustain the experts motivation to commit time in outreach activities. This is why a major part of the proposal will consist in implementing incentives to this end. Since publications represent the currency of the academic world, the proposal will offer more opportunities for researchers to publish their work and value their efforts towards making software tools more reusable; this traditionally means organizing new workshops as well as journal issues devoted to topics related to the proposal goal: reproducibility, software engineering issues in NLP research, outreach contributions, etc. The organization of these scientific events must take a special care at establishing their evaluation criteria; in particular, instead of the traditional evaluation criteria focused on scientific originality, they could give more importance to software-related aspects such as documentation quality, usability, openness. This would for instance encourage the submission of work which reproduces existing methods, or provides the community with better quality software tools. In the long term, the recognition of the workshop or journal by the scientific community would make authors more willing to be published in such reputable references, hence increasing efforts of the community towards software tools and outreach.

The case of data resources is slightly different, because it is already very common to publish those in academic venues. Thus, for data resources, the focus should be put on developing methods to involve non-experts in their creation; this would directly benefit the research community because data resources are costly to produce.

These direct academic benefits are not the only advantages experts can gain from making their software tools more available and usable, in particular for non-experts. The potential popularity of their software can induce secondary benefits: other NLP researchers, students or technology enthusiasts can provide feedback and give them ideas to improve their software, which might in turn open new scientific perspectives. The visibility offered this way to valuable research software would also benefit their authors, in the form of further citations, feedback, collaborations and project funding.

8.3.2.3 Involving Non-experts Communities

Clearly the "non-expert" group includes people who might be interested in AI technology for various reasons. Everyone can have their own motivations: curiosity, an
appetite for technology advances, the desire to acquire knowledge or learn new skills,
the intellectual challenge, etc. AI technology in particular can have a strong appeal to
the general public from this perspective. Maybe one of the main reasons why people
would want to contribute is simply the satisfaction of participating to an interesting collaborative project; when properly organized, benevolent contributors can be a
great help to a project, as seen in many examples of great and valuable achievements
over the Internet, e.g. Wikipedia, Stack Exchange, and many open-source software
projects. The conditions of success for collaborative projects can be analyzed, in
order to structurally maximize the chances of success.

First, potential contributors must be aware of the existence of the project, and
know that anyone can contribute. To our knowledge, there are nowadays very few
research projects really open to non-expert contributions, if any. Inviting the general
public to participate in a significant way to scientific projects is likely to be seen as
an opportunity by many people, as opposed for example to some quite misguided
attempts at involving the public in some extra-scientific part of the proposal, with
the risk of people not taking it seriously.[34] Then it is important to make people feel
welcome and feel that their input is taken into account. Clearly this can happen only
if the experts on the proposal are ready to spend some time answering questions,
improving the documentation, etc.

However, it is also important that people feel that they are not exploited. This
perception is likely if others obtain wealth on the basis of contributions beyond their
own while they obtain, at best, recognition. This perception was widely attested in
the aftermath of the crowd-sourced localization of Facebook's interfaces into multiple languages [39]. A perception of exploitation may be an explanation for evidently
malicious behaviours that have been witnessed in calls for general public participation in crowd-sourcing in the private sector [24]. Nonetheless, one may notice that
members of the public continue to demonstrate willingness to volunteer for charities
that they perceive in having good causes, even when the volunteers know that executives who run the charities are paid handsome salaries, and even when executives have
been discovered to have been obtaining unwarranted personal gifts at the expense of
the charities. Further, as suggested (footnote 25), if there is a means for members of
the community to see accurate (yet individual privacy preserving) accounts of the
extent to which the rest of the community is contributing to an endeavour, as well as
the recompense correspondingly accrued, then one may reasonably expect greater
levels of participation than if community-wide participation rates are not visible.

Technology enthusiasts often have programming skills and are familiar with the
environment and tools around software projects. Thus they are more likely to be
interested and to contribute, but it is also important not to exclude people who do
not have a technology background; for example, students (not necessarily following

---

[34]https://www.en.wikipedia.org/wiki/Boaty_McBoatface—last verified February 2018.

computer science studies) might explore opportunities to participate and this can be a motivation for some of them to pursue AI studies. Finally, there can also be a interest from companies which see a professional use for the technology. Importantly, small and medium-sized businesses which do not have the same access to technology as big companies might give a valuable input to direct the technology in a direction that fits their needs. Many kinds of contributions do not require any particular skills:

- Providing feedback: ask questions, fuel discussions on usage, shortcomings, etc.;
- Testing, identifying issues, propose improvements or ideas;
- Providing or annotating data;
- Identifying documentation problems, translating documentation;
- People with programming skills can contribute to the software itself, but also code extensions or interfaces, create packages for specific systems (e.g. as a plugin, add-on, app …);
- Imagine and propose ideas about potential applications, combinations with other tools, etc.;

It should be emphasized that these interactions are meant to happen at the level of a specific project, where a community can progressively gather together around a shared interest for this proposal in particular; the dynamics of such a community depend only on the people who feel that they belong to it, whether they are experts or not. Nevertheless, various steps can be taken in the perspective of encouraging such communities to grow. This would include identifying projects which have a potential for attracting contributions from non-experts and giving them some visibility; this would involve publicize them on social media, but depending on the project more specific targets can be considered, like specialized websites and forums in the open-source world, or technology media interested in recent AI trends. Contents from the latter can occasionally trigger articles in the mainstream media, thus giving potentially extensive visibility to the projects. Some projects can also be highlighted by creating demos or applications based on their software.

The appeal to the general public is to be considered as an essential goal of the proposal. This means that a significant part of the work in delivering on our proposal will focus on how to make some specific NLP projects appealing (whether through entertainment value, through the self-satisfaction that arises from acts of altruism, or appeals to pure altruism), beyond making them open to the general public, for instance:

- Building demonstrations of what AI can do with language; famous examples include IBM Watson Jeopardy Challenge [5] and Personality Insights [29].
- "Serious Games" can be designed to entertain and educate, but games can even be used to collect data resources in original ways [26].
- Determining how to ensure that data obtained from crowd sourcing is reliable; see e.g. [19].

- Establishing through public discussion and debate the extent to which participation in citizen science is a public duty (in the spirit of recycling or avoiding littering).[35]

8.3.2.4   Design: Flexibility and Sustainability

The proposal is oriented towards leveraging existing software tools and encouraging the production of new ones. This perspective entails a few strong characteristics which differ significantly from more traditional software-oriented projects. The core part of the development does not take place in-house but in many independent projects with their own goals, design choices and coding practices. Thus the governance of the proposal aims at generating added value from multiple independent projects, by focusing on helping them grow a sustainable community of contributors as well as encouraging potential combinations and applications of projects (these two aspects being deeply interdependent). Consistently with this philosophy, the proposal favours a broad-range strategy based on the diversity of tools (in type, quality, level of development, etc.), as opposed to a centralized strategy concentrated on a closed set of tools.

An important consequence of this diversity strategy is that it does not enforce any framework or compatibility requirements, in order not to add constraints to the projects. This unusual feature deserves a particular explanation: the lack of strong standards has been seen as an issue in NLP for a long time; despite regular attempts based on various software frameworks (e.g. Gate,[36] Stanford CoreNLP,[37] Apache UIMA,[38] etc.), the NLP research community has consistently kept working for the most part with simple file formats[39] (frequently raw text or CSV files, xml when needed), on an ad-hoc basis depending on the task at hand. In other words, the community tends to favour simplicity with lightweight task-oriented components over inter-component compatibility (inter-operability); we argue that this is not by lack of standardization effort or lack of good standard, but on the contrary that this is a meaningful choice motivated by the complexity of language-related tasks. As a consequence, we choose to adopt this lack of standard as a feature for the proposal, especially since this fits particularly well with our goal of maximal flexibility and design choices on a project basis. This does not preclude the adoption of a particular standard, framework or platform by some software components. Instead, in this perspective we argue that standardisation should happen downstream, i.e. after the "raw" component has been developed and tested in various contexts; the adaptation of the component to some framework is seen as being part of the late phase of developing applications for it: at this stage, the component can be integrated into

---

[35] One may anticipate a robust multi-disciplinary debate on this topic given a public perception that industrial pollution is at times treated with greater laxity than public littering.

[36] https://www.gate.ac.uk/—last verified: February 2018.

[37] https://www.stanfordnlp.github.io/CoreNLP/—last verified: February 2018.

[38] https://www.uima.apache.org/—last verified: February 2018.

[39] As seen for example in the vast majority of the shared tasks organized by the community.

various forms which fit its potential uses. Decoupling the development of the core from the integration into a final larger system has the additional advantage of allowing different people to do these different tasks, i.e. not leaving the developer of the core necessarily in charge of the integration of their component. Again this fits well with the philosophy of the proposal, where a community of contributors can participate in different parts of the proposal. Of course, optional guidelines can still be proposed to help project developers increase the compatibility of their project and consequently its potential for reuse.

This voluntarily agnostic approach follows the principle that the general ecosystem should be very flexible about the kind of resources which are accepted, as long as they can be evaluated, explained, examplified through reproducible experiments; it is crucial that the process does not add constraints over these basic criteria. Similarly in terms of technology development level, any software starting from TRL 3 (proof of concept software, see Sect. 8.3.1.5) is a valid candidate, since the goal is to promote cutting-edge research methods, not final-product applications.

8.3.2.5 Core Components of the Proposal

Due to its evolving nature and broad scope, the proposal cannot easily be broken down into clearly defined subtasks.[40] However one can roughly classify the nature of the actual tasks of the proposals into the four following categories (the order of which is not relevant):

- **Referencing Projects**. It is of course unavoidable for this project to progressively build a collection of open-source projects related to AI research. The exact referencing process is going to be defined progressively over time. It involves identifying projects (active or not) which fall inside the scope of the project of course, but also organizing the information in a way which facilitates the exploration of potential connections between projects. Scientific literature is naturally the main sources of pointers to software resources; some existing resources might also help in this task, e.g.:

  - ELRA[41] proposes a catalogue of language resources[42] identified through an "International Standard Language Resource Number" (ISLRN).[43]
  - Github,[44] the largest open-source repository, offers various means to explore projects: highlighted projects grouped into collections, topic tags, user ratings, etc.

---

[40]This belongs to the next stage of designing and implementing the proposed plan, taking into account available resources and specific targets.

[41]http://www.elra.info—last verified: February 2018.

[42]http://www.catalog.elra.info/—last verified: February 2018.

[43]http://www.islrn.org—last verified; February 2018.

[44]http://www.github.com—last verified: February 2018.

- **Projects Development**. Development takes place in relation with one or several projects in accordance with the objectives of the project. Naturally a participant's role can be to engage in the core development of a particular project, but it can also involve developing extensions or enhancements to make the software usable in some a particular context, for example:

  - Testing the component for a particular task (which may or may not have been the task for which the component was originally designed);
  - Using the component to run or reproduce an experiment;
  - Combining the component with some other component(s) to achieve a different task;
  - Implementing an interface for a new usage, e.g. demonstrating the component in a webpage;
  - Make the component more user-friendly, improve the documentation, write a tutorial.

- **Outreach**. Outreach can take many different forms:

  - Social media, open-source discussion lists, online specialized press, casual conversation during human interaction in social activities, etc. The outreach strategy typically depends on the project or technology being advertised;
  - Documentation, tutorials and software demonstrations are essential to attract interest and encourage people to use and/or get involved with a project;
  - Community management: helping interested people finding a project and/or a task that they like is important for the projects to grow; sometimes it can be useful to help old or even abandoned projects finding new contributors by advertising them;
  - Community feedback: providing means for members of the community to see accurate (yet individual privacy preserving) accounts of the extent to which the rest of the community is contributing;
  - The organization of scientific events (conferences, journals) is clearly also an important part of the outreach strategy (see Sect. 8.3.2.2).

- **Research Contributions**. As explained in Sect. 8.2.2, the project is expected to produce scientific contributions in the field of NLP, in particular by improving reproducibility and providing the field with better research instrument (both in quantity and quality).

8.4 Impact

The impact of the social and educational side of the proposal can be measured with indicators based on the participation in its constituent projects, such as number of participants, number of projects, etc. However it is important to bear in mind that this project goes beyond short-term quantifiable results, with potentially long-term benefits in many areas.

8.4.1 Long-Term Benefits of Involving Society with AI

8.4.1.1 Education by Exploration

A few decades from now, the proportion of people with programming skills will be much higher. It is even not absurd to imagine a future in which basic algorithmic skills have become as important as reading and writing skills.[45] Not everybody will be willing to participate to open-source AI projects, but it seems reasonable to assume that some people will be interested. In a long-term perspective, enabling and supporting this change now can only amplify its positive effects in the future, by making more and more people aware of this possibility. The educational side of the project is characterized by hands-on experience, i.e. it is clearly not designed as a course at all but rather encourages people to participate to discussions and to contribute in the way they feel the most interested. People are offered the possibility to open the "black box" of AI in an open and flexible way, to the extent of their choice depending on their motivations.

8.4.1.2 Creativity

AI is currently being designed and developed mostly by scientists and industrials, often with commercial applications in mind. The inclusion of a more diverse public in the development process of AI might open new perspectives of applications. By providing access to a broad range of diverse tools as well as their code, the project empowers people to develop their own variants of AI tools as well as to combine existing tools into new complex tools for any application. This could give rise to entirely new uses; it is important to emphasize that as opposed to more standard approaches (e.g. [23, 30]), our approach does not provide a limited collection of pre-made tools, but is meant to give access to every aspect of a much larger range of tools; this entails more effort on the part of the user, but it also offers much more freedom for them to build any customized AI system. This approach fits in a much wide perspective where building an AI system can be seen as a creative process which consists in the combination of smaller building blocks; the goal does not have to be utilitarian and is bound only by imagination (for instance in arts [18]).

8.4.1.3 Multilinguality

Major progress has been achieved in multilinguality in the recent years. In NLP applications, scalability in terms of data size has been addressed for the most part, but

---

[45]It is important to keep in mind that the level of education of a population can evolve drastically over a few generations, as history shows: only one third of the world population was literate in 1950, and this proportion raised to 85% over sixty years [46]. Similarly, the proportion of a generation achieving tertiary education level has doubled in most western countries in the past 30 years [42].

scalability in terms of language diversity is still a significant challenge. As of version 2.1, the Universal Dependencies corpus [40] includes 102 annotated datasets and 59 distinct languages,[46] thanks to the authors' and contributors' great effort. Packaging such a diversity of languages in a uniform format is a major step towards the ability to process multiple languages in an homogeneous way, which is the cornerstone of language-wise scalability. Researchers [6, preface] have claimed that *"Previously, to build robust and accurate multilingual natural language processing (NLP) applications, a researcher or developer had to consult several reference books and dozens, if not hundreds, of journal and conference papers."* One might add that said researcher or developer would also have to find, test and integrate multiple language-specific software tools. Thus, language scalability also requires a more streamlined engineering process: it becomes impractical to find a specific software tool for every language to process, let alone the best tool for every specific language. Instead, evaluating software tools is progressively shifting from accuracy in a specific language to robustness and adaptability to a wide range of languages.

The NLP community as well as institutions[47] are actively addressing multilinguality, since this is one of the main challenges for NLP technology to become widespread and really useful to society. In the approach that we propose, multilinguality is implied as it is a valuable consequence of the broad diversity of NLP tools that we promote and of the opening to general public contributions. In particular, the adaptation of NLP tools to various languages generally requires annotated data in the target language; given the chance, many people would certainly be happy to help preserving and promoting their own language by contributing to building linguistic resources. This would in turn help the NLP community provide them with better quality tools adapted to their language.

8.4.1.4 Opening the AI Black Box

As explained in Sect. 8.2.1.1, societal and ethics questions around AI are becoming crucially important for the future of the society. Many aspects of AI (e.g. [2, 22, 49]) require informed choices from the society; the general public should be able to question AI research blind spots. Thus it is vital that AI methods are made open to public scrutiny, and to provide the general public with means to get an understanding and even a say in the evolution of AI.

From an ethical point of view, the lack of transparency of modern ML techniques is questionable: e.g. [8, 10]; but simplistic expectations about total visibility are also unhelpful [1]. This proposal offers new kinds of actor-network configurations that could produce different starting points for understanding how transparency relates to accountability. AI experts are experts in their field, most notably in Machine Learning (ML), but can be subject to unconscious biases like anyone (see e.g. [16, 20, 28]).

---

[46]http://www.universaldependencies.org/—last verified: February 2018.

[47]For instance: http://www.mlp.computing.dcu.ie/, http://www.meta-net.eu—last verified: February 2018.

Incorporating non-expert and diverse participants in NLP research helps to deepen the potential for ethical reflection from the outset while encouraging the development of alternative evaluation strategies around the values embedded in research. Also, by facilitating a better understanding of AI, it allows researchers to take responsibility for the way research is presented to and understood by public [21].

8.4.2 Indirect Benefits to Research and Industry

The approach that we propose can be seen as a seed for exploring alternative research opportunities, and diversifying/enriching innovation in new economic areas.

The availability of a well tested and maintained collection of tools is an advantage for the research community and the whole innovation chain. The process of testing, using and extending NLP tools can lead to discoveries or improvements. For instance, testing a tool on some benchmark dataset may be published in a paper comparing the results to existing methods. Combining tools for some new task or comparing tools against each other are also likely to lead to research contributions. Overall, the richness and diversity of the tools makes it easier and faster to test different ideas or methods. This contributes to a more efficient process of selecting the best approach for a task, e.g. by filtering out unsuccessful ideas sooner. This means that the good ideas can be developed faster, hence a gain in research and development productivity.

This is true in particular for companies which do not have the resources of the technology giants: by making software tools accessible, such companies could afford to test new methods without requiring the expensive in-house knowledge of AI experts; this means that the approach we propose can lower the skill barriers for SMEs. A similar argument can be made for companies/organizations which deal with sensitive data and therefore cannot make their data available to external experts; by providing them with open software tools, they can experiment in-house instead of long and costly IP processes and/or confidentiality agreements.

8.5 Final Remarks

The proposal for advancing NLP described above in some sense suggests a return to basics. In a simplifying analogy, we propose systematizing the reporting of knowledge and applications of NLP advances in an accessible manner in the way that Wikipedia enables contributing to archives of specific slices of general knowledge. Through this, we also propose revisiting some of the "received wisdom" in the field to ensure that assumptions, theories and technologies retain internal and external validity on further inspection, and to make them available for exaptation to other problems. While we ourselves have contributed modestly to the field's advances (for example, in parsing [44], in grammatical inference [34], etc.), we have also participated in "revisiting" activities (for example, among other results, in assessing syntactic expressivity/complexity results [55], combining string similarity measures [38], assessing quality estimation methods [35, 36], etc., and perhaps most importantly to our argument in this paper, a revisiting of the tokenization problem [37][48]) our proposal is much bigger than our own lab, and requires collaboration as a "big team" along the lines that we have described above: we hereby invite others who share productive interests in this domain (from whatever sector of activity) to make contact with us towards coordinating the proposed endeavour. More NLP components than any individual can master require advancing and revisiting in the manner that we propose. Further, additional AI domains merit analysis using the same paradigm.

Acknowledgements The ADAPT Centre for Digital Content Technology is funded under the SFI Research Centres Programme (Grant 13/RC/2106) and is co-funded under the European Regional Development Fund.

References

1. Ananny, M., Crawford, K.: Seeing without knowing: limitations of the transparency ideal and its application to algorithmic accountability. New Media Soc. **20**(3), 973–989 (2018)
2. Anonymous. AI image recognition fooled by single pixel change (Nov 2017). https://bbc.com. http://www.bbc.com/news/technology-41845878
3. Bardzell, J.: Interaction criticism and aesthetics. In: Proceedings of the SIGCHI Conference on Human Factors in Computing Systems, CHI '09, pp. 2357–2366. ACM, New York, NY, USA (2009)
4. Benkler, Y.: Freedom in the commons: towards a political economy of information. Duke Law J. **52**, 1245–1276 (2003). https://scholarship.law.duke.edu/dlj/vol52/iss6/3
5. Best, J.: IBM watson: the inside story of how the jeopardy-winning supercomputer was born, and what it wants to do Next. TechRepublic. https://www.techrepublic.com/article/ibm-watson-the-inside-story-of-how-the-jeopardy-winning-supercomputer-was-born-and-what-it-wants-to-do-next
6. Bikel, D., Zitouni, I.: Multilingual Natural Language Processing Applications: From Theory to Practice, 1st edn. IBM Press (2012)
7. Bonnefon, J., Shariff, A., Rahwan, I.: Autonomous vehicles need experimental ethics: are we ready for utilitarian cars? CoRR (2015). arXiv:1510.03346
8. Bornstein, A.M.: Is artificial intelligence permanently inscrutable? Nautilus (2016). http://nautil.us/issue/40/learning/is-artificial-intelligence-permanently-inscrutable
9. Brekke, K.A., Kipperberg, G., Nyborg, K.: Social interaction and responsibility ascription: the case for household recycling. Land Econ. **86**(4), 766–784 (2010)
10. Castelvecchi, D.: Can we open the black box of AI? Nature **1**(538), 20–23 (2016). http://www.nature.com/news/can-we-open-the-black-box-of-ai-1.20731
11. Cookson, C.: DeepMind computer teaches itself to become world's best Go player. Financ. Times (2017). https://www.newscientist.com/article/2132086-deepminds-ai-beats-worlds-best-go-player-in-latest-face-off
12. European Commission. Key Enabling Technologies. https://ec.europa.eu/growth/industry/policy/key-enabling-technologies_en
13. European Commission. Horizon 2020—Work Programme 2014–2015, General Annexes, G. Technology readiness levels (TRL), July 2014. https://ec.europa.eu/research/participants/data/ref/h2020/wp/2014_2015/annexes/h2020-wp1415-annex-g-trl_en.pdf

---

[48]This is arguably most important because tokenization, individuation of the linguistic atoms to be analyzed in a text is the first step in the prototypical NLP pipeline.

14. Evans, N., Levinson, S.C.: The myth of language universals: language diversity and its im portance for cognitive science. Behav. Brain Sci. **32**, 429–492 (2009)
15. Fogel, K.: Producing Open Source Software: How to Run a Successful Free Software Project, 2nd edn. O'Reilly Media (Jan 2017). http://www.producingoss.com/
16. Fokkens, A., van Erp, M., Postma, M., Pedersen, T., Vossen, P., Freire, N.: Offspring from reproduction problems: what replication failure teaches us. In: Proceedings of the 51st Annual Meeting of the Association for Computational Linguistics, pp. 1691–1701, Sofia, Bulgaria. Association for Computational Linguistics, August 2013
17. Friedman, B.: Value-sensitive design. Interactions **3**(6), 16–23 (1996)
18. Gayford, M.: Robot Art Raises Questions about Human Creativity. MIT Technology Review, Feb 2016. https://www.technologyreview.com/s/600762/robot-art-raises-questions-about-human-creativity
19. Graham, Y., Ma, Q., Baldwin, T., Liu, Q., Parra, C., Scarton, C.: Improving evaluation of document-level machine translation quality estimation. In: Proceedings of the 15th Conference of the European Chapter of the Association for Computational Linguistics, pp. 356–361, Valencia, Spain. Association for Computational Linguistics (2017)
20. Hand, D.J.: Classifier technology and the illusion of progress. Stat. Sci. **21**(1), 1–14 (2006). https://doi.org/10.1214/088342306000000060
21. Johnson, D.G., Verdicchio, M.: Reframing AI discourse. Minds Mach. **27**(4), 575–590 (2017)
22. Kelly, K.: The Myth of a superhuman AI. Wired (2017). https://www.wired.com/2017/04/the-myth-of-a-superhuman-ai
23. Knight, W.: Google's Self-Training AI Turns Coders into Machine-Learning Masters. MIT Technology Review (Jan 2018). https://www.technologyreview.com/s/609996/googles-self-training-ai-turns-coders-into-machine-learning-masters
24. Kroulek, A.: Crowd-Sourced Translation Goes Awry For Facebook. k International: The Language Blog. (Aug 2010). https://www.k-international.com/blog/wrong-translation-for-facebook/
25. Kuhn, T.S.: The Structure of Scientific Revolutions. University of Chicago Press (1962)
26. Lafourcade, M., Joubert, A., Le Brun N.: GWAPs for Natural Language Processing, pp. 47–72. Wiley (2015)
27. Le Dantec, C., Poole, E., Wyche, S. (2009) Values as lived experience: evolving value sensitive design in support of value discovery. In: Conference on Human Factors in Computing Systems—Proceedings, pp. 1141–1150 (2009)
28. Levy, O., Goldberg, Y., Dagan, I.: Improving distributional similarity with lessons learned from word embeddings. Trans. Assoc. Comput. Linguist. **3**, 211–225 (2015)
29. Lewis T.: IBM's Watson says it can analyze your personality in seconds—but the results are all over the place. Business Insider UK (July 2015). http://uk.businessinsider.com/ibms-supercomputer-can-now-analyze-your-personality-based-on-a-writing-sample-heres-how-you-try-it-2015-7
30. Lunden, I.: AWS ramps up in AI with new consultancy services and Rekognition features. TechCrunch (Nov 2017). https://techcrunch.com/2017/11/22/aws-ai/
31. Masterman, M.: The nature of a paradigm. In: Lakatos, I., Musgrave, A. (eds.), Criticism and the Grown of Knowledge, pp. 59–89. Cambridge University Press (1970)
32. McMillan, R.: AI has arrived, and that really worries the world's brightest minds. Wired (2015). https://www.wired.com/2015/01/ai-arrived-really-worries-worlds-brightest-minds
33. Metcalf, J., Keller, E.F., Boyd, D.: Perspectives on Big Data, Ethics, and Society, White Paper (2017)
34. Moreau, E.: Identification of natural languages in the limit: exploring frontiers of finite elasticity for general Combinatory Grammars. In: 12th Conference on Formal Grammars (FG 2007), page Online Proceedings, Dublin, Ireland, France, Aug. 2007. CSLI Publications Online Proceedings (2007)
35. Moreau, E., Vogel, C.: Weakly supervised approaches for quality estimation. Mach. Trans. **27**(3), 257–280 (2013)

36. Moreau, E., Vogel, C.: Limitations of MT quality estimation supervised systems: the tails prediction problem. In: Proceedings of COLING 2014, the 25th International Conference on Computational Linguistics, Dublin, Ireland, Aug 2014, pp. 2205–2216. Dublin City University and Association for Computational Linguistics (2014)
37. Moreau, E., Vogel, C.: Multilingual word segmentation: training many language-specific Tokenizers Smoothly thanks to the universal dependencies corpus. In: Proceedings of the Eleventh International Conference on Language Resources and Evaluation (LREC 2018), Miyazaki, Japan (May 2018)
38. Moreau, E., Yvon, F., Cappé, O.: Robust similarity measures for named entities matching. In: COLING 2008, Manchester, UK, pp. 593–600. ACL (Aug 2008)
39. Newman, A.A.: Translators Scoff at LinkedIn's Offer of $0 an Hour. New York Times (June 2009). http://www.nytimes.com/2009/06/29/technology/start-ups/29linkedin.html
40. Nivre, J., Agić, Ž., Ahrenberg, L., Aranzabe, M.J., Asahara, M., Atutxa, A., Ballesteros, M., Bauer, J., Bengoetxea, K., Bhat, R.A., Bick, E., Bosco, C., Bouma, G., Bowman, S., Candito, M., Cebiroğlu Eryiğit, G., Celano, G.G.A., Chalub, F., Choi, J., Çöltekin, Ç., Connor, M., Davidson, E., de Marneffe, M.-C., de Paiva, V., de Ilarraza, A.D., Dobrovoljc, K., Dozat, T., Droganova, K., Dwivedi, P., Eli, M., Erjavec, T., Farkas, R., Foster, J., Freitas, C., Gajdošová, K., Galbraith, D., Garcia, M., Ginter, F., Goenaga, I., Gojenola, K., Gökörmak, M., Goldberg, Y., Gómez Guinovart, X., Saavedra, B.G., Grioni, M., Grūzītis, N., Guillaume, B., Habash, N., Hajič, J., Hà, L., Haug, D., Hladká, B., Hohle, P., Ion, R., Irimia, E., Johannsen, A., Jørgensen, F., Kaşıkara, H., Kanayama, H., Kanerva, J., Kotsyba, N., Krek, S., Laippala, V., Hng, P.L., Lenci, A., Ljubešić, N., Lyashevskaya, O., Lynn, T., Makazhanov, A., Manning, C., Mărăduc, C., Mareček, D., Martínez Alonso, H., Martins, A., Mašek, J., Matsumoto, Y., McDonald, R., Missilä, A., Mititelu, V., Miyao, Y., Montemagni, S., More, A., Mori, S., Moskalevskyi, B., Muischnek, K., Mustafina, N., Müürisep, K., Nguy Th, L., Nguy Th Minh, H., Nikolaev, V., Nurmi, H., Ojala, S., Osenova, P., Øvrelid, L., Pascual, E., Passarotti, M., Perez, C.-A., Perrier, G., Petrov, S., Piitulainen, J., Plank, B., Popel, M., Pretkalniņa, L., Prokopidis, P., Puolakainen, T., Pyysalo, S., Rademaker, A., Ramasamy, L., Real, L., Rituma, L., Rosa, R., Saleh, S., Sanguinetti, M., Saulīte, B., Schuster, S., Seddah, D., Seeker, W., Seraji, M., Shakurova, L., Shen, M., Sichinava, D., Silveira, N., Simi, M., Simionescu, R., Simkó, K., Šimková, M., Simov, K., Smith, A., Suhr, A., Sulubacak, U., Szántó, Z., Taji, D., Tanaka, T., Tsarfaty, R., Tyers, F., Uematsu, S., Uria, L., van Noord, G., Varga, V., Vincze, V., Washington, J.N., Žabokrtský, Z., Zeldes, A., Zeman, D., Zhu, H.: Universal dependencies 2.0. LINDAT/CLARIN Digital Library at the Institute of Formal and Applied Linguistics, Charles University (2017)
41. Nuzzo, R.: How scientists fool themselves—and how they can stop. Nature (2015). http://www.nature.com/news/how-scientists-fool-themselves-and-how-they-can-stop-1.18517
42. OECD Data.: Population with tertiary education (2017). https://data.oecd.org/eduatt/population-with-tertiary-education.htm
43. Pearson, J.: Why an AI-Judged beauty contest picked nearly all white winners. Motherboard (2016). https://motherboard.vice.com/en_us/article/78k7de/why-an-ai-judged-beauty-contest-picked-nearly-all-white-winners
44. Popowich, F., Vogel, C.: A logic based implementation of head-driven phrase structure grammar. In: Brown, C., Koch, G. (eds.) Natural Language Understanding and Logic Programming, vol. III, pp. 227–246. Elsevier, North-Holland (1991)
45. Reinhardt, A.: There's Sanity Returning. Bus. Week **3579**, 62–64 (1998)
46. Roser, M., Ortiz-Ospina, E.: Literacy (2017). https://ourworldindata.org/literacy/
47. Sample, I.: Artificial intelligence risks GM-style public backlash, experts warn. https://www.theguardian.com (Nov 2017). https://www.theguardian.com/science/2017/nov/01/artificial-intelligence-risks-gm-style-public-backlash-experts-warn
48. Schmid, H.: Probabilistic part-of-speech tagging using decision trees. In: Proceedings of International Conference on New Methods in Language Processing, Manchester, UK (1994)
49. Searle, J.R.: Minds, brains, and programs. Behav. Brain Sci. **3**(3), 417–424 (1980)
50. Sengers, P., Boehner, K., David, S., Kaye, J.J.: Reflective design. In: Proceedings of the 4th Decennial Conference on Critical Computing: Between Sense and Sensibility, CC '05, pp. 49–58, New York, NY, USA. ACM (2005)

51. Shilton, K., Anderson, S.: Blended, not bossy: ethics roles, responsibilities and expertise in design. Interact. Comput. **29**(1), 71–79 (2017)
52. Sneddon, J.: Why linux users make the most valuable customers. OMG! Ubuntu! (July 2017). http://www.omgubuntu.co.uk/2017/07/linux-users-are-more-valuable-customers
53. Titcomb, J.: 'Facebook is listening to me': why this conspiracy theory refuses to die (Oct 2017). http://www.telegraph.co.uk. http://www.telegraph.co.uk/technology/2017/10/30/facebook-listening-conspiracy-theory-refuses-die
54. van Wynsberghe, A., Robbins, S.: Ethicist as designer: a pragmatic approach to ethics in the lab. Sci. Eng. Ethics **20**(4), 947–961 (2014)
55. Vogel, C.M., Hahn, U., Branigan, H.: Cross-serial dependencies are not hard to process. In Proceedings of the 16th International Conference on Computational Linguistics, pp. 157–162. COLING'96, Copenhagen, Denmark (1996)
56. Yuste, R., Goering, S., y Arcas, B.A., Bi, G., Carmena, J.M., Carter, A., Fins, J.J., Friesen, P., Gallant, J., Huggins, J.E., Illes, J., Kellmeyer, P., Klein, E., Marblestone, A., Mitchell, C., Parens, E., Pham, M., Rubel, A., Sadato, N., Sullivan, L.S., Teicher, M., Wasserman, D., Wexler, A., Whittaker, M., Wolpaw, J.: Four ethical priorities for neurotechnologies and AI. Nature **1**(551), 159–163 (2017). https://www.nature.com/news/four-ethical-priorities-for-neurotechnologies-and-ai-1.22960

Chapter 9
Big Data and Multimodal Communication: A Perspective View

Costanza Navarretta and Lucretia Oemig

Abstract Humans communicate face-to-face through at least two modalities, the auditive modality, speech, and the visual modality, gestures, which comprise e.g. gaze movements, facial expressions, head movements, and hand gestures. The relation between speech and gesture is complex and partly depends on factors such as the culture, the communicative situation, the interlocutors and their relation. Investigating these factors in real data is vital for studying multimodal communication and building models for implementing natural multimodal communicative interfaces able to interact naturally with individuals of different age, culture, and needs. In this paper, we discuss to what extent big data "in the wild", which are growing explosively on the internet, are useful for this purpose also in light of legal aspects about the use of personal data, comprising multimodal data downloaded from social media.

9.1 Introduction

Humans communicate face-to-face through at least two modalities, the auditive modality, speech, and the visual modality, gestures, and therefore face-to-face communication is said to be multimodal. In the present work, we use gesture as a general term covering all kinds of unobtrusive co-speech communicative body behaviour [1] such as gaze direction, facial expressions, head movements, body postures, arm and hand gestures. Other modalities, such as haptics, can also be involved in communication. Speech and gesture are related on many levels and various gesture types have different temporal and/or semantic relation to speech [25, 34]. Moreover, gestures are multi-functional, and the same gesture can have disparate meanings depending on

C. Navarretta (✉)
Department of Nordic Studies and Linguistics, Centre for Language Technology,
University of Copenhagen, Emil Holms Kanal 2, 2300 Copenhagen, Denmark
e-mail: costanza@hum.ku.dk

L. Oemig
Agency for Digitisation, Ministry of Finance, Landgreven 4, 1017 Copenhagen, Denmark
e-mail: lucoe@digst.dk

© Springer Nature Switzerland AG 2019
A. Esposito et al. (eds.), *Innovations in Big Data Mining and Embedded Knowledge*,
Intelligent Systems Reference Library 159,
https://doi.org/10.1007/978-3-030-15939-9_9

the context. The relation between speech and gesture is influenced by multiple factors, e.g. the culture, the language, the communicative settings, the relation between the individuals which communicate, their degree of familiarity, and their personal characteristics.

Understanding and being able to formalise the complex relation between speech, gestures and communicative context is not only essential for theories of language and cognition, but has also practical uses because it provides formal models for the implementation of natural communicative interfaces comprising communicative software agents and social robots. These interfaces should be able to interact in a natural way with individuals of different age, culture, and needs. Therefore, formal models of multimodal behaviour in communication and, more generally in various types of interaction, must take into account how humans respond to different situations and stimuli given their background and personality.

Large amounts of multimodal interactions can already be found on the internet, and people without technical background are recording multimodal interactions via simple and common devices such as mobile phones. The technological development which allows not only the collection and analysis of video- and audio-recorded communication, but also the implementation of advanced multimodal interfaces has haft a great impact on research. Multimodal communication is currently addressed by a growing number of disciplines such as linguistics, psychology, computer science, anthropology, engineering, philosophy, and sociology. The aims of these disciplines are different, but since multimodal communication is a complex phenomenon, it is often addressed interdisciplinarily.

Even though there are numerous videos of recorded communicative situations on the web, only a minimal part of them has been collected or uploaded according to pre-defined criteria, and annotated multimodal corpora are few. Therefore, a lot of effort has been done in order to find the best methodologies for extracting information from raw multimodal data. The issue is whether all information about multimodal communication can be extracted from raw data "in the wild", that is data which have been uploaded on the internet for different purposes, and whether all data can be used freely since it has been published on the internet. Moreover, multimodal audio- and video-recordings are by nature personal data and physical, physiological or behavioural characteristics of the persons recorded can also be used to extract biometric data. In this article, we will discuss the use of big data for studying multimodal communication providing some answers to the following questions:

- Can multimodal big data "in the wild" provide sufficient knowledge about multimodal communication to build formal models for designing intelligent multimodal communicative interfaces, that is interfaces which take into account cultural and individual differences and needs of the users?
- To what extent can big data give more insight on the way humans communicate?
- What are the general European regulation for processing personal data in research and how these rules relate to the type of knowledge which can be extracted from multimodal data?

Our fields are computational linguistics applied to the field of multimodal communication and law dealing with the treatment of personal data.

The paper is organised as follows. First in Sect. 9.2, we present studies which show that factors such as the language and the degree of familiarity of the participants influence communication and more specifically the relation between speech and gesture. Secondly in Sect. 9.3, we shortly describe examples of annotated multimodal data and discuss the impact of big data on research investigating multimodal communication. In Sect. 9.4, we account for how the European legislation addresses the use of personal data, with particular focus on multimodal data, and finally in Sect. 9.5, we discuss to what extent big data can be useful for modelling multimodal communication including contextual knowledge also in light of the legislation on personal data.

9.2 Studies on Multimodal Communication

In this section, we present examples of research addressing the production and/or perception of multimodal behaviour and their relation to contextual factors such as the type of language, the culture, the physical and communicative settings, the reciprocal relation between the conversation participants, their age and personality.

The rhythmic relation between speech and beat gestures has been studied inter alia by Kendon [24] and Loehr [32, 33]. They have noticed that there is a temporal alignment between pitch accents and the stroke of hand gestures, head movements and eye blinks. Since intonation is language specific, the results of these studies confirm the strong dependence of gesture and language. This is also the case for perception experiments showing that people are sensible to changes in the natural temporal alignment of speech and co-speech hand gestures [17, 28]. Other studies have addressed speech and gestural pauses and have concluded that the two modalities are coordinated at many different levels including the syntactic, semantic and pragmatic ones [14, 19, 23].

The form of co-speech hand gestures and the syntactic characteristics of languages are also related. For example, Özyürek et al. [45] show that the syntactic framing of motion verbs in English and Turkish, which are a satellite-framed and a verb-framed language respectively, influences the form of co-speech iconic gestures. Similar studies have been performed in many other languages. Researchers have also pointed out that the form of iconic gestures does not only depend on the semantic characteristics of the object referred to, but also by pragmatic and linguistic factors such as the verb action form [13]. Following this line of studies, Lis [29] finds that the form of iconic gestures is influenced by the preceding hand gestural context and Lis and Navarretta [30] succeed in training classifiers on linguistic features of the verbs, such as *Aktionsart*, in order to predict the movement type of co-occurring Polish iconic gestures. They also show that classifiers can use information about the linguistic characteristics of verbs as training data for predicting the viewpoint of the iconic gestures co-occurring with the verbs, confirming that language specific characteristics influence the shape and movement of iconic gestures.

Feedback head movements were compared in three Nordic multimodal corpora of dyadic first encounters involving young people [44]. The study revealed differences in the type and frequency of head movements in the three datasets, even though the three countries in which the corpora were collected are very near culturally, and Swedish and Danish belong to the same language family. One of the most interesting differences in the three corpora was that up-nods are significantly more frequent in the Swedish encounters than in the Danish and Finnish ones, while feedback head tilts and side head movements only occur in the Danish conversations. As expected, feedback speech expressions in the Danish and Swedish encounters are similar.

Navarretta and Lis [41] found both similarities and differences in the way feedback is expressed through words and head movements in Danish and Polish comparable spontaneous triadic conversations. The most noticeable difference was that repeated multimodal feedback expressions are significantly more frequent in the Polish conversations than in the Danish ones.

A comparison of feedback head movement and facial expressions in two different types of Danish conversation showed that the degree of familiarity of the participants, the number of the participants involved in the conversations, and the physical settings influence how the participants express feedback [43]. More particularly, participants in first encounters use more often feedback facial expressions than participants in conversations with near friends and family members. The analysis of the data also showed that feedback head movements are more frequent in the corpus with participants who know each other well than in the first encounters. The study also found that the type and frequency of head movements and facial expressions partially depend on the physical settings, and more precisely whether the participants are sitting or standing, and whether they are facing each other or not. The number of participants involved in the conversation also has an impact on the frequency and type of feedback.

Even though differences in feedback behaviour have been found in various languages or in different conversation types in the same language, the data also show that there are common features in the ways subjects use head movements when giving and eliciting feedback. For this reason, machine learning algorithms were trained on the annotations of a corpus in order to automatically classify feedback head movement in other corpora [38, 42]. The application of machine learning models on conversations of different languages or types in these experiments was facilitated by the fact that all the corpora were annotated according to the same annotation framework and only the granularity of the annotations varied.

Other studies have addressed differences in gesturing which are related to the age of the participants. For example, Feyereisen and Havard [16] compared the production of iconic gesture in adults of different age. They found that older adults produce less frequently representational gestures than younger adults when gestures are activated by visual images, while the two groups do not differ in gesture production when the gestures are activated by motion images.

Individual differences in the place in which conversation participants give feedback in the same conversational context have been discovered in an experiment performed by de Kok and Heylen [11], while Hostetter and Potthoff [20] have showed a correlation between two self-reported personality traits, extroversion and neurocity, and the production of representational gestures. They have also noticed that the frequency of gesturing and the physical context are related since speakers produce more representational gestures when the interlocutors can see them clearly, than when this is not the case. Individual differences in the shape of gestures were exploited by Navarretta who used information about the physical characteristics of gestures as training data for classifiers in order to identify the gesturers [37].

The gestures of the people with whom we interact also influences our gestures, according to the mirror theory [52, 53]. In line with this theory, Navarretta [40] has found a positive correlation between the gestures of a participant and the gestures produced by the interlocutor in first encounters in which the participants face each other. The gestures which are mirrored more often in the encounters are facial expressions. Moreover, information about the gestures produced by a subject has been used to predict automatically the presence and type of the gestures of the interlocutors [40]. Kramer et al. [26] also discovered mirroring effects in human-machine interactions. Their experiments showed that subjects interacting with a software agent smile more frequently when the agent produces numerous smiles than when it smiles less frequently, even when the participants are not aware of whether the agent smiles frequently or not. It must be noticed that studies addressing the relation between e.g. syntactic or discourse structure and gestures as well as those investigating the temporal relation between speech and gestures have often been based on small datasets in one language, e.g. [18, 24, 32, 39].

Not only the production, but also the perception of gestures has been found to depend on the culture. Riviello et al. [51] have compared how people of different cultures (U.S. Americans, French and Italians) perceive emotions expressed by actors in unimodal and multimodal data. The results of the study show that the perception of emotions is not only affected by the modalities involved, but also by the culture of the participants and their familiarity with the language in which the emotions are expressed.

A branch of linguistic research which has addressed multimodality in dialogues for decades is Conversation Analysis (CA). Conversation analysts have performed qualitative studies of social interactions and presented theories about e.g. turn taking and repairs [54] which have also inspired computational linguistic research in e.g. the field of dialogue processing [60] and of multimodal behaviour generation in software agents [9]. Recently, CA researchers have focused on the spatiality and embodiment of interactions and on human interaction with technological devices [35, 58, 59].

Summing up, most of the research investigating factors which influence the production and/or the perception of multimodal behaviour consists of experimental studies and qualitative analysis of specific types of interactions. Qualitative and quantitative analyses of annotated corpora of multimodal communication have also been conducted and machine learning algorithms have been applied on annotated corpora in order to test hypotheses about language as well as the appropriateness of the annotations of the corpora. Information about the participants and the context of the conversations was known in all these studies that looked at the relation about language, gesture and interaction context.

9.3 Big Data and Multimodality

Processing multimodal communication data not only requires the automatic identification and the interpretation of the data in the different modalities, but also capturing the relation between them in the context in which they were produced. This knowledge is also useful for generating multimodal behaviour.

Large text corpora of different types, domains and languages have been collected the past fifty years, and many of them are annotated with linguistic, semantic and pragmatic information. Most computational linguistic research has used these corpora for e.g. analysis, information extraction and automatic annotation of new data, often through machine learning algorithms. Applications such as machine translation, summarisation and abstracting have also been built using some of these corpora. Similarly, transcribed dialogues have been used for building query-answering systems, and have been the basis for the definition of a dialogue act standard [6].

The collection and annotation of multimodal corpora is more recent and freely available large scale annotated multimodal corpora are few. Some examples are shortly presented in the following section.

9.3.1 Multimodal Corpora

The AMI corpus [8] is an English scenario-based 100 h corpus in which conversations between three or four participants are audio- and video-recorded by multiple cameras and microphones. The interactions are fictive or real project meetings during which the participants use various devices. The conversations have been transcribed and linguistic features at several levels of complexity have been manually or automatically annotated. For example, the annotations comprise part-of-speech, syntax and dialogue acts. Only a few gesture types have been annotated in a subset of the interactions and the granularity of these annotations is very coarse. The AMI corpus has been extensively used for training and testing machine learning algorithms to automatically annotate linguistic information on dialogues.

The HuComTech corpus is a 50 h Hungarian multimodal corpus [21]. It consists of 111 formal and 111 informal conversations between male and female university students. The corpus has been transcribed and enriched with numerous linguistic features, and gestures have also been annotated.

The RECOLA French multimodal corpus consists of approx. 10 h of audio and video-recordings of dyadic interactions between participants, who were solving a task in collaboration. The corpus has been collected for detecting affective states [50] and it also contains physiological recordings (electrocardiogram, and electrodermal activity). The affective and social behaviours in the corpus have been self-reported by the participants. A large part of the database is freely available for research and has been used in Audio/Visual Emotion challenges.

Other multimodal annotated corpora, such as the Nordic multimodal first encounters corpora (NOMCO) [44, 46] are also available for research. The NOMCO corpora are not very large (between one hour and three hours of recordings), but they are comparable, that is the physical and communicative settings and the type of participants are similar and they are annotated according to the same multimodal annotation framework, MUMIN [2]. First encounters have also been collected in other languages. Recently, a multimodal corpus annotated with opinions in online videos has been presented [65]. The corpus has been coded with subjectivity and sentiment intensity labels both in the audio and visual modalities.

One problem with annotated multimodal data is that various research groups use different annotation schemes and systems. Furthermore, the annotation of gestures and their relation to speech is not as formalised as the annotation of e.g. the syntax and semantics of texts. However, as far as the annotations are consistent and the semantics of the schemes is well-explained, annotated data are extremely useful, and can be used as golden standards against which to evaluate theories, models and algorithms. Another problem related to multimodal corpora is their availability, since they can often only be used for a specific research project. Nonetheless non-annotated multimodal data on the WEB are growing and they also comprise human-machine interactions and collections of sensor data of different types, and it is therefore interesting to use them in research on communication, as far as this is possible. Sensors comprise e.g.eye tracking, motion capture devices, heart rate and sweat monitors. Neuroimaging and EEG data are also available, and they give insight in aspects of human behaviour which have not earlier been available to communication research. Examples of collections containing these types of big data are the previously mentioned RECOLA corpus, the CMU Graphics Lab Motion Capture Database,[1] and eye tracking and neuroimaging data of people looking at the film *Forrest Gump* at http://studyforrest.org.

In the following, we shortly present some of the methods that have been used for processing big data and, successively, we address how data in the wild can be used for research according to the European legislation.

9.3.2 Processing Big Data

The most commonly applied technique for treating big data of different types is machine learning. Unsupervised machine learning is applied on raw data in order to discover patterns in them, while supervised machine learning refers to algorithms that are trained on data enriched with various annotations in order to classify new data with

---

[1]http://mocap.cs.cmu.edu/.

the same annotation labels. Hybrid forms, such as semi-supervised classification, are also common.

Recently, researchers have begun reusing a particular type of machine learning algorithms, so called neural networks, and some of these algorithms have performed surprisingly well on certain raw data types. Neural networks have been most successful when applied to signal processing, providing generalised representations for clusters of pixels or acoustic signals. More specifically, deep learning algorithms have been successfully applied to computer vision data [27], such as astrophysical data [4] and facial pictures [36, 48] while so-called recurrent neural networks have improved speech processing significantly [12, 62].

Even though tools for the automatic identification and interpretation of gestures from unrestricted videos are not available yet, research act to recognising e.g. facial expressions from still images or frontal videos [31] and human actions from videos [22] is continuously improving. In some cases, annotation schemes such as the Facial Action Coding System (FACS) or manual annotations of data have been used as a framework for training or interpreting the data. Machine learning has also been applied on large textual data in the field of computational social media [57, 63] and computational linguistics [10]. Good results in computational linguistics have been achieved when deep learning has been used in particular applications such as machine translation even with few parallel data as training material [49, 56].

Open source tools, such as OpenSMILE [15], OpenPose [7, 55, 61], and OpenFace [3], which use machine learning to extract features and classify speech and/or body movements are also useful for processing speech and body behaviour from audio and videos, respectively.

Concluding, multimodal annotated corpora are extremely useful for research, but they are few because of privacy issues and since the manual or semi-automatic annotation of multimodal data is time consuming. However, other types of data, such as sensor and neuroimaging data can also be used for research on multimodal interaction. Furthermore, the success of machine learning algorithms in recognising patterns in speech and video sequences is promising since it will provide at least semi-automatic identification of gesture occurrences and, possibly, improve the automatic recognition of speech in many languages.

9.4   Using Big Data "in the Wild" Lawfully

In the following, we present the main principles and rules for treating personal data when we deal with big data, also when these are downloaded from the internet.

The general European data protection legislation is based on Regulation (EU) 2016/679 of the European Parliament and of the Council of 27 April 2016 on the protection of natural persons with regard to the processing of personal data and the free movement of such data also called General Data Protection Regulation (GDPR in what follows).

Big data are covered by the GDPR when they contain personal data. Personal data cover any information relating to an identified or identifiable person, cf. GDPR article 4(1). This means that if a person can be identified from big data, the GDPR is applied, while the GPDR does not apply if the data is anonymized.

In order to process personal data, a legal basis is needed. The GDPR distinguishes between two categories of personal data: sensitive personal data and personal data. The legal basis for processing personal data and sensitive data are different. Personal data is any information about a natural person, such as the subject's name, address, phone number, email address and IP address. The definition of "personal data" is broad, cf. Judgement of 20 December 2017 (C-434/16, EU:C:2017:994, paragraph 27), where the European Court of Justice found that even the written answers submitted by a candidate at a professional examination and any examiner's comments with respect to those answers constitute personal data.

Sensitive personal data is "racial or ethnic origin, political opinions, religious or philosophical beliefs, or trade union membership, and the processing of genetic data, biometric data for the purpose of uniquely identifying a natural person, data concerning health or data concerning a natural person's sex life or sexual orientation", cf. GDPR article 9(1). Particularly relevant when working with visual and audio data, it is the concept of biometric data. In the definition of biometric data in the GDPR article 4(11) facial images are in fact given as an example, but the GDPR's recital 51 explains that processing of photographs should not automatically be considered as processing of sensitive personal data. Photographs will only be covered by the definition of "biometric data" when they are processed through specific technical means, which allow the unique identification or authentication of an individual as a natural person. Videos can be biometric data if specific technical processing relating to the physical, physiological or behavioural characteristics of a natural person allows or confirms the unique identification of that person.

In the following, we will focus on the legal basis for processing personal data which is particularly important for research. Processing personal data and sensitive personal data is lawful if the data subject has given consent to the processing of her or his data for one or more specific purposes, cf. GDPR article 6(1)(a) for personal data and GDPR article 9(2)(a) for sensitive personal data. The consent has to be: "freely given, specific, informed and unambiguous indication of the data subject's wishes by which he or she, by a statement or by a clear affirmative action, signifies agreement to the processing of personal data relating to him or her", cf. GDPR article 4(11).

Processing personal data and sensitive personal data is also allowed if the personal data involved are manifestly made public by the data subject, cf. GDPR article 9(2)(e). Therefore, if the personal data have been published on the internet for example on Youtube, a public profile on Facebook or Twitter, it's allowed to process these data without any permission from the data subject. This article can only be applied, if the personal data have been published by the data subject and no one else. Personal data are considered to be published, if they have been made available by the data subject to a large group of people.

Moreover, personal data can be processed if the processing is necessary in order to perform a task which has public interest, or in the exercise of official authority vested by the data controller, cf. GDPR article 6(1)(e).

Processing of sensitive personal data is also allowed if it is needed for scientific or historical research purposes, for statistical purposes or for archiving purposes in the public interest, cf. GDPR article 9(2)(j). For using this paragraph as the legal basis for processing sensitive personal data, four conditions must be fulfilled. First, processing is necessary for scientific or historical research purposes, for statistical purposes or for archiving purposes in the public interest. Secondly, the process has to be based on Union or Member State Law. Thirdly, the law has to be proportionate to the aim pursued. Finally, the law must respect the essence of the right to data protection and provide for suitable and specific measures to safeguard the fundamental rights and the interests of the data subject.

Besides having a legal basis to process data, researchers and other actors who work with big data also need to comply with the general principles relating to processing of personal data of the GDPR. The GDPR article 5 defines general principles which apply to all forms of personal data processing. These principles are not the legal basis for processing the data, but their compliance is a pre-condition for processing the data. The noncompliance with any of the general principles will make data processing unlawful. The principles are *lawfulness, fairness, transparency, purpose limitation, data minimisation, accuracy, storage limitation, integrity* and *confidentiality*. Besides these principles, there is a new principle in the GDPR regarding accountability. The controller[2] shall be responsible for, and be able to demonstrate the compliance with all the above mentioned principles.

The principle lawfulness, fairness and transparency indicates that personal data must be processed lawfully, fairly and in a transparent manner in relation to the data subject.

The purpose limitation refers to the fact that specified and legitimate purposes must justify the collection of personal data and further processing of these data must be in line with the initial purposes. Moreover, further processing of the data for historical or scientific research as well as for archiving in the public interest and for statistical purposes cannot be considered incompatible with the initial purposes, cf. GDPR article 5(1)(b). However, this further processing of personal data for archiving, statistical purposes or research must be subject to appropriate safeguards for the rights and freedoms of the data subject. Data can for example be pseudo anonymized if the research or archiving purposes can be fulfilled this way. Moreover, in cases where the purposes of research can be achieved by further processing the data, and the processing does not permit or no longer permits the identification of data subjects, those purposes must be fulfilled in that way, cf. GDPR article 89(1).

---

[2]In the GDPR article 4(7) the controller is defined as "the natural or legal person, public authority, agency or other body which, alone or jointly with others, determines the purposes and means of the processing of personal data; where the purposes and means of such processing are determined by Union or Member State law, the controller or the specific criteria for its nomination may be provided for by Union or Member State law".

The data minimisation principle means that personal data must be relevant, adequate and limited to what is necessary in relation to the purposes for which they are processed. Therefore the controller has to limit the collection of personal data to what is really relevant and necessary to accomplish a given task.

The principle of accuracy states that data has to be accurate and kept up to date. The controller has to make sure that inaccurate data are erased or rectified without delay.

The storage limitation implies that data cannot be kept for a longer time than it is necessary for the purposes for which the personal data are processed. Personal data may be stored for longer periods if the personal data only is processed for archiving purposes in the public interest, scientific or historical research purposes or statistical purposes in accordance with the GDPR article 89(1) and subject to implementation of appropriate technical and organisational measures.

The principle about integrity and confidentiality means that personal data shall be processed in a manner that ensures appropriate security of the personal data, including protection against unauthorised or unlawful processing and against accidental loss, destruction or damage, using appropriate technical or organisational measures.

When the controller starts collecting data directly from the data subject, the controller has to inform the subject about the processing of data when the personal data are obtained and all the information required in GDPR article 13.[3]

The controller shall not give all this information if the data subject already has it, but since the data subject is usually not aware of all this information, this exception rarely applies.

Where the data is not collected directly from the data subject, cf. GDPR article 14, for example if it is collected from the internet or other public sources, the controller shall give the same information as described in the footnote, and also inform the data subject where from the personal data originate, and if applicable, whether they came from publicly accessible sources.

As in the case of data collected directly from the data subject, the controller shall not give the information if the data subject already has it. However, the data subject will not know about the processing of data in most cases, and therefore the controller should provide all this information.

---

[3]This includes all information about (i) the identity and the contact details of the controller, (ii) the contact details of the data protection officer (where applicable), (iii) the purposes of the processing for which the personal data are intended and the legal basis for the processing, (iv) the recipients or categories of recipients of the personal data (if any), (v) the fact (when relevant) that the controller intends to transfer personal data to a third country or international organisation, (vi) the period for which the personal data will be stored, or if that is not possible, the criteria used to determine that period, (vii) the existence of the right to request from the controller access to and rectification or erasure of personal data or restriction of processing concerning the data subject or to object to processing as well as the right to data portability, (viii) in cases when the processing is based on consent, the existence of the right to withdraw consent at any time, without affecting the lawfulness of processing based on consent before its withdrawal, (ix) the right to lodge a complaint with a supervisory authority, and (x) the existence of automated decision-making, including profiling (if any).

Another exception to the information requirement, is if the provision of such information proves impossible or would involve a disproportionate effort, in particular when processing for archiving purposes in the public interest, scientific or historical research purposes or statistical purposes. In such cases the controller shall take appropriate measures to protect the data subject's rights and freedoms and legitimate interests, including making the information publicly available.

Summing up, researchers need a legal basis in order to process personal data and sensitive personal data. The general principles for processing personal data must be fulfilled and finally, the data must be processed in a secure and appropriate way.

9.5   Discussion

The availability of large amounts of both visual and auditory data on the internet and the development of algorithms and methodologies for processing big data will continue to improve the state-of-the-art of the automatic treatment of multimodal data. Especially, the automatic identification of types of action from videos and the recognition of speech is reaching good accuracy and we expect the results of these tasks to become even better in the future. Moreover, research has shown that multimodal processing in many cases is more robust than unimodal processing of the data. This is for example the case for the identification of emotions [5] or sentiment analysis [64].

Furthermore, there are some types of information, such as the marking of the temporal span of both visual and auditive signals, which can be accomplished with more precision by machines than humans. In fact the time stamp annotation of e.g. gestures or speech tokens and even more of their components, such as gesture phases or phonemes is hard and time consuming. Therefore, one important contribution of the automatic processing of multimodal data will be the automatic extraction of the temporal relation between speech segments and gesture types.

The detection of semantic and pragmatic relations between speech and gesture, on the contrary, often requires human judgement. E.g. in the Danish NOMCO corpus [47], the annotators linked gestures to speech segments when they judged gestures to be related to the gesturer's speech and/or the interlocutor's speech. These segments do not always correspond to syntactic categories, and the temporal alignment of gesture and speech varies in the data depending on the speech content, the co-speech gesture and their semantics. For these reasons, it is important to build more corpora annotated with these types of knowledge and make the corpora available for analysis and as training and testing data for machine learning algorithms.

Regularities in the way people communicate multimodally despite differences in culture, age etc. will also be discovered or confirmed using big data in the wild. More specifically, studies performed on data of limited size can be in some cases scaled up on larger data. We also expect that in the future, it will be possible to extract automatically information about the language used in the interactions, the gender and at least a general indication of the age of individuals involved in the interactions.

Research will also address the automatic classification of conversation types with respect to e.g. language, content of the speech tokens, types of gesture and number of participants. Automatic extraction of the number and co-occurrence of e.g. gesture types and linguistic expressions in big data will also be addressed.

However, information about the social context, the background of and the relation between the various participants will in many cases not be available, unless this information is explicitly given as metadata or part of the uploader's personal profile. Nowadays, on some types of social media such as Facebook, it is already possible to extract a large amount of personal data from the users' profiles. Nevertheless according to the European legislation, personal data cannot be used and processed unless subjects have made their profile public or given their consent to the use of the data. It must also be noted, that a closed profile on Facebook can also be considered public if the subject has made the data available to a large group of persons. This is a consequence of a judgement by the European Court of Justice which determined that data made visible to an indefinite number of people cannot be considered private.[4] However, deciding how many people form a large group, requires individual estimation.

The fact that only public profiles can be used restricts the population covered by the data, since persons who have open profiles or have many "friends" are often extrovert or use social media because they have professions which require visibility, such as politicians or musicians. Furthermore, it can be difficult, or even impossible, to determine automatically who has uploaded audio- and video-recordings available under a public profile. In the case of interactions involving more individuals, it is also necessary to control that all the individuals have agreed upon sharing their personal data on the internet, and this can also be difficult. Therefore, most of the videos on the internet that can be used together with personal data about the participants will be videos of monologues of the data subject or public debates.

A part from the legal and practical problems related to the extraction of contextual information about subjects appearing in multimodal interactions on the internet, big data "in the wild" is also problematic because of their representativeness. Data on the social media are usually uploaded by companies, organisations or people who have access and interest in social media. Even though the number of users of social platforms increases every day and the spectrum of population covered by the social media becomes larger, especially in countries with a good digital infrastructure, these subjects are not representative of the world population. Elder, impaired or economically weak individuals are not prototypical users of social media. Since much research on e.g. social robots has the aim of supporting elder or impaired subjects, there is a need for data which show their specific multimodal behaviour. Therefore, it is important to collect, annotate and process data which show how different types of people communicate with each other, and how they interact with machine interfaces the first time they are presented for them and after they have got acquainted with them.

---

[4]Cf. Judgement of 6 November 2003, Lindqvist (C 101/01, EU:C:2003:596, paragraph 47).

It must also be noted that the lack of annotated data in different languages and domains is still a challenge in the field of computational linguistics even when the domain of research is unimodal (texts and spoken resources) since most language models work satisfactorily in the domain of the training data, while they only partially succeed in other domains. Moreover, processing dialogues is more difficult than processing written data. Additionally, individuating sufficiently large multimodal data about specific phenomena, such as types of hand gesture co-occurring with specific verb constructions in certain languages, is nearly impossible. Collecting these data specifically in e.g. narrative settings as proposed by McNeill for studying specific hand gestures [34] or setting up experiments as it is common in psycholinguistic or psychologic research can still be the best solution in many cases, even though experimental settings are not natural and the data used to solicit multimodal behaviours might to some extent influence them.

So, on the one hand, big data are extremely useful for training and testing models act to the automatic identification of gestures and speech independently from the peculiarities of individuals or particular groups. Big data are also useful to test hypotheses and models as well as to identify general aspects in human-human or human-machine interaction. On the other hand, it is also necessary to process data whose contextual knowledge is known and where implicit knowledge is made explicit, and to inform the involved data subjects about the processing of personal data. Moreover, it is important to study communication between all types of people, independently from whether they are active or not on the internet. Therefore, researchers should go on individuating both general aspects of human communication, and individual behaviours in order to be able to design applications for specific groups of users.

References

1. Allwood, J., Nivre, J., Ahls'en, E.: On the semantics and pragmatics of linguistic feedback. J. Semant. **9**, 1–26 (1992)
2. Allwood, J., Cerrato, L., Jokinen, K., Navarretta, C., Paggio, P.: The MUMIN coding scheme for the annotation of feedback, turn management and sequencing. Multimodal corpora for modelling human multimodal behaviour. Spec. Issue Int. J. Lang. Resour. Eval. **41**(3–4), 273–287 (2007)
3. Amos, B., Ludwiczuk, B., Satyanarayanan, M.: Openface: A general-purpose face recognition library with mobile applications. Technical report, CMU-CS-1 6-118, CMU School of Computer Science (2016)
4. Batalha, N.M., Rowe, J.F., Bryson, S.T., Barclay, T., Burke, C.J. et al.: Planetary candidates observed by Kepler. III. Analysis of the first 16 months of data. Astrophys. J. Suppl. Ser. **204**(2), 24 (2013)
5. Bourbakis, N., Esposito, A., Kavraki, D.: Extracting and associating meta-features for understanding people's emotional behaviour: face and speech. Cogn. Comput. **3**, 436–448 (2011)
6. Bunt, H., Alexandersson, J., Carletta, J., Choe, J.W., Fang, A.C., Hasida, K., Lee, K., Petukhova, V., Popescu-Belis, A., Romary, L., Soria, C., Traum, D.: Towards an ISO standard for dialogue act annotation. Proc. LREC **2010**, 2548–2555 (2010)

7. Cao, Z., Simon, T., Wei, S.E., Sheikh, Y.: Realtime multi-person 2D pose estimation using part affinity fields. In: CVPR (2017)
8. Carletta, J., Ashby, S., Bourban, S., Flynn, M., Guillemot, M., Hain, T., Kadlec, J., Karaiskos, V., Kraaij, W., Kronenthal, M., Lathoud, G., Lincoln, M., Lisowska, A., McCowan, I., Post, W., Reidsma, D., Wellner, P.: The AMI meeting corpus: a pre-announcement. In: Renals, S., Bengio, S. (eds.) Machine Learning for Multimodal Interaction, Second International Workshop, vol. 10. Lecture Notes in Computer Science, pp. 28–39. Springer, Berlin (2006)
9. Cassell, J., Pelachaud, C., Badler, N., Steedman, M., Achorn, B., Becket, T., Douville, B., Prevost, S., Stone, M.: Animated conversation: rule-based generation of facial expression, gesture & spoken intonation for multiple conversational agents. In: Proceedings of the 21st Annual Conference on Computer Graphics and Interactive Techniques, pp. 413–420. ACM (1994)
10. Collobert, R., Weston, J.: A unified architecture for natural language processing: deep neural networks with multitask learning. In: Proceedings of the 25th International Conference on Machine Learning, ICML '08, pp. 160–167. ACM, New York, NY, USA (2008)
11. de Kok, I., Heylen, D.: The MultiLis corpus dealing with individual differences in nonverbal listening behavior. Toward Autonomous, Adaptive, and Context-Aware Multimodal Interfaces. Theoretical and Practical Issues- Third COST 2102 International Training School, Caserta, Italy, March 15–19, 2010, Revised Selected Papers, pp. 362–375. Springer, Berlin (2010)
12. Deng, L., Li, J., Huang, J.T., Yao, K., Yu, D., Seide, F., Seltzer, M., Zweig, G., He, X., Williams, J., Gong, Y., Acero, A.: Recent advances in deep learning for speech research at Microsoft. In: 2013 IEEE International Conference on Acoustics, Speech and Signal Processing, pp. 8604–8608 (2013)
13. Duncan, S.: Gesture, verb aspect, and the nature of iconic imagery in natural discourse. Gesture 2(2), 183–206 (2002)
14. Esposito, A., Esposito, A.M.: On speech and gesture synchrony. In: Esposito, A., Vinciarelli, A., Vicsi, K., Pelachaud, C., Nijholt, A. (eds.) Communication and Enactment - The Processing Issues. LNCS, vol. 6800, pp. 252–272. Springer, Berlin (2011)
15. Eyben, F., Weninger, F., Gross, F., Schuller, B.: Recent developments in opensmile, the munich open-source multimedia feature extractor. In: Proceedings of the 21st ACM International Conference on Multimedia, MM '13, pp. 835–838. ACM, New York, NY, USA (2013)
16. Feyereisen, P., Havard, I.: Mental imagery and production of hand gestures while speaking in younger and older adults. J. Nonverbal Behav. 23(2), 153–171 (1999)
17. Giorgolo, G., Verstraten, F.A.: Perception of 'speech-and-gesture' integration. Proc. Int. Conf. Audit.-Vis. Speech Process. 2008, 31–36 (2008)
18. Hadar, U., Steiner, T.J., Grant, E.C., Rose, F.C.: The relationship between head movements and speech dysfluencies. Lang. Speech 27(4), 333–342 (1984)
19. Hadar, U., Steiner, T.J., Grant, E.C., Rose, F.C.: The timing of shifts of head postures during conservation. Hum. Mov. Sci. 3(3), 237–245 (1984)
20. Hostetter, A.B., Potthoff, A.L.: Effects of personality and social situation on representational gesture production. Gesture 12(1), 62–83 (2012)
21. Hunyadi, L., Bertok, K., Nemeth, T., Szekrenyes, I., Abuczki, A., Nagy, G., Nagy, N., Nemeti, P., Bodog, A.: The outlines of a theory and technology of human-computer interaction as represented in the model of the HuComTech project. In: 2011 2nd International Conference on Cognitive Infocommunications, CogInfoCom 2011 (2011)
22. Ikizler-Cinbis, N., Sclaroff, S.: Object, scene and actions: combining multiple features for human action recognition. In: Daniilidis, K., Maragos, P., Paragios, N. (eds.) Computer Vision - ECCV 2010: 11th European Conference on Computer Vision, Heraklion, Crete, Greece, September 5–11, 2010, Proceedings, Part I, pp. 494–507. Springer, Berlin (2010)
23. Kendon, A.: Some relationships between body motion and speech. In: Seigman, A., Pope, B. (eds.) Studies in Dyadic Communication, pp. 177–216. Pergamon Press, Elmsford, New York (1972)

24. Kendon, A.: Gesture and speech: two aspects of the process of utterance. In: Key, M.R. (ed.) Nonverbal Communication and Language, pp. 207–227. Mouton (1980)
25. Kendon, A.: Gesture - Visible Action as Utterance. Cambridge University Press, Cambridge (2004)
26. Krämer, N., Kopp, S., Becker-Asano, C., Sommer, N.: Smile and the world will smile with you—the effects of a virtual agent's smile on users' evaluation and behavior. Int. J. Hum.-Comput. Stud. **71**(3), 335–349 (2013)
27. Krizhevsky, A., Sutskever, I., Hinton, G.E.: ImageNet classification with deep convolutional neural networks. In: Pereira, F., Burges, C.J.C., Bottou, L., Weinberger, K.Q. (eds.) Advances in Neural Information Processing Systems, vol. 25, pp. 1097–1105. Curran Associates Inc. (2012)
28. Leonard, T., Cummins, F.: The temporal relation between beat gestures and speech. Lang. Cogn. Process. **26**(10), 1457–1471 (2010)
29. Lis, M.: Multimodal representation of entities: a corpus-based investigation of co-speech hand gesture. Ph.D. thesis, University of Copenhagen (2014)
30. Lis, M., Navarretta, C.: Classifying the form of iconic hand gestures from the linguistic categorization of co-occurring verbs. In: 1st European Symposium on Multimodal Communication (MMSym'13), pp. 41–50 (2013)
31. Liu, M., Li, S., Shan, S., Chen, X.: AU-inspired deep networks for facial expression feature learning. Neurocomputing **159**(Supplement C), 126–136 (2015)
32. Loehr, D.P.: Gesture and intonation. Ph.D. thesis, Georgetown University (2004)
33. Loehr, D.P.: Aspects of rhythm in gesture and speech. Gesture **7**(2), (2007)
34. McNeill, D.: Hand and Mind: What Gestures Reveal about Thought. University of Chicago Press, Chicago (1992)
35. Mondada, L.: Emergent focused interactions in public places: a systematic analysis of the multimodal achievement of a common interactional space. J. Pragmat. **41**, 1977–1997 (2009)
36. Mou, D.: Automatic Face Recognition, pp. 91–106. Springer, Berlin (2010)
37. Navarretta, C.: Individuality in communicative bodily behaviours. In: Esposito, A., Esposito, A.M., Vinciarelli, A., Hoffmann, R., Müller, V.C. (eds.) Behavioural Cognitive Systems. Lecture Notes in Computer Science, vol. 7403, pp. 417–423. Springer, Berlin (2012)
38. Navarretta, C.: Transfer learning in multimodal corpora. In: IEEE (ed.) In Proceedings of the 4th IEEE International Conference on Cognitive Infocommunications (CogInfoCom2013), pp. 195–200. Budapest, Hungary (2013)
39. Navarretta, C.: Fillers, filled pauses and gestures in Danish first encounters. In: Abstract proceedings of 3rd European Symposium on Multimodal Communication, pp. 1–3. Speech Communication Lab at Trinity College Dublin, Dublin (2015)
40. Navarretta, C.: Mirroring facial expressions and emotions in dyadic conversations. In: Chair, N.C.C., Choukri, K., Declerck, T., Goggi, S., Grobelnik, M., Maegaard, B., Mariani, J., Mazo, H., Moreno, A., Odijk, J., Piperidis, S. (eds.) Proceedings of the Tenth International Conference on Language Resources and Evaluation (LREC 2016), pp. 469–474. European Language Resources Association (ELRA), Paris, France (2016)
41. Navarretta, C., Lis, M.: Multimodal feedback expressions in Danish and polish spontaneous conversations. In: NEALT Proceedings. Northern European Association for Language and Technology, Proceedings of the Fourth Nordic Symposium of Multimodal Communication, pp. 55–62. Linköping Electronic Conference Proceedings (2013)
42. Navarretta, C., Lis, M.: Transfer learning of feedback head expressions in Danish and polish comparable multimodal corpora. In: Proceedings of 9th Language Resources and Evaluation Conference (LREC 2014), pp. 3597–3603. Reykjavik, Island (2014)
43. Navarretta, C., Paggio, P.: Verbal and non-verbal feedback in different types of interactions. In: Proceedings of LREC 2012, pp. 2338–2342. Istanbul Turkey (2012)

44. Navarretta, C., Ahlsn, E., Allwood, J., Jokinen, K., Paggio, P.: Feedback in Nordic first-encounters: a comparative study. In: Proceedings of LREC 2012, pp. 2494–2499. Istanbul Turkey (2012)
45. Özyürek, A., Kita, S., Allen, S., Furman, R., Brown, A.: How does linguistic framing of events influence co-speech gestures? Insights from crosslinguistic variations and similarities. Gesture **5**(1–2), 219–240 (2005)
46. Paggio, P., Ahlsén, E., Allwood, J., Jokinen, K., Navarretta, C.: The NOMCO multimodal Nordic resource - goals and characteristics. In: Proceedings of LREC 2010, pp. 2968–2973. Malta (2010)
47. Paggio, P., Navarretta, C.: The Danish NOMCO corpus: multimodal interaction in first acquaintance conversations. Lang. Resour. Eval. **51**(2), 463–494 (2017). https://doi.org/10.1007/s10579-016-9371-6
48. Parkhi, O.M., Vedaldi, A., Zisserman, A.: Deep face recognition. In: Proceedings of the British Machine Vision Conference (BMVC) (2015)
49. Raina, R., Madhavan, A., Ng, A.Y.: Large-scale deep unsupervised learning using graphics processors. In: Proceedings 26th Annual International Conference on Machine Learning, pp. 873–888 (2009)
50. Ringeval, F., Sonderegger, A., Sauer, J., Lalanne, D.: Introducing the RECOLA multimodal corpus of remote collaborative and affective interactions. In: Proceedings of the 2nd International Workshop on Emotion Representation, Analysis and Synthesis in Continuous Time and Space (EmoSPACE 2013), Held in Conjunction with the 10th International IEEE Conference on Automatic Face and Gesture Recognition (FG 2013). Shanghai, China (2013)
51. Riviello, M.T., Esposito, A., Vicsi, K.: A cross-cultural study on the perception of emotions: how Hungarian subjects evaluate American and Italian emotional expressions. In: Esposito, A., Esposito, A.M., Vinciarelli, A., Hoffmann, R., Müller, V.C. (eds.) Cognitive Behavioural Systems: COST 2102 International Training School, Dresden, Germany, February 21–26, 2011, Revised Selected Papers, pp. 424–433. Springer, Berlin (2012)
52. Rizzolatti, G.: The mirror neuron system and its function in humans. Anat. Embryol. **210**, 419–421 (2005)
53. Rizzolatti, G., Craighero, L.: The mirror-neuron system. Annu. Rev. Neurosci. **27**, 169–192 (2004)
54. Sacks, H., Schegloff, E., Jefferson, G.: A simplest systematics for the organization of turn-taking for conversation. Language **50**(4), 696–735 (1974)
55. Simon, T., Joo, H., Matthews, I., Sheikh, Y.: Hand keypoint detection in single images using multiview bootstrapping. In: CVPR (2017)
56. Singh, S.P., Kumar, A., Darbari, H., Singh, L., Rastogi, A., Jain, S.: Machine translation using deep learning: an overview. In: 2017 International Conference on Computer, Communications and Electronics (Comptelix), pp. 162–167 (2017)
57. Stieglitz, S., Dang-Xuan, L., Bruns, A., Neuberger, C.: Social media analytics. Bus. Inf. Syst. Eng. **6**(2), 89–96 (2014)
58. Streeck, J.: Gesturecraft - The Manufacture of Meaning. John Benjamins Publishing Company (2009)
59. Streeck, J., Goodwin, C., LeBaron., C. (eds.): Embodied Interaction: Language and Body in the Material World. Cambridge University Press, Cambridge (2011)
60. Traum, D.R.: A computational theory of grounding in natural language conversation. Ph.D. thesis, Computer Science Department, University of Rochester (1994)
61. Wei, S.E., Ramakrishna, V., Kanade, T., Sheikh, Y.: Convolutional pose machines. In: CVPR (2016)
62. Weninger, F., Erdogan, H., Watanabe, S., Vincent, E., Roux, J.L., Hershey, J.R., Schuller, B.: Speech enhancement with LSTM recurrent neural networks and its application to noise-robust ASR. In: Vincent, E., Yeredor, A., Koldovský, Z., Tichavský, P. (eds.) Latent Variable Analysis and Signal Separation: 12th International Conference, LVA/ICA 2015, Liberec, Czech Republic, August 25–28, 2015, Proceedings, pp. 91–99. Springer International Publishing, Cham (2015)

63. You, Q., Luo, J., Jin, H., Yang, J.: Robust image sentiment analysis using progressively trained and domain transferred deep networks. In: Proceedings of the Twenty-Ninth AAAI Conference on Artificial Intelligence, AAAI'15, pp. 381–388. AAAI Press (2015)
64. Zadeh, A., Chen, M., Poria, S., Cambria, E., Morency, L.: Tensor fusion network for multimodal sentiment analysis. CoRR (2017). arXiv:abs/1707.07250
65. Zadeh, A., Zellers, R., Pincus, E., Morency, L.P.: MOSI: Multimodal corpus of sentiment intensity and subjectivity analysis in online opinion videos. IEEE Intell. Syst. **31**(6), 81–88 (2016)

Chapter 10
A Web Application for Characterizing Spontaneous Emotions Using Long EEG Recording Sessions

Giuseppe Placidi, Luigi Cinque and Matteo Polsinelli

Abstract Emotions are important in daily life and in several research fields, especially in Brain Computer Interface (BCI) and Affective Computing. Usually, emotions are studied by analyzing the brain activity of a subject, monitored by Electroencephalography (EEG), functional Magnetic Resonance Imaging (fMRI) or functional Near Infrared Spectroscopy (fNIRS), after some external stimulation. This approach could lead to characterization inaccuracies, due to the secondary activations produced by the artificial elicitation and to the subjective emotional response. In this work, we design a web application to support spontaneous emotions characterization. It is based on a database for EEG signals where a large amount of data from long recording sessions, collected from subjects during their daily life, are stored. In this way, EEG signals can be explored to characterize different spontaneous emotional states felt by several people. The application is also designed to extract features of specific emotions, and to compare different emotional states. Researchers all over the world could share both raw data and classification results. Since large datasets are treated, the application is based on strategies commonly used in big data managing. In particular, a column-oriented database is used to store a huge amount of raw EEG signals, while a relational database is employed to keep metadata information. A web application interface allows the user to communicate with the repository and a computational module performs the features extraction.

Keywords Spontaneous emotions · Electroencephalography (EEG) · Big data · Brain computer interface (BCI) · Emotion characterization · Human-computer interaction · Affective computing

G. Placidi (✉) · M. Polsinelli
A2VI-Lab, C/O Department of Life, Health and Environmental Sciences, University of L'Aquila, L'Aquila, Italy
e-mail: giuseppe.placidi@univaq.it

M. Polsinelli
e-mail: matteo.polsinelli@graduate.univaq.it

L. Cinque
Department of Computer Science, Sapienza University of Rome, Rome, Italy
e-mail: cinque@di.uniroma1.it

10.1   Introduction

Brain Computer Interfaces aim at translating the Central Nervous System (CNS) activity into artificial output that replaces, restores, enhances, supplements or improves the natural CNS output [1]. Since brain activity includes many electrophysiological, chemical and metabolic processes, a BCI could be driven by one of these phenomena through different (or combined) acquisition methods, such as fMRI [2–4], fNIRS [5–7], EEG [8], etc. Between them, EEG is the most popular acquisition technique due to its low cost, low invasiveness, great portability and easiness of use.

The natural application of BCIs, in which a lot of scientists have produced big efforts over the last years, is to provide an alternative communication and interaction channel towards the external environment for people with severe disabilities [9]. Moreover, the knowledge of the patient's mental state is fundamental to improve the effectiveness of rehabilitation procedures without boring and/or disturbing it [10–18]. However, BCIs can also be useful for healthy people if applied in the field of human-computer interaction [19, 20] mainly used in Affective Computing [21], a new discipline that has gained great attention in recent years.

Research on emotions grows continuously because they are important in daily life and have to be considered in the design of a BCI. Since they affect brain activity, BCIs should adapt internal classification and interpretation algorithms to the signal variability. Moreover, subjects could use emotions to drive BCIs [22, 23], by modulating some emotional states.

Several studies were addressed in providing emotions classification algorithms [24–26]. The most common practice to study emotions is to record EEG signals during the elicitation of emotional states by means of external stimuli like pictures, videos, virtual reality interaction and sounds [27–32]. Though this procedure is the most used, it has some drawbacks: (1) the response to a stimulation is subjective (for example, a given picture can produce "fear" in a subject and leave completely "indifferent" another subject); (2) the activation of a given emotional process is subjective; (3) external stimulations produce brain activity that could obscure the signal produced by genuine emotional states (for example, a visual stimulation activates the visual cortex that, as a consequence, activates memory and some other mental process that, finally, can activate an emotional state); (4) the emotional states induced by external stimulations are difficult to use for driving a BCI and leave no room to the independence of the user.

In recent works [33–36], communication protocols and classification strategies that allowed to control a BCI, through the "disgust" induced by remembering an unpleasant odor, have been presented. The subject elicits an emotion by remembering it, being free on "when" activate the considered emotional state. This makes the protocol suitable for assessing the consciousness of neurological patients [37] and for driving a BCI [38–40]. The proposed classifiers were tailored on specific emotions but could be generalized and adapted to other affective states [41, 42]. One of the future developments, indeed, was aimed to find and classify different emotional states spontaneously activated by the users (self-induced emotional states). However, though the self-induced emotional states are the future for driving BCIs, the study regarding the effective pattern localization, of different emotional states, remains. Moreover, the application of different emotional states in affective computing requires to study "genuine" emotions (the EEG signal has to be generated just by emotions spontaneously generated by a subject). In this paper, we propose the design of a Web Application to manage data collected from long EEG acquisition sessions to study EEG signals from spontaneous emotions and to characterize the resulting emotional states.

To obtain EEG datasets of different emotions that are as much as possible spontaneous, our proposal is to acquire and study extended EEG recordings of daily life by different subjects. This approach involves that an examined subject wears a portable EEG headset with storing capability to record brain activity for a long period during which he/she has to take note about the felt emotions, including the occurrence of relaxing situations. These notes have to be converted in metadata and attached to the acquired data, thus allowing to label portions of signal with the proper emotional information. This long-term EEG approach [43], in which some emotionally meaningful signals are surrounded by a lot of background data, drives us to deal with challenging scenarios. In fact, we have to store and quickly manage huge datasets and we are interested in supplying the storage with feature extraction capabilities to synthetize signals in lighter and more understandable forms. Moreover, we need several sessions from a large number of subjects to collect a considerable quantity of significant information which allow us to separate objective (inter-subject invariant) features of each emotion from subjective ones. This involves heavy workload for data collection which forces to collaborate with other research groups worldwide, especially for contributing to collect data from different subjects, with different life experiences and cultures from all around the world. Lastly, always in a collaboration outlook, sharing data, metadata and extracted features could support the work of other research groups.

The manuscript is organized as follows: Sect. 2 defines the actors, showing their interactions with the application and user requirements; Sect. 3 illustrates the system design, proposing models and implementation guidelines for each of the main components of the application; Sect. 4 presents the conclusions and future developments.

10.2   Actors and Requirements

In recent years, the cloud-computing infrastructure has proven to be suited for managing "big-data" [44], as those we intend to collect. Cloud computing is one of the most significant technologies in the modern ICT industry and business and service for enterprise applications has become a powerful architecture to perform large-scale storage and complex computing. The cloud computing infrastructures, and the requirements of collaboration and sharing, involve the employment of a Web Application instead of a stand-alone software.

In the design of the proposed Web Application, different actors, with their own requirements, have to be considered.

The simplest form of interaction is for data consultation carried out by the user called Consulter. For each stored session, Consulter has to be able to visualize and download any combination of these elements:

- Meta-Data: Information regarding sessions (date, setup, study in which the session was recorded, etc.) and subject (gender, age, health information, etc.);
- Raw Data: EEG signals enriched with emotional tags. Should be either the entire session or parts of it, for example defined time intervals or specific emotions occurrence;
- Features: Synthetized information about emotional content of sessions. The web application is furnished with a generalized version of the algorithm described in [33] and has to be able to recognize automatically if and how signals belonging to a specific emotion have significant dissimilarities in frequency (computed by means of the r^2 coefficient [45]) with respect to the recorded relaxing situation signals [33].

In particular, "Raw Data" could be useful for researchers that are experts of EEG signals and their classification techniques and, more generally, of signal analysis techniques. In the same time, "Features" could be a practical way, for researchers who are not skilled with the classification topics, to gather information about the meaningfulness of channels or frequencies related to a specific emotion.

Another, more complex, form of interaction with the system is through a "contributor user": the user has both the possibility of data consultation (the Consulter privileges) and of inserting data into the database. Researchers who want to upload data have to follow a standardized procedure of data collection, defined by the following rules:

- Subject materials: contributors have to furnish to each subject a wireless EEG headset with storing capabilities, and an equipment to report emotional annotations. The headset has to be easily wearable, since the subject should autonomously perform this operation at home. Moreover, the hardware should be not cumbersome to avoid limitations of movement. Equipment for emotional jotting could be a notebook, a voice recorder or a digital device like a smartphone;
- Subject session briefing: contributors have to explain to the subject how to record a session correctly. The subject has to record the longest possible session, while living its daily life. When a specific emotion is felt, the user should record, with the higher possible accuracy, time of occurrence, type of emotion (chosen between a list of pre-defined), intensity (using an integer value between 1 to 5) and the cause (like an event or a sensorial input) of this emotional state. Moreover, the subject should spend some time to record at least one relaxing situation, in which he/she should calm down, thus avoiding to focus on specific emotional states. Signals from relaxing situations are necessary for comparison when extracting features regarding other emotional states;

- File format and additional information: signals can be uploaded if conformed to a common data format. To the best of our knowledge, EDF+ [46, 47] is the most used standard for exchange and store multichannel biological and physical signals. Contributors also should provide information about the acquisition hardware, the study and the subject. Some of these information have reserved fields in the EDF+ header. Therefore the application is expected to be able to parse the uploaded file header (corresponding to the first 256 bytes) and to extract them. In case of lack of information, the application should force the Contributor to insert missing data manually (by means of a specific form). Some essential additional information do not have reserved space in the EDF+ header: for this reason, Contributors should provide them by means of manual insertion, by using a dedicated wizard;
- Emotional metadata. Contributors should transcribe emotional annotations by filling an appropriate form to add a tag within the raw data.

When a session is successfully uploaded, the application is expected to synthetize and store information about differences, in frequency, calculated between signals belonging to any emotional state with the relaxing states.

Lastly, a super-user class of users, the Owner, has to be considered. Owner has the typical role of system administrator (managing, for example, users' privileges) and is authorized to upload data analysis results different from those the system performs automatically. The Owner class is tailored for researchers that are familiar with the algorithm in [33] and allows them to download raw data, process them locally (with the original, customizable version of the algorithm) and, finally, to upload the results as part of metadata.

10.3   System Design

Since our primary needs are the storage and the management of big-data and their complex analysis, we adopted a Cloud Computing Architecture [48]. In the last years the Cloud services have become dominant architectures for scientific applications. Indeed, voluminous experiments are deployed in the cloud to exploit the available parallel computing facilities in remote specific servers and to store the increasing volume of data produced by the experiments [49]. In order to make available to the community a simple, fast and intuitive system we decided to design a web application. The architecture of the proposed cloud system, reported in Fig. 1, is composed by three main components:

- "Application", that allows the user, through a web browser or through external API requests, to handle data by using the operations allowed by his/her privileges. These features are implemented in the Web Application component. Application should also execute the signals analysis algorithm. Since this operation requires a huge computational effort (with respect to a standard web application's operations) it is developed as a dedicated module (Computational Module) following the approaches for distributed algorithms.

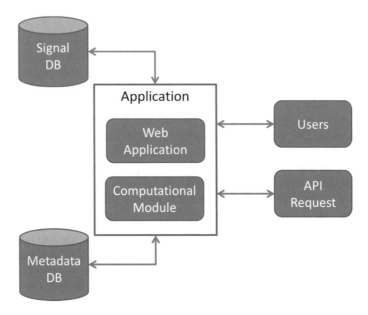

Fig. 1 Architecture diagram

- "Signals Database", to store the EEG signals from long sessions and to produce efficient queries from a large amount of data.
- "Metadata Database", to contain the information needed to manage the system and the metadata corresponding to the uploaded sessions.

We plan to store data into two different databases due to the fact that signals, session metadata and system information are characterized by different features requiring different storage policies.

10.3.1 Signal Database

Since our primary needs are the storage and the management of big-data and their complex analysis, we adopted a Cloud Computing Architecture [48].

Signal Database is designed to store raw signals. This kind of data could be represented by a simple structure, without relationship between entities, but is composed by a large data volume to be accessed quickly. We use a non-relational database (NoSQL) [50, 51] because raw EEG data especially fit the characteristics of a column oriented database structure. Signals are represented as a sequence of samples (rows) through time, each reporting the signal amplitudes of a given set of channels (columns).

We take advantage by three traits of the column oriented database model:

- Sorting: rows represent a large set of key value maps that are maintained ordered, allowing fast queries to the data, in particular blocks of contiguous rows. By coding keys, containing both the session and the sample's index, an efficient access to signal intervals is allowed.
- Sparsity: the main difference with a relational database lies in the management of null values. By considering a series of different attributes, a relational database would store a null value where information is absent. This may cause a significant waste of space, in case of heterogeneous data. Conversely, a column-oriented database could adapt the space occupation to the size of data effectively present in a row. This is suitable for our representation since EEG data sources could be different (for example, the number of channels could range from some units to hundreds).
- Structure Flexibility: column oriented databases dynamically manage the columns. Column families are static and have to be declared when the table is created, but columns can be added to an existing column family at any time. This allows to store efficiently data from channel locations that are not already present in the database.

We use Hbase [52] as column oriented database implementation. As shown in Fig. 2, the Signal Database consists of a single Table, named Signal. We use an 8-byte row-key obtained by the concatenation of two values (4-byte length each):

- Id_session: this value is used as a link between the two data storage systems;
- Sample_index: each session is represented by a sequence of sorted samples. The index is used to keep track of the order.

The 4-byte size allows to represent 232 sessions, each containing at most 232 samples (corresponding to about 48 days of continuous recording at a sampling rate of 1024 Hz). We use a single column family (Channels), a set of columns representing the signal amplitude for a specific channel (in Fig. 2 channels are called CH1, CH2, etc. but in a real scenario they will have the names corresponding to the 10–20 system [53]).

This structure serves to maintain sorted the data, by considering first the session identifier and, then, the sample's index. In that way, requests for signals from a time interval or related to a specific emotion occurrence could be easily satisfied. Indeed, for a given session, the application translates time or annotated information in the

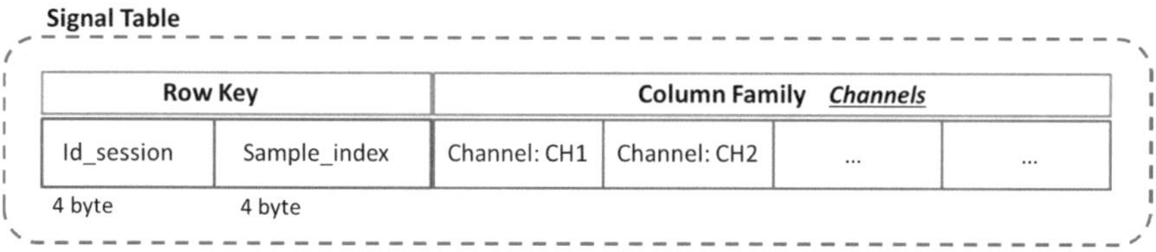

Signal Table

Row Key		Column Family *Channels*			
Id_session	Sample_index	Channel: CH1	Channel: CH2	...	...
4 byte	4 byte				

Fig. 2 Signal Database structure: the names of channel are just indicative (original names of the 10–20 system can be used)

corresponding samples indexes pair by using the metadata stored in the relational database (e.g. the sampling rate and the time interval of an emotion occurrence) and, then, extracts target data from the Signal Database by means of a native and efficient HBase function scan.

10.3.2 Metadata Database

Metadata and system information are suited to be represented by traditional relational model and take advantage by Relational Database Management System (RDBMS) features, since they have a consistent structure, multiple relationships between entities and relative by small size. We opt for the Relational Oracle Database [54].

Figure 3 illustrates the structure of the Metadata Database (in this figure, some system tables are not reported, e.g. the user role authentication and management tables, that are developed following the Spring Authentication standard [55]).

Consulters, Contributors and Owners are represented by the table *User*. In each *Session*, *Subject* indicates the type of an *Emotion* and its *Cause*. Session represents

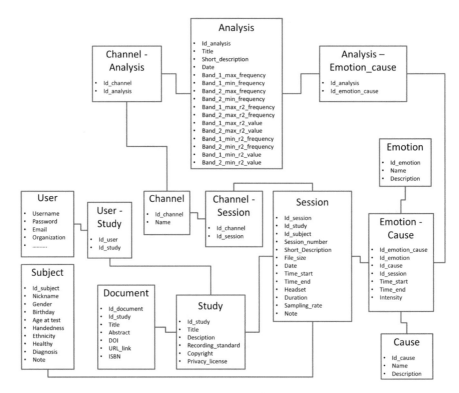

Fig. 3 Metadata Database structure: a relational database is used

an experiment belonging to a more articulated *Study* (that can be linked with one or more *Documents*, like papers, reports or books). *Analysis* represents an instance of comparison between the occurrences of two *Emotions* in the same *Session* and contains information about setup options and results. *Analysis* could be performed by considering different *Channels* combinations between those that have been used for recording. Emotion contents are fixed and cannot be modified: it contains nine elements (relax, joy, sadness, fear, anger, anticipation, surprise, disgust, and trust), that are those described by Plutchik [56], plus the relax status. Due to its meaning of absence of emotions, relax is not related to a cause and is not assessed by an intensity value. Any *Emotion-Cause* row corresponding to a relaxing situation should contain just temporal information.

10.3.3   Web Application

The Web Application is designed, with Java Enterprise solution, on a three-layer architecture based on the Model View Controller (MVC) pattern [57]. The persistence layer is ensured by the use of the Hibernate framework [58] mapped on the relational Metadata Database. Application and Presentation layers are designed by using Spring technology [55]. To obtain a fluid layout, that automatically fits into the browser window, the Web Application User Interface is developed in HTML5 [59], CSS3 [60], jQuery library [61] and by using the technique of responsive design [62].

Since the web application is shared by a large scientific community, it is necessary to establish users role, classified according to some privileges:

Anonymous: is not registered in the system. This kind of user has access only to basic information resources, available in the system homepage, regarding the type of information stored and accessible from the application.

Consulter (Cs): has a registered account in the system. This user can visualize and download EEG raw-data and metadata of experiments stored in the system by others members.

Contributor (Ct): has a registered account in the system. This kind of user has the same basic set of privileges as the Consulter. In addition, Contributor can create, remove and manage new EEG studies and sessions (inserted by himself/herself).

Owner (Ow): is the super administrator of the system. This user inherits the privileges of Contributor, can upload his/her locally elaborated analysis results (computed with customized instances of the algorithm in [33]) and manage all the operations necessary for the maintenance of the system.

From the functionalities point of view, there are three main possible scenarios:

- *Data and Analysis, Visualization and Download* (Cs, Ct, Ow). The web application should provide the possibility of browsing and visualizing all the stored studies with related sessions and for each session the associate metadata. In particular, for each session, the system provides the visualization of the EEG signals into a Charting Area. The emotional tags annotated for the represented session are highlighted into the chart. In order to properly display the signals and the related information into the chart, the system uses an asynchronous loading and chunks of the signals into the window to be displayed. Moreover, to avoid repeated requests to the server, the system uses a caching request system. In the Charting area page it is possible to filter the visualized signals. From the dashboard, or from the menu, the user can reach the emotions query page. The page is arranged with the following sections: (1) query form (2) filters section (3) query result visualization (4) details information. The user can perform simple or advanced queries to retrieve a specific emotion, groups of emotions or combination of cause-emotions and the research can be refined through a series of predefined filters. The system returns a navigable tree view containing all the pairs study/session for the searched emotion: also in this case, the user can refine the research by mixing a series of filters. By clicking on an entry tree, the system updates the details panel with all the information related to the selected study. In the details information panel, a button serves to redirect the user to the charting area page to visualize the selected session and to link the data analysis results (those automatically computed after the session upload and/or eventually inserted by an Owner). The analysis results, organized in a table form, are open in a dedicated page.

A button is used to download the desired information (e.g. the signals and additional information of a session or a portion of session, signals and metadata of an emotion from the repository, signals and related metadata of an emotion from a single session) in every data visualization page (charting area, session or study page, analysis presentation page).

With this function, users can download a compressed package containing:

- Signals Raw-data in EDF+ format: the system should create new EDF+ files retrieving the signals information from the Signal table and fill the EDF+ header fields with the associated metadata.
- XML [63] files and corresponding JSON [64] files, containing the metadata information about sessions, studies, tags and analysis results (if available),
- A HTML file containing descriptions and meta-data (plus analysis results, if available) encapsulated into a graphic format readable with any browser.

- Study/Session Upload (Ct, Ow). The dashboard contains a link to the "New Study" page that drives the user through a wizard procedure:

1. The system displays the form for data entry (several fields are mandatory). The user must insert information such as a title, a short description of the study, recording protocol standards and information relative to the copyright and privacy policies.
2. If the user needs to attach documents regarding the study, can do this in two ways:
   2.1 Automatic retrieve from DOI or ISBN. The user inserts the DOI or the ISBN into the specific field and the system automatically searches the associated information, then checks and validates the retrieved data.

     2.2 Manual insertion, by filling a dedicated form or by loading a digital document.

3. The system displays the form for the insertion of the new session data information.

     3.1 Subject setup: the user can select an existing subject (from previous studies) from a drop down list or add a new subject.

         3.1.1 New Subject: the system proposes the form to add the subject information. The user must insert the nickname that he/she wants to associate to the subject necessary to research it for future insertions: gender, birthday, handedness, ethnicity. The user can select the checkbox for healthy subjects, or provide the diagnosis description for patients and, if necessary, insert additional information into the note text area.

     3.2 Automatic information extraction from EDF+ files. The user selects the EDF+ file and the application extracts and parses the available information from the header (first 256 byte of the file), e.g. number of channels, sampling rate, file size.

     3.3 Session data entry: several fields in the form are filled by the previous step. The user must validate or update data and add information like session number, description, channels positions and notes.

     3.4 Emotional Tagging: The wizard proposes to the user a series of mandatory fields to report the list of annotated emotions in electronic way (through the selection of five attributes: Emotion, Cause, Intensity, Start Time, End Time). The user can choose the Emotion from a drop-down list and the system automatically proposes a list of causes from those inserted in previously stored sessions. If the user does not find the appropriate cause, he/she can add a new one and associate it. In case of a relax tag, only temporal information (without the cause or the intensity value) must be attached.

     3.5 Upload the file EDF+: The system accepts only EDF+ files. After the conclusion of the upload operation the system runs an automatic function that parse the EDF+ file and executes the queries to insert the raw-data into the Signal Database. The original EDF+ file uploaded from the user is not stored into the web application but is used in a temporary folder until parsing is terminated.

     3.6 The user, from the dashboard or from the menu, can add a new Session into his/her own previously inserted studies. The system shows to the user a search box to search the desired study. The procedure to add information regarding new session is a portion of that used to add a new study (step 3).

- Analysis Results Upload (Ow). Owners can select, from the dashboard, the link "Add Result". In this page the user can reach, using the associate identifier, the desired session and attach to it the analytical result, computed separately in an offline mode:

1. The analysis title, description and date;
2. The set of channels involved in the analysis;
3. The emotions occurrences composing the two compared sets (the comparison couple can be Emotion-Relax or Emotion-Emotion, but each set has to contain instances of the same emotion);
4. The limits of the frequency bands in which the comparison is performed;
5. The analysis results: maximum and minimum r^2 values inside both bands and spatial location.

To create an open architecture for sharing content and data between communities and application, the application provides a RestFul Web Service Application Programming Interface (API) [65, 66] that offers Consulter and/or Contributors functionalities. The API requests are based on the URI HTTP transfer protocol and the returned values are encapsulated into a JSON file.

10.3.4 Computational Module

The algorithm proposed in [33] was tailored to identify the emotion of disgust produced by an unpleasant odor memory. In particular, neuroscience literature drove us through the selection of frequency bands (8–12 and 32–42 Hz) and scalp locations (P4, C4, T8 and P8) that are supposed to be involved in brain activity this specific emotion [67–71].

The Computational Module arises from the need to equip the Application with a generalized version of the calibration phase of this algorithm by considering the annotated emotions with intensity greater than or equal to 3, when a session is uploaded. With respect to the original strategy, it searches the most significant frequency dissimilarities inside less focused bands, 1–25 and 26–50 Hz. For each signal related to an emotion, the algorithm divides it in partially overlapping segments and applies the Short Time Fourier Transform to analyze frequency components. The r^2 function is used to select (and average) only the spectra of the segments that are, in pairs, more similar.

The r^2 function is also evaluated to identify the frequencies (in the considered bands) where the differences between the emotion and relax are larger and smaller.

This module:

- manages signals containing emotional tags to be compared each other;
- performs signal manipulations and domain transformations.

Moreover, having supposed that the set up and the execution of some online analysis operations could be added to the users privileges, we implement this module by using a Map/Reduce strategy [72].

The Map/Reduce was introduced by Google [73, 74] for manipulation and generation of large datasets. The idea is to split the application logic in two basic functions: Map and Reduce (Fig. 4). A Map/Reduce process divides a data-set into independent

Fig. 4 The Map/Reduce
approach graphic
representation

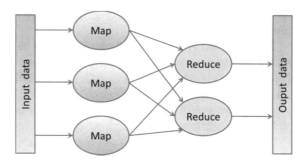

blocks, organizes them in key-value pairs using the Map function and splits them on different nodes for parallel processing. Subsequently, results from each node are recovered and, using the Reduce function, reprocessed for the final outcome. This approach could be applied to the described algorithm, where several computations could be executed in the same time. For example:

- Each channel and, inside it, the signal related to each emotion occurrence is separately considered;
- The STFT is implemented by computing the Fourier Transform of each segment independently;
- The mutual comparison between segments and, then, between emotions occurrences is divided into several independent tasks, one for each pair of signals.

In order to take advantage from this approach, we develop this component using the Hadoop framework [75, 76]. Hadoop is an open-source framework, (Apache Software Foundation) developed in Java to support the processing of large data sets in a distributed computing environment. Hadoop consists of two main parts: a module that manages the scheduling, the distribution and the execution of calculations on different nodes (implementing the Map/Reduce approach), and a module that deals with the management of the distributed file-system, called Hadoop Distributed File System (HDFS). The HDFS is designed to store huge files on all cluster machines and Hadoop is also tolerant to node failure. Hadoop splits the computation in separated jobs, at the logical level, without taking care of their physical distribution on nodes.

10.4   Conclusion

We described the design of a web application to store long sessions EEG signals from daily life experiences of different subjects to gather information regarding spontaneous emotions. By means of a web application, researchers are able to store, manage and share EEG raw data with corresponding metadata information and emotional tags. The Web application is also equipped with a module designed to analyze automatically all the occurrences of a given emotion by studying them in the frequency domain.

The described cloud architecture allows both to store and to compute a large volume of signals by employing emerging paradigms and technologies from the big data research field. The proposed application uses a column-oriented database to store the produced heterogeneous EEG signals, data sorting capabilities and sparse data managing approach of this architecture. A relational database supports the application by storing metadata regarding sessions, studies and emotions.

Using the Hadoop framework and the Map/Reduce paradigm to parallelize the jobs, the system is able to optimize and reduce the computational time necessary to process the big data, and makes possible to add nodes, to increase storage space and/or computational resources. The use of Hadoop also allows the organization of the internal computation algorithm in parallel jobs. The integration of the system with data visualization and query search form allows the researchers to visualize and download whole sessions, segments of interest or information regarding features of specific emotions.

Next steps are the implementation of the proposed web-application (the name proposed for the repository is EMOTIONET) and its test. Initially, tests will be conducted by using data collected only by our research group. After that, the application will be open to other groups for data usage and insertion.

References

1. Wolpaw, J., Wolpaw, E.: Brain-Computer Interfaces. Oxford University Press, Oxford (2011)
2. Yoo, S., Fairneny, T., Chen, N., Choo, S., Panych, L., Park, H., Lee, S., Jolesz, F.: Brain–computer interface using fMRI: spatial navigation by thoughts. NeuroReport **15**(10), 1591–1595 (2004)
3. DeCharms, R., Maeda, F., Glover, G., Ludlow, D., Pauly, J., Soneji, D., Gabrieli, J., Mackey, S.: Control over brain activation and pain learned by using real-time functional MRI. Proc. Nat. Acad. Sci. **02**(51), 18626–18631 (2005)
4. Ruiz, S., Lee, S., Soekadar, S., Caria, A., Veit, R., Kircher, T., Birbaumer, N., Sitaram, R.: Acquired self-control of insula cortex modulates emotion recognition and brain network connectivity in schizophrenia. Hum. Brain Mapp. **34**(1), 200–212 (2013)
5. Sitaram, R., Zhang, H., Guan, C., Thulasidas, M., Hoshi, Y., Ishikawa, A., Shimizu, K., Birbaumer, N.: Temporal classification of multichannel near-infrared spectroscopy signals of motor imagery for developing a brain–computer interface. NeuroImage **34**(4), 1416–1427 (2007)
6. Schurholz, M., Rana, M., Robinson, N., Ramos-Murguialday, A., Cho, W., Rohm, M., Rupp, R., Birbaumer, N., Sitaram, R.: Differences in hemodynamic activations between motor imagery and upper limb FES with NIRS. In: IEEE Engineering in Medicine and Biology Society, Annual International Conference of the IEEE, pp. 4728–4731 (2012)
7. Rea, M., Rana, M., Lugato, N., Terekhin, P., Gizzi, L., Brotz, D., Fallgatter, A., Birbaumer, N., Sitaram, R., Caria, R.: Lower limb movement preparation in chronic stroke: a pilot study toward an fNIRS-BCI for gait rehabilitation. Neurorehabilitation Neural Repair **28**(6), 564–575 (2014)
8. Hwang, H., Kim, S., Choi, S., Im, C.: EEG-based brain-computer interfaces: a thorough literature survey. Int. J. Hum. Comput. Interact. **29**(12), 814–826 (2013)

9. Daly, J., Huggins, J.: Brain-computer interface: current and emerging rehabilitation applications. Arch. Phys. Med. Rehabil. **96**(3), S1–S7 (2015)
10. Placidi, G.: A smart virtual glove for the hand telerehabilitation. Comput. Biol. Med. **37**(8), 1100–1107 (2007)
11. Franchi, D., Maurizi, A., Placidi, G.: A Numerical Hand Model for a Virtual Glove Rehabilitation System, pp. 41–44 (2009)
12. Franchi, D., Maurizi, A., Placidi, G.: Characterization of a simmechanics model for a virtual glove rehabilitation system. In: 2nd International Symposium on Computational Modeling of Objects Represented in Images, Fundamentals, Methods and Applications 2010. Lecture Notes in Computer Science (including subseries Lecture Notes in Artificial Intelligence and Lecture Notes in Bioinformatics), pp. 141–150 (2010)
13. Spezialetti, M., Avola, D., Placidi, G., De Gasperis, G.: Movement analysis based on virtual reality and 3D depth sensing camera for whole body rehabilitation. In: 3rd International Symposium on Computational Modelling of Objects Represented in Images: Fundamentals, Methods and Applications, Rome, Sept 2012, pp. 367–372 (2012)
14. Bisconti, S., Spezialetti, M., Placidi, G., Quaresima, V.: Functional near-infrared frontal cortex imaging for virtual reality neuro-rehabilitation assessment. In: 3rd International Symposium on Computational Modelling of Objects Represented in Images: Fundamentals, Methods and Applicationsp, Rome, Sept 2012, pp. 187–192
15. Placidi, G., Avola, D., Iacoviello, D., Cinque, L.: Overall design and implementation of the virtual glove. Comput. Biol. Med. **43**(11), 1927–1940 (2013)
16. Placidi, G., Avola, D., Ferrari, M., Iacoviello, D., Petracca, A., Quaresima, V., Spezialetti, M.: A low-cost real time virtual system for postural stability assessment at home. Comput. Methods Programs Biomed. **117**(2), 322–333 (2014)
17. Petracca, A., Carrieri, M., Avola, D., Basso Moro, S., Brigadoi, S., Lancia, S., Spezialetti, M., Ferrari, M., Quaresima, V., Placidi, G.: A virtual ball task driven by forearm movements for neuro-rehabilitation. In: 2015 International Conference on Virtual Rehabilitation Proceedings (ICVR), IEEE, pp. 162–163 (2015)
18. Moro, S.B., Carrieri, M., Avola, D., Brigadoi, S., Lancia, S., Petracca, A., Quaresima, V.: A novel semi-immersive virtual reality visuo-motor task activates ventrolateral prefrontal cortex: a functional near-infrared spectroscopy study. J. Neural Eng. **13**(3), 036002 (2016)
19. Avola, D., Cinque, L., Placidi, G.: A novel multimodal framework to support human-computer interaction. In: Proceedings of the 6th International Symposium of GIRPR, pp. 1–12 (2012)
20. Avola, D., Spezialetti, M., Placidi, G.: Design of an efficient framework for fast prototyping of customized human–computer interfaces and virtual environments for rehabilitation. Comput. Methods Programs Biomed. **110**(3), 490–502 (2013)
21. Picard, R.: Affective Computing. MIT Press, Cambridge, Mass. (1997)
22. Molina, G., Tsoneva, T., Nijholt, A.: Emotional brain-computer interfaces. Int. J. Auton. Adapt. Commun. Syst. **6**(1), 9–25 (2013)
23. Spezialetti, M., Cinque, L., João Manuel Tavares, R.S., Placidi, G.: Towards EEG-based BCI driven by emotions for addressing BCI-illiteracy: a meta-analytic review. Behav. Inf. Technol. **37**(8), 855–871 (2018)
24. Bos, D.O.: Eeg-based emotion recognition: the influence of visual and auditory stimuli. [Online]. http://hmi.ewi.utwente.nl/verslagen/capita-selecta/CS-Oude_Bos-Danny.pdf
25. Wang, X., Nie, D., Lu, B.: Emotional state classification from EEG data using machine learning approach. Neurocomputing **129**, 94–106 (2014)
26. Stikic, M., Johnson, R., Tan, V., Berka, C.: EEG-based classification of positive and negative affective states. Brain-Comput. Interfaces **1**(2), 99–112 (2014)
27. Lang, P.J., Bradley, M.M., Cuthbert, B.N.: International affective picture system (IAPS): affective ratings of pictures and instruction manual. Univ. of Florida, Gainesville, FL, Tech. Rep. A-8 (2008)
28. Bradley, M.M., Lang, P.J.: International affective digitized sounds (IADS): stimuli, instruction manual and affective ratings. The Center for Research in Psychophysiology, Univ. of Florida, Gainesville, FL, Tech. Rep. B-2 (1999)

29. Avola, D., Cinque, L., Levialdi, S., Petracca, A., Placidi, G., Spezialetti, M.: Time-of-flight camera based virtual reality interaction for balance rehabilitation purposes. In: International Symposium Computational Modeling of Objects Presented in Images Springer, Cham, pp. 363–374 (2014)
30. Ferrari, M., Bisconti, S., Spezialetti, M., Basso Moro, S., Di Palo, C., Placidi, G., Quaresima, V.: Prefrontal cortex activated bilaterally by a tilt board balance task: a functional near-infrared spectroscopy study in a semi-immersive virtual reality environment. Brain Topogr. **27**(3), 353–365 (2014)
31. Basso Moro, S., Bisconti, S., Muthalib, M., Spezialetti, M., Cutini, S., Ferrari, M., Placidi, G., Quaresima, V.: A semi-immersive virtual reality incremental swing balance task activates prefrontal cortex: a functional near-infrared spectroscopy study. NeuroImage **85**, 451–460 (2014)
32. Avola, D., Cinque, L., Levialdi, S., Petracca, A., Placidi, G., Spezialetti, M.: Markerless hand gesture interface based on LEAP motion controller. In: DMS 2014, pp. 260–266 (2014)
33. Placidi, G., Avola, D., Petracca, A., Sgallari, F., Spezialetti, M.: Basis for the implementation of an EEG-based single-trial binary brain computer interface through the disgust produced by remembering unpleasant odors. Neurocomputing **160**, 308–318 (2015)
34. Iacoviello, D., Petracca, A., Spezialetti, M., Placidi, G.: A real time classification algorithm for EEG-based BCI driven by self induced emotions. Comput. Methods Programs Biomed. **122**(3), 293–303 (2015)
35. Iacoviello, D., Petracca, A., Spezialetti, M., Placidi, G.: A classification algorithm for electroencephalography signals by self-induced emotional stimuli. IEEE Trans. Cybern. **46**(12), 3171–3180 (2015)
36. Placidi, G., Petracca, A., Spezialetti, M., Iacoviello, D.: Classification strategies for a single-trial binary brain computer interface based on remembering unpleasant odors. In: IEEE Engineering in Medicine and Biology Society (2015) 37th Annual International Conference the IEEE, pp. 7019–7022 (2015)
37. Pistoia, F., Carolei, A., Iacoviello, D., Petracca, A., Sacco, S., Sarà, M., Spezialetti, M., Placidi, G.: EEG-detected olfactory imagery to reveal covert consciousness in minimally conscious state. Brain Inj. **29**(13–14), 1729–1735 (2015)
38. Placidi, G., Petracca, A., Spezialetti, M., Iacoviello, D.: A modular framework for EEG web based binary brain computer interfaces to recover communication abilities in impaired people. J. Med. Syst. **40**(1), 34 (2015)
39. Placidi, G., Cinque, L., Di Giamberardino, P., Iacoviello, D., Spezialetti, M.: An affective BCI driven by self-induced emotions for people with severe neurological disorders. International Conference on Image Analysis and Processing, pp. 155–162. Springer, Cham (2017)
40. Di Giamberardino, P., Iacoviello, D., Placidi, G., Polsinelli, M., Spezialetti, M.: A brain computer interface by EEG signals from self induced emotions. Lect. Notes Comput. Vis. Biomech. **27**, 713–721 (2018)
41. Iacoviello, D., Pagnani, N., Petracca, A., Spezialetti, M., Placidi, G.: A poll oriented classifier for affective brain computer interfaces. In: NEUROTECHNIX 2015—Proceedings of the 3rd International Congress on Neurotechnology, Electronics and Informatics, Lisbon, pp. 41–48 (2015)
42. Placidi, G., Di Giamberardino, P., Petracca, A., Spezialetti, M., Iacoviello, D.: Classification of emotional signals from the DEAP dataset. In: Proceedings of the 4th International Congress on Neurotechnology, Electronics and Informatics (NEUROTECHNIX 2016), Porto 7–8/11/2016, pp. 15–21 (2016)
43. Michel, V., Mazzola, L., Lemesle, M., Vercueil, L.: Long-term EEG in adults: sleep-deprived EEG (SDE), ambulatory EEG (Amb-EEG) and long-term video-EEG recording (LTVER). Neurophysiol. Clin./Clin. Neurophysiol. **45**(1), 47–64 (2015)
44. Agrawal, D., Das, S., El Abbadi, A.: Big data and cloud computing: current state and future opportunities. In: Proceedings of the 14th International Conference Extending Database Technology—EDBT/ICDT (2011)
45. Draper, N., Smith, H.: Applied Regression Analysis. Wiley, New York (1998)

46. Kemp, B., Olivan, J.: European data format 'plus' (EDF+), an EDF alike standard format for the exchange of physiological data. Clin. Neurophysiol. **114**(9), 1755–1761 (2003)
47. Kemp, B.: European Data Format (EDF) (2015) Edfplus.info, [Online]. http://www.edfplus.info/
48. Talia, D.: Clouds for scalable big data analytics. Computer **46**(5), 98–101 (2013)
49. Liu, H.: Big data drives cloud adoption in enterprise. IEEE Internet Comput. **17**(4), 68–71 (2013)
50. Han, J., Haihong, E., Guan, L., Du, J.: Survey on NoSQL database. In: Pervasive Computing and Applications (ICPCA), 6th International Conference on IEEE 2011, pp. 363–366 (2011)
51. Hecht, R., Jablonski, S.: NoSQL evaluation: a use case oriented survey. In: International Conference on Cloud and Service Computing (CSC). IEEE 2001, pp. 336–341 (2011)
52. Apache Team.: Apache HBase™ Reference Guide. Hbase.apache.org (2015). [Online]. https://hbase.apache.org/book.html
53. Jurcak, V., Tsuzuki, D., Dan, I.: 10/20, 10/10, and 10/5 systems revisited: their validity as relative head-surface-based positioning systems. NeuroImage **34**(4), 1600–1611 (2007)
54. Oracle.com.: Oracle | Hardware and Software, Engineered to Work Together (2015). [online] http://www.oracle.com
55. Spring.io.: spring.io (2015). [Online]. https://spring.io/
56. Camras, L., Plutchik, R., Kellerman, H.: Emotion: theory, research, and experience. Vol. 1. Theories of emotion. Am. J. Psychol. **94**(2), 370 (1981)
57. Leff, A., Rayfield, J.: Web-application development using the model/view/controller design pattern. In: Proceedings of the 5th IEEE International Enterprise Distributed Object Computing Conference (2001)
58. Hibernate.org.: Hibernate. Everything data. – Hibernate (2015). [Online]. http://hibernate.org/
59. W3.org.: HTML5 (2015). [Online]. http://www.w3.org/TR/html5/
60. W3.org.: Cascading Style Sheets (2015). [Online]. http://www.w3.org/Style/CSS/
61. jquery.org, "jQuery", Jquery.com.: (2015). [Online]. https://jquery.com/
62. Marcotte, E.: Responsive Web Design (2010). Alistapart.com [Online]. http://alistapart.com/article/responsive-web-design
63. W3.org.: "Extensible Markup Language (XML)," (2015). [Online]. http://www.w3.org/XML/
64. Tools.ietf.org, "RFC 7159 - The JavaScript Object Notation (JSON) Data Interchange Format", 2015. [Online]. http://tools.ietf.org/html/rfc7159
65. Richardson, L., Ruby, S.: RESTful Web Services. O'Reilly, Farnham (2007)
66. Fielding, R., Taylor, R.: Principled design of the modern web architecture. In: Proceedings of the 22nd International Conference on Software Engineering (2000)
67. Henkin, R., Levy, L.: Lateralization of brain activation to imagination and smell of odors using functional magnetic resonance imaging (fMRI): left hemispheric localization of pleasant and right hemispheric localization of unpleasant odors. J. Comput. Assist. Tomogr. **25**(4), 493–514 (2001)
68. Niemic, C.: Studies of emotion: a theoretical and empirical review of psychophysiological studies of emotion. J. Undergrad. Res. **1**(1), 15–18 (2002)
69. Balconi, M., Lucchiari, C.: Consciousness and arousal effects on emotional face processing as revealed by brain oscillations. A gamma band analysis. Int. J. Psychophysiol. **67**(1), 41–46 (2008)
70. Li, M., Lu, B.: Emotion classification based on gamma-band EEG. In: Engineering in Medicine and Biology Society, 2009. EMBC 2009. Annual International Conference of the IEEE, pp. 1223–1226 (2009)
71. Kassam, K., Markey, A., Cherkassky, V., Loewenstein, G., Just, M.: Identifying emotions on the basis of neural activation. PLoS ONE **8**(6), e66032 (2013)
72. Wang, L., Chen, D., Ranjan, R., Khan, S., KolOdziej, J., Wang, J.: In: 2012 IEEE 18th International Conference on Parallel and Distributed Systems (ICPADS). IEEE, pp. 164–171 (2012)
73. Dean, J., Ghemawat, S.: MapReduce: simplified data processing on large clusters. Commun. ACM **51**(1), 107 (2008)

74. Dean, J., Ghemawat, S.: MapReduce: a flexible data processing tool. Commun. ACM **53**(1), 72 (2010)
75. Hadoop.apache.org.: "Welcome to Apache™ Hadoop®!" (2015). [Online]. https://hadoop. apache.org/
76. Berrada, G., van Keulen, M., Habib, M.: Hadoop for EEG storage and processing: a feasibility study. Brain Informatics and Health, pp. 218–230 (2014)

Chapter 11
Anticipating the User: Acoustic Disposition Recognition in Intelligent Interactions

Ronald Böck, Olga Egorow, Juliane Höbel-Müller, Alicia Flores Requardt, Ingo Siegert and Andreas Wendemuth

Abstract Contemporary technical devices obey the paradigm of naturalistic multi-modal interaction and user-centric individualisation. Users expect devices to interact intelligently, to anticipate their needs, and to adapt to their behaviour. To do so, companion-like solutions have to take into account the affective and dispositional state of the user, and therefore to be trained and modified using interaction data and corpora. We argue that, in this context, big data alone is not purposeful, since important effects are obscured, and since high-quality annotation is too costly. We encourage the collection and use of *enriched data*. We report on recent trends in this field, presenting methodologies for collecting data with rich disposition variety and predictable classifications based on a careful design and standardised psychological assessments. Besides socio-demographic information and personality traits, we also use speech events to improve user state models. Furthermore, we present possibilities to increase the amount of enriched data in cross-corpus or intra-corpus way based on recent learning approaches. Finally, we highlight particular recent neural recognition approaches feasible for smaller datasets, and covering temporal aspects.

R. Böck (✉) · A. Wendemuth
Cognitive Systems Group and Center for Behavioral Brain Sciences,
Otto von Guericke University Magdeburg, Universitätsplatz 2, 39106 Magdeburg, Germany
e-mail: ronald.boeck@ovgu.de

A. Wendemuth
e-mail: andreas.wendemuth@ovgu.de

O. Egorow · J. Höbel-Müller · A. F. Requardt · I. Siegert
Cognitive Systems Group, Otto von Guericke University Magdeburg, Universitätsplatz 2,
39106 Magdeburg, Germany
e-mail: olga.egorow@ovgu.de

J. Höbel-Müller
e-mail: juliane.hoebel@ovgu.de

A. F. Requardt
e-mail: alicia.requardt@ovgu.de

I. Siegert
e-mail: ingo.siegert@ovgu.de

© Springer Nature Switzerland AG 2019
A. Esposito et al. (eds.), *Innovations in Big Data Mining and Embedded Knowledge*,
Intelligent Systems Reference Library 159,
https://doi.org/10.1007/978-3-030-15939-9_11

11.1 Introduction

Considering an interaction of at least two interlocutors can be seen as a challenging issue since each communication is afflicted with ambiguities (cf. e.g. [138]). So, the famous four-sides model developed by Schulz von Thun shows different message layers which have to be interpreted properly by each interlocutor to guide a communication towards a success—this model was inspired by the five axioms of communication by Watzlawick (cf. [106, 146]). In particular, the interpersonal or interindividual relationship is important. This aspect also includes the intrapersonal perspective of each communication partner—anticipating what the message's sender means and "feels" is a characteristic of a vital interaction.

Besides the issues based on contextual information, personal aspects also enhance and complicate communication at the same time. Therefore, humans developed strategies to interpret and evaluate the additional intra- and interpersonal aspects of communication. As we know, this works—with some restrictions—quite well in human-human interaction (HHI). But does it also work in the relation to machines or, in particular, in human-computer interaction (HCI)? Is this question generally valid in this context? As Carroll discusses in [22], the way of interaction with a technical system or a machine has changed during the last decades. We see a development from a purely artificial controlling and input in a mouse-and-keyboard setting towards a more naturalistic interaction with technical devices using speech and gestures. But in order to shape this development according to the human needs, it is necessary to discuss what "more naturalistic" means in the context of HCI.

Naturalistic interaction is defined as a voluntary and willing, yet unforced, unscripted and not acted dialogic interaction with the freedom of choice of modalities (e.g. gestures, expressions, movements, speech, technical interaction devices), allowing the interlocutors to interact with each other and the environment as they please (cf. [7, 8, 135]). In this sense, it is directly connected to technical devices, since HHI—in comparison to HCI—can be most likely assumed as a naturalistic interaction by definition. For this, it was and still is important to analyse HHI to identify and transfer the underlying mechanisms used in such communication. Based on these concepts, technical systems need to be able to work in the aforementioned multi-modal manner (cf. [135]), especially in terms of input sensors and output channels (cf. Sect. 11.7). In fact, today's devices tend to use multiple modalities like touch, gestures, and speech (cf. [22]). On the one hand, we can find many examples for such devices, most prominent smartphones and tablet computers, and also applications dealing with a user-friendly interaction (cf. e.g. [55]). On the other hand, even with this more intuitive way of controlling, the step towards a naturalistic communication is still not done, yet. Therefore, the idea of companion-like systems or assistant systems arose (cf. [10, 150]), and lead to the development of devices or providing technical methods that are more adaptive towards the interlocutor's behaviour (cf. [10, 11, 54, 150]). These companion-like devices will be systems which are highly adapted to the users and their needs. But they need information beyond the common signals received and interpreted in the contemporary HCI. As already discussed, HHI is rich on subtle information (cf. [146]), which has to be transferred to HCI as well. One particular cue is the use of emotional or *dispositional* events to enable technical systems to interpret the human interlocutor's behaviour in an advanced manner. Unfortunately, it is still not enough; a more holistic view of the user is necessary including personality traits and special speech events like Discourse Particles (DPs) (cf. Sect. 11.5.1).

Considering such a holistic user observation, we can state that besides the *dispositional* interpretation defined as a rather generalised view on the user's emotions and feelings towards the current state of the interaction (cf. [12]), a representation of the user's current situation and behaviour should be regarded, based on the previously mentioned ideas of Schulz von Thun and Watzlawick. This can—more generalised— be subsumed by the term *user state* which integrates the situation of the user as well as of the environment and the behaviour afflicted with dispositional reactions triggered by either an interaction or a circumstance. These investigations are also related to two aspects: (1) interpretation of social signals in HHI and HCI (cf. [88, 138, 139]), and (2) handling multiple information sources, which is directly connected to aspects of multimodal fusion (cf. e.g. [28, 132]). Regarding social signals, we can state that a multitude of aspects has to be considered to draw correct conclusions during an interaction. Therefore, it is an "emerging domain" for new research (cf. [139]) but is beyond the topics to be discussed in this chapter. In fact, we concentrate more on the second issue related to the interpretation and integration of multiple sensor inputs (partially) available in a communication.

Especially in the context of learning approaches used in establishing user observations in HCI, data related to the particular domain is absolutely important. Most learning approaches applied in such a context are known to be greedy in terms of necessary data (cf. e.g. [105, 128]), in particular to achieve a good generalisation performance. Therefore, the use of "Big Data" is in the focus of various researchers and research communities related to affective computing (cf. e.g. [6, 35, 57, 86, 105]). Big data is characterised by a large collection of data in a certain context. On the one hand, such enormous amount of material can help to generate detection or classification systems in a generalised manner. On the other hand, being focussed on data only is dangerous: Can meaningful characteristics extracting fundamental knowledge be identified for a particular domain? Is the material useful for the particular issue or is it contaminated by side-effects?

In this regard, user state analysis is highly depending on the situation of HCI and the communication partners—it is obvious that this is also valid for HHI. Therefore, it is too trivial to say "the more data, the better data", since the variations between several tasks are sometimes important. As discussed by Siegert et al., each user has specific characteristics which should be modelled if possible—at least particular groupings should be covered (cf. [120]). Therefore, additional information beyond a pure data collection is necessary. In some situations, the credibility and match of the data can automatically be assessed through context information (cf. e.g. [88]). Better still, it is highly desirable to analyse corpora providing "important" information rather than a massive number of samples. One example for such additional information are questionnaires providing a ground truth related to personality features like the "Big Five" (agreeableness, conscientiousness, extraversion, neuroticism, openness; cf. [45]). As we will see in Sect. 11.5, a modelling incorporating speaker characteristics as well as correlations with personality traits can improve the detection and assessment of acoustic events like DPs (cf. [117]). Of course, for each analysis, a feasible collection of data also in terms of the samples' number is essential. Therefore, the general idea of big data is a comprehensible, and, in fact, feasible solution to implement automatic tools. Nevertheless, we highly argue for a more specific view which is not only relying on an enormous amount of data but rather on specific data. For this, we will use the term *enriched data*. With this term, we imply that we have a suitable number of samples allowing to develop technical devices but the data is also rich in information (cf. e.g. linguistically rich). Given this additional information, a smaller set of material can counterbalance the advantages of massive data collections. By interpretation, such corpora are not only a mere collection from any source like YouTube or Instagram; They usually are well-selected and well-elaborated regarding a particular task or domain. In Sects. 11.4 and 11.5, we will demonstrate how such enriched data can be used to develop classification systems for detecting user states as well as changes of user characteristics. This is mainly discussed on corpora which are rather small—compared to big data—but balance this aspect by a well-defined study design (cf. e.g. [40]). In Sect. 11.3 we introduce two corpora that can be seen as enriched data in the context of disposition recognition.

In this chapter, we introduce our perspective on enriched data—to be seen as a counterpart of big data—which is grounded on analyses presented in the remainder of the chapter. Since our research is embedded in the community of affective computing, a brief overview is given on the state of the art in terms of corpora collection and recognition techniques. This also includes a brief description of methods feasible for classification and combination of multiple information sources. Finally, we will discuss future trends on enriched data derived and extended based on literature as well as work explained in the current chapter (cf. Sect. 11.7).

11.2 State of the Art of Corpora Collection and Recognition Techniques

Recent literature shows a substantial shift from research on acted speech in full-blown emotions [16, 82], incorporating few meta-data, to research on naturalistic and more subtle dispositions [26, 73] incorporating enriched data and approximating affective user states "in the wild". It can be argued that acted emotional speech must be similar to natural expressions to be identified correctly by human recognisers. However, Bachorowski et al. suppose that there is affective information in human voice, which is not being under conscious control [4]. Moreover, acted emotions may be exaggerated and prototypical in contrast to natural emotions [98], whereas full-blown emotions are rarely found in normal daily conversations [73]. Instead, complex vocalisations of milder and more subtle dispositions like (negative) irritation, frustration, resignation, or (positive) well-being and interest, frequently occur in HHI as well as in HCI [26, 73].

Acoustic disposition recognition can be realised by supervised learning models such as Support Vector Machines (SVMs) by constructing a hyperplane in a high-dimensional space (cf. [32, 102]), Hidden Markov Models (HMMs) assuming speech to be a Markov process with unobserved states corresponding to a disposition (cf. [16, 107]), Linear Discriminant Classifiers assuming each class has a Gaussian probability density (cf. [75]), and by unsupervised learning models such as k-Nearest Neighbours Clustering (cf. [75]). Especially biologically inspired techniques are gaining attention (cf. Sect. 11.6), such as Segmented-Memory Recurrent Neural Networks (SMRNNs) (cf. [43]) and Extreme Learning Machines (ELMs) (cf. [42]). Compared to the trend of applying deep learning in the big data community, a similar trend can be observed in the community related to disposition recognition on enriched data utilising Convolutional Neural Networks (CNNs) (cf. [5, 77, 148]) which are used for automatic feature extraction as well as for classification. A prerequisite for acoustic disposition recognition is represented by extracting vocal parameters, such as pitch, intensity, speaking rate, and voice quality (cf. [4, 87]. Other features derived from a speech segment include formants, Mel-Frequency Cepstral Coefficients (MFCCs), pauses, energy-based features, log frequency power coefficients, and Perceptual Linear Prediction cepstral coefficients (PLPs) [142]. Additionally, statistics from time series are computed to capture their behaviour over time. A variety of feature sets is available, such as the "emobase" feature set [37] or the Geneva minimalistic acoustic parameter set [36] provided by openSMILE [38], a popular audio feature extraction toolkit.

When using the aforementioned feature sets, one has to be aware of features being influenced by not only oral variations, personality traits, and the temporal development of the emotional state [17], but also by extrinsic factors, such as recording environments or compression techniques [21, 118]. Depending on intrinsic and extrinsic experimental conditions, features can be highly correlated so that they cannot ensure a favourable impact on the system. Consequently, the following question arises in the context of disposition recognition on enriched data dealing with feature dimensionality issues: What is the most representative and most compact feature set to be used for the disposition recognition? To answer this question, studies on speech-based disposition recognition were conducted applying popular features. Böck et al. applied a Leave-One-Speaker-Out (LOSO) cross-corpus HMM-based emotion classification using three benchmark corpora: Berlin emotional database (emoDB) comprising expressive speech, eNTERFACE comprising naturalistic speech, and SmartKom comprising naturalistic but low-quality speech to determine handcrafted suitable features such as MFCCs and PLPs [16]. Sidorov et al. extracted optimal feature sets with an advanced feature selection technique using a self-adaptive multi-objective genetic algorithm and a probabilistic neural network for classification purposes regarding different corpora, such as emoDB and Vera-am-Mittag corpus (VAM), comprising speech from a German talk-show [109]. A Principal Component Analysis (PCA)-based feature selection approach was proposed by Siegert et al. [112] (cf. Sect. 11.6). Despite the presence of benchmark corpora being an adequate basis for research, determining an optimal feature set based on literature is challenging, as state-of-the-art studies are based on different fundamental aspects regarding the experiment characterised by a speaker-dependent or a speaker-independent design, and regarding the lack of a consistent applying of performance measures (cf. e.g. [36, 125]).

To tackle these issues specific corpora providing related information are necessary. Consequently, a future trend on creating corpora for disposition recognition is characterised by incorporating multimodal information sources (cf. the LAST MINUTE Corpus (LMC) [93] and the integrated Health and Fitness Corpus (iGF) described in Sects. 11.3.1 and 11.3.2, respectively). A full exploitation of multimodal data can be enabled by providing metadata and annotations ensuring information enrichment. However, obtaining such detailed information results in subjectivity and time consumption issues, like the resource-expensive labelling process (transcription or annotation) of data, which needs to be performed by several trained or (considerably more) untrained raters (cf. [97, 114]). Especially if the data is multi-modal, or if time alignment is required, this can only be performed with the aid of software tools (cf. e.g. [27]). Inter-rater reliability and time issues are addressed by Siegert et al. in the context of the annotation of emotions from audio and/or video [114]. By using an active learning technique (cf. Sect. 11.6), Schels et al. mitigate the trade-off between the amount of coded material and the accuracy of automatic classifiers trained on the EmoRec corpus [144, 145], comprising a naturalistic HCI in a gaming scenario [97]. Hence, a small number of speech samples can be created representing a representative set of enriched data contrasting big data.

Alongside subjectivity and time effort, there are performance issues of supervised learning techniques, as a classification system's performance relies strongly on the variability and the amount of the available annotated multimodal data. Although annotated data is available, it represents a sparse set compared to the large amount of unlabelled data. As the time exposure required for the annotation process increases with data size, this raises the question whether a reduced subset of a fully annotated dataset would be sufficient for supervised classification purposes. Hence, Thiam et al. propose an active learning technique combining outlier detection methods with uncertainty sampling in order to select the most informative samples. To enhance the capabilities of a disposition recognition system, the authors focus on the recognition of speech events such as laughter and heavy breathing. This approach has been successfully tested on a subset of the Ulm University Multimodal Affective Corpus, using solely the speech modality, as the system remains stable once the most interesting samples are annotated [129].

Analysing multimodal corpora including enriched data is important to discover multimodal information sources related to each other, which can be beneficial in a fusion approach. Multimodal fusion combines various modalities. There are mainly two levels of fusion, namely (1) feature-level or early fusion[1] combining features such as visual and audio features into one feature vector, and (2) decision-level fusion or late fusion examining and classifying features of each modality independently to

---

[1] Poria et al. state that some authors define early fusion as fusion directly on signal level, introducing mid-level fusion as fusion on feature level [89].

combine the results in a decision vector (cf. [89]). Applying decision-level fusion, Kindsvater et al. show the relation between body gestures and posture, vocal changes (e.g. pitch), and facial expressions when completing tasks with varying degree of difficulty. First, unimodal classifiers are built which are afterwards being combined to a multimodal mental load estimation by implementing Markov Fusion Networks (MFN) and Kalman Filter Fusion. While the outcomes of unimodal classifiers were very unstable, unreliable, and characterised by missing values, the fusion with MFNs and Kalman filter stabilises the result, especially compared to those of the unimodal classifiers [67]. Further studies on disposition recognition using multiple modalities surveyed by Poria et al. can be found in [89].

11.3 Generating Enriched, Naturalistic Interaction Data

While planning the data collection and experimental design, there is a variety of factors to be considered. Besides obvious factors such as the participants' sex, age and experience, there are also other factors that can possibly influence HCI—for example personality traits, user states and their changes. To collect an appropriate amount of diverse, enriched data covering all features in question, we need to carefully plan the experimental design of the recorded interactions and fully control the conditions. At the same time, it should not be forgotten that HCI is evolving, leaving the laboratory environment and moving into the wild—therefore, the scenarios must be close-to-real-life. In this section, we present two examples for naturalistic yet highly controlled corpora. An overview of the datasets is given in Table 11.1. The first dataset is the LMC that shows naturalistic interaction with a personalised assistant, and the second corpus is the iGF, displaying interaction in a health care context.

11.3.1 The LAST MINUTE Corpus

The LMC is a collection of 133 multimodal HCI experiments recorded using a Wizard-of-Oz (WoZ) setup. A detailed overview on the collecting, storing and annotating the data as well as privacy issues can be found in [93]. The scenario revolves around an imaginary journey to "Waiuku", a place unknown to the participants. The participants are asked to prepare the trip and therefore to accomplish close-to-real-life tasks with different levels of complexity. Each experiment takes about 30 min and consists of three modules with two different conversation styles: a personalisation module (free speech), a problem solving module (command-like speech), and a closure module (free speech), which are explained in the following. A detailed overview of the experimental stages is shown in Fig. 11.1.

The first part of each interaction is the personalisation module, intended to introduce the users to the system and to generate a more natural behaviour: The users are asked to introduce themselves, to recall recent emotional moments and their previ-

Table 11.1 Overview of the recorded modalities and enriched data of LMC and iGF

	LMC	iGF
Participants	130	41
Video data	Face (4-directions), 3D-data (bumblebee), system screen	Face (3-directions), body posture, system screen
Audio data	Participant (shotgun, neckband), system output	Participant (shotgun, neckband), system output
Physiological data	GSR, Respiration, ECG[a]	EMG, ECG, GSR
Transcripts	GAT, dialog success measure, dialog function of hesitations	–
Annotation	Experimental stages, hesitations, Filled Pauses (FPs)	Experimental stages
Questionnaires	ASF-E, NEO-FFI, IIP-C, SVF, ERQ, BIS/BAS, AttrakDiff, TA-EG	TAT/PSE, PRF, PMPN, TA-EG, PANAS, HAKEMP, NEO-FFI, SSI-K, ERQ, self-defined measure of walking problems

[a] only for some of the participants

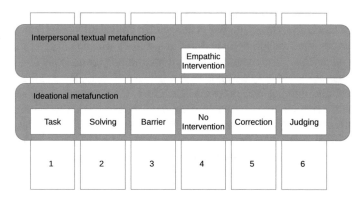

Fig. 11.1 Design of the LMC presenting the structure and function of each stage. The figure is taken from [40] referring to Halliday's model called "systemic functional grammar"

ous experience with technical systems. Here, the users are encouraged to talk freely and provide longer statements. After this part, the task is presented and the problem solving module begins: The users are asked to pack a suitcase for an imaginary journey to Waiuku. This part of the dialogue follows a specific structure: The users perform certain actions (e.g. choosing an item and requesting a list of items), whereas the system confirms or denies these actions. Since this part of the conversation is task-focussed, the users talk in a different, more command-like style. During this part, the difficulty of the task increases over time: On the one hand, the users have a strict time constraint from the very beginning, and on the other hand, there are also two major barriers, the weight limit and the destination change, intended to alter the course of interaction at specific points. After a stage of untroubled, "normal" interaction, the system points out that the airline's suitcase weight limit is reached and therefore, some items need to be removed. After this re-organisation stage the second barrier appears: The users learn that the destination is in the Southern Hemisphere and hence the journey is not a summer trip, but a winter trip, and therefore, they need to re-organise the suitcase again under time constraint and even more challenging weight constraints because of heavier winter gear. After this "nerve-jangling" experience, each experiment is completed with a closure module, where the system asks further questions about the task in general and users satisfaction with their solution of the task, encouraging them to talk freely again.

The recorded data consists of audio and video recordings for almost all participants as well as physiological data such as blood volume pressure and skin conductance for some of them. The data was transliterated and richly annotated by trained annotators, including not only the stages of the experiment but also parts of the interaction where trouble occurred. There are also annotations of special speech events like FPs, speech overlaps, off-talk, and laughter.

Besides the interaction itself, also other data is collected. Apart from general information like the participants' age and sex, socio-biographic (educational level, experience with computers) and psychometric parameters are collected using validated questionnaires, such as the NEO-FFI covering the "Big Five" factors [25], the inventory of interpersonal problems (IIP) focussing on interpersonal relationships [56], and the stress-coping questionnaire (SVF) [63] (cf. Table 11.1). It should also be mentioned that the participating users were chosen at random, but aiming at an equal distribution of sex and age groups, with a younger group's age ranging from 18 to 28, and the elder group consisting of participants elder than 60.

In addition to the LMC experiments, nearly half of the participants also took part in in-detail interviews conducted after the experiments intended to evaluate their experience. Therein, they were asked to described their individual experience of the experimental interaction and the simulated system [71, 72]. The interviews focussed on the participants' emotions and dispositions occurring during the interaction, their subjective ascriptions to the system, and the overall evaluation of the system.

This dataset can be seen as enriched data as well as "big data" for naturalistic HCI research, since it comprises 56 h of audio, video, and biophysiological HCI data, and over 87 h of interview data, both transliterated manually, as well as data on socio-biographic and psychometric parameters for each of the 133 participants. Therefore, it is a magnificent example for a corpus providing enriched data.

As the creation of such material demands careful planning and efficient interdisciplinary cooperation, this endeavour has been pursued in a large collaborative research framework on companion technology [11]. Furthermore, the process of annotating and enriching the data takes magnitudes longer than the pure planning and recording time.

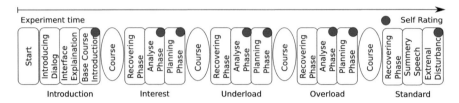

Fig. 11.2 Design of iGF presenting the structure and function of each stage. The figure is adapted from [132]

11.3.2 The Integrated Health and Fitness Corpus

Another example of a carefully planned data collection is the iGF, which comprises WoZ-interaction experiments with 65 participants (Min: 50 years, Max: 80 years, Mean: 66 years), with a duration of approximately 100 min each. Details on the collection, data storage and further information can be found in [132]. The corpus contains data of 10 modalities recorded using 20 sensory channels (cf. Table 11.1), and is aimed to provide data for the user dispositions of interest, cognitive overload, and cognitive underload as well as HCI-related emotional reactions of fear, frustration, and joy. The experimental phases, including indicators for self-ratings, are depicted in Fig. 11.2.

The health-and-fitness-scenario used for the data collection revolves around a self-explanatory gait training system and comprises the interaction data during the planning of a training course. The experiment is divided into five distinct modules. After each of the first four, a gait course is placed in order to retain the users' involvement, also including a pause for resting. Furthermore, each module is concluded by a self-rating of the user (cf. [132]).

The experiments start with an introduction module, where the gait training system and its usage are explained. This module is intended to induce natural behaviour by questioning the subjects on personal details like age, body size, profession, and technical affection. The first module is followed by three cognitive elicitation modules for interest, underload, and overload. In the interest module, the users receive information on gait exercises and are asked to create a personalised gait course. In the cognitive underload module, the system analyses the gait course in a simplified way, asking the users to read barely varying information on the aforementioned exercises, which leads to boredom and cognitive underload. In the cognitive overload module, the users receive negative feedback regarding their previous gait course and overall training progress, and then are asked to plan a new course, now with restrictions on exercises as well as a time constrain, which leads to cognitive overload. After this, the experiments are concluded by an emotion induction module, where the users are asked to summarise their impressions during a short talk.

The corpus comprises material of long HCI sessions (mean duration: 97 min) located in a non-ideal recording environment (a gym) and recorded in 20 synchronised modalities. Therefore, iGF is continuously being post-processed, including a rich annotation and transliteration. Primary results on iGF are presented in Sect. 11.7.

11.4 Using Socio-Demographic Information to Improve Disposition Modelling

The importance of certain psychometric parameters in HCI has already been in the focus of several investigations, (cf. [29, 100]). Considering the task of disposition modelling, it can be stated that the users' socio-demographic and psychometric parameters influence the users' dispositions and therefore can be used to improve their modelling. In this context, a remarkable number of characteristics can be collected through questionnaires like age, sex, personality, and stress-coping abilities. Therefore, the experimental design for the data collection has to be well-elaborated (cf. Sect. 11.3) relying on the knowledge of psychological experts.

11.4.1 The Influence of Age and Sex Information

Besides obvious differences in vocal characteristics of male and female (cf. [20]) as well as younger and elder users (cf. [78]), also differences in emotional behaviour patterns can be observed considering age and sex (cf. [39]). There are several investigations focussing on these influences in HCI. For example, Vogt et al. applied a sex differentiation to improve speech emotion recognition noticing an approximately 5% improvement using the correct sex information [143]. A two-stage approach for emotion classification was implemented by Shahin et al.: The first step is an automatic detection of the speaker's sex; the second step is the emotion classification itself. Given this method, the classification performance improves by approximately 11% absolute [107]. Dobrisek et al. utilised an adaptation approach based on Universal Background Models to perform emotion recognition on the eNTERFACE'05 database, and achieved an unweighted average recall (UAR) increase of about 3% by incorporating sex information [30].

The influence of other factors on disposition recognition on naturalistic material has not yet been thoroughly investigated—although it is known that age has an impact on the vocal characteristics (cf. [78]), the emotional response (cf. [48]), and the semantics (verbosity, politeness) (cf. [90, 94]). The age influence was analysed by Sidorov et al. in several experiments without achieving significant improvements [110]. This might be related to the utilised material with its limited age coverage. The LMC covers coverage of different age groups, as shown in Table 11.2 and therefore can be used to investigate the age influence in more detail.

As discussed in Sect. 11.3.1, the LMC provides two predefined barriers invoking a remarkable difference in expressed dispositions. Therefore, a two-class disposi-

Table 11.2 Distribution of speaker groups in LMC

	Male	Female	Total
Young	16	21	70
Elderly	18	24	60
Total	34	45	79

Fig. 11.3 Recognition results showing the advantage of including age and sex information for disposition recognition. The stars denote the significance level: **($p < 0.01$) using ANOVA

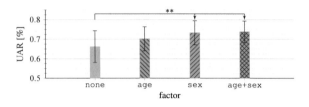

tion recognition problem (high/low expressivity) can be established. To analyse the effects of age and sex, all participants are grouped accordingly (cf. Table 11.2). Applying Gaussian Mixture Models and mainly spectral features—in particular, MFCCs, Zeroth Cepstral Coefficient, Fundamental Frequency, and Energy—in a LOSO validation experiment, a significant improvement in disposition recognition can be obtained, incorporating age and sex modelling and none (no speaker group modelling). In Fig. 11.3 the classification results are presented in terms of UAR. According to Hattie, the improvement is significant for none to sex (F = 8.706, p = 0.0032, dCohen = 0.492) and none to age+sex (F = 10.358, p = 0.0013, dCohen = 0.526) [51]. Comparing the achieved UARs utilising either age or sex groups, it can be seen that the sex grouping outperforms age grouping. A further discussion on details can be found in [120].

11.4.2 The Influence of Personality Traits in Human-Computer Interaction

Although age and sex are the most obvious and well-known characteristics, they are not the only ones influencing HCI. Personality traits also have an effect on interaction behaviour (cf. [62, 126]) as well as on emotion expressions (cf. [127, 131]). So, Matsumoto et al. found strong evidence that the recognition of emotional expression in faces is related to *openness to experience* and, to a lesser extent, *conscientiousness* [84]. Mairesse et al. analysed linguistic cues and performed experiments to automatically recognise personality traits [83]—although generally, text-based features are found to be inappropriate for this task [41]. In contrast, speech characteristics show capabilities for personality trait recognition [134], demonstrated in the INTERSPEECH 2012 Speaker Trait Challenge (cf. [101]). Furthermore, Siegert et al. presented models to link the temporal evolution of emotions to user characteristics such as mood and extraversion [113].

Although it is known that the user's personality is important for the interaction and for the overall communication success in HCI (cf. [100]), only a few studies analysed this issue. For example, Weinberg et al. identified personality traits as well as interpersonal relationships as relevant aspects in the field of HCI [147]. Van der Veer et al. found evidence suggesting that *extraversion* is related to computer aptitude and achievement [136].

The impact of the users' personality on the interaction itself is less investigated. To the best of our knowledge, the most prominent analysis was reported by Resseguier et al. considering the relation between facial expressions while playing a video game and the user's personality [91]. Larger studies on the influence of personality on HCI could not be performed since suitable datasets were missing. With LMC and iGF (cf. Sect. 11.3), two corpora are available that can be used for such detailed analyses. One example of personality traits influencing the interaction course is the use of particular speech events—namely DPs. These remarkable effects are described in Sect. 11.5.1.

11.5 Speech Events Indicating User States and Changes

Beyond mere words, speech uses a variety of channels to carry information, such as the prosody itself, speech events like DPs and FPs, overlaps, etc. In order to recognise the user's current dispositional state and adapt the interaction course, it is necessary to understand the meaning of these speech phenomena and—of course—automatically detect them. Since the prosody and general acoustic characteristics have been used as an important information source in a variety of disposition recognition applications by now (cf. [16, 43, 61, 77, 116]), we want to focus on special speech events and their influence on HCI.

11.5.1 Special Speech Events in Interactions

In a natural conversation, there are various speech events mainly intended to control and influence the communication rather than to submit content. Enriched corpora like the LMC provide the opportunity to analyse such events and the relations between their functions in the general interaction course.

Feedback Particles and Filled Pauses

The first speech event we are focussing on is related to feedback mechanisms. In HHI, different feedback paradigms contribute to the progress of the conversation by fulfilling different speech organising functions, the most important are managing the speakers' own linguistic contributions in the interactional flow. A special focus is on linguistic feedback signals as described by Allwood et al. and used to enable the exchange of information via the four basic communicative functions: contact, perception, understanding, and other attitudinal reactions [2]. One way to express these feedback signals is by using so-called Discourse Particles (DPs). They serve as discourse connectors, confirmation-seekers, intimacy signals, topic-switchers, hesitation markers, boundary markers, or fillers (cf. [65]). DPs are not only limited to English and German speech, but do also occur in other languages like Italian [9], French [52, 53], Dutch and Norwegian [1], and Japanese [149]. Pronouncing DPs, interaction partners are able to share information on their behavioural or affective state without interrupting the flow of the conversation [99].

DPs from daily, natural conversations are elements like *yeah, okay, right, uhm*, or *aha*. These monosyllabic words demand particular attention, as they are among the most frequent feedback words in spoken language (cf. [2]). Since they usually have very little semantic content, we need to analyse their prosody—depending on their intonation, these interjections have different meanings (cf. [99]). Furthermore, they show the same intonational characteristics as whole spoken sentences. One of the most diverse DPs is the element *hm*, which is also referred to as the "pure intonation carrier" [18, 99]. This feedback signal deserves special attention since it can be used to evaluate the user's emotional reaction (cf. [116]).

For the German language, Schmidt was able to identify seven form-function relations on the isolated DP *hm* : attention, thinking, finalisation, confirmation, decline, positive assessment, and request to respond [99]. Lotz et al. investigated the automatic evaluation of these sub-DPs in several consecutive publications, using the F_0 contour as a feature for the correlation with Schmidt's mathematically modelled prototypes [79] within non-isolated speech [80], and develop a pre-processing for further reducing the classification error [81].

Investigations on the LMC (cf. Sect. 11.3.1) by Siegert et al. showed that the use of DPs can be related to personality traits and certain corpus' stages. During the personalisation phase, there were no significant differences in the usage of DPs between participants with certain low and high traits (below and above median, respectively)— but in the problem solving phase, differences for certain personality traits arised. A detailed analysis of the data revealed that the participants having a positive stress coping mechanism trait below the median uttered significantly more DPs than those above the median [117]. This observation shows that participants having lower skills in stress management with regard to positive distraction express substantially more DPs. For all other traits, no significant differences and thus no influence was detected. A detailed overview over the results of the significance analysis for several traits is shown in Table 11.3.

The personalisation phase of the LMC was not intended to cause mental stress, it was aimed at making the user familiar with the system—therefore, it can be assumed that this was not a situation raising the user's stress level, and thus this personality trait would have no influence. In contrast, in the problem solving phase of LMC, stressful situations were generated. Siegert et al. analysed the user's behaviour directly at crucial interaction points and compared the behaviour after the weight limit barrier (WLB) to the smooth and non-disturbing baseline interaction (BSL) showing that the influence of the user's personality became more apparent. In Fig. 11.4, it can

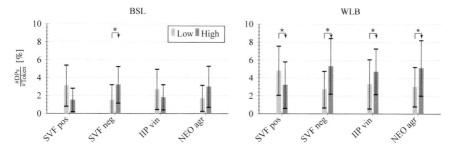

Fig. 11.4 Mean and standard deviation for the DPs of the two barriers regarding different groups of user characteristics. The stars denote the significance level: *($p < 0.05$), ⋆ denotes close proximity to significance level. The figure is taken from [117]. The abbreviations are given in Table 11.3

be seen that users having better skills in stress management with regard to positive distraction uttered substantially less DPs (cf. SVF neg. values). A detailed discussion on the particular personality traits and corresponding questionnaires is given in [117].

Besides DPs, there are also another noteworthy types of feedback signals, above all the so-called Filled Pauses (FPs) such as *uh* in English or *ähm* in German. In the same way as DPs, they occur in different languages (cf. [50, 96, 137]) and are used in HHI as well as in HCI (cf. [122]). FPs have several communicative and affective functions, such as turn-holding and indicating cognitive load [47]. The importance of this interaction phenomenon is well-known, and their detection has been in focus for some time, including the INTERSPEECH 2013 Computational Paralinguistics Challenge (cf. [104]). The increasing interest resulted in several available detection approaches: From simple SVM-based approaches with additional filtering (cf. [90]) to deep learning methods with sophisticated optimisation (cf. [46]).

Overlapping Speech in Natural Interactions

Speech overlaps can also provide information on the users' (dispositional) state [111]. An overlap is a part of a conversation where two or more participants speak simultaneously and can be grouped into four categories: simple confirmation signals, inter-

Table 11.3 Achieved level of significance of DP usage between LMC's personalisation and problem solving phase regarding selected personality traits. The F and p values provide information on the statistical significance

Trait	Personalisation		Problem solving	
	F	p	F	p
SVF positive distraction strategies (SVF pos)	2.015	0.156	3.546	0.058
SVF negative strategies (SVF neg)	1.271	0.260	3.515	0.061
IIP vindictive competing (IIP vin)	2.315	0.128	3.735	0.053
NEO-FFI agreeableness (NEO agr)	1.777	0.183	3.479	0.062

ruptions or crosstalk, premature turn-taking, and a simultaneous start after a pause (cf. [121]). But overlaps can also be distinguished with regard to their behavioural function, since they can be of cooperative or competitive nature. Whereas the former category is used to express support for the current speaker (cf. [152]), the latter is related to competition towards the current speaker and the wish to change the course of interaction (cf. [44]). Therefore they might be of interest for assessing the user state during an interaction—and even for more distinct dispositions, such as aggression (cf. [76]).

In order to use this information, it is necessary to detect overlaps and distinguish between their different types automatically. This task is not an easy one, with ongoing research concentrating on finding the right features—for example acoustic in different combinations with lexical or emotional features (cf. [34, 108]), as well as the right classification methods, from decision trees (cf. [108]) to deep learning architectures (cf. [24]).

11.5.2 Detection of User Dispositions Changes

Usually, in HCI dispositions are interpreted and classified, being understood as "static" events embedded in an interaction. For this, they are generally handled as being stable during an utterance—especially in utterance-level classification (cf. e.g. [102, 103]). Since phonemes of vowels provide a remarkable power to discriminate dispositions, frame-level classification is also considered, but usually the results are aggregated to generate a decision on larger segments (cf. [102, 140, 141]). In fact, dispositions are affected by various circumstances and thus tend to change during an interaction, especially if it is lasting for longer periods of time and covering different topics (cf. Sect. 11.3). This can be seen in neural mechanisms as well (cf. [64]), using brain imaging during an interaction (cf. [151]). Therefore, it is stringent to observe changes of user dispositions during communication as an additional source of information (cf. [17, 19, 61]). This is also reflected in the design of annotation tools—such as FeelTrace, a tool allowing continuous marking of dispositions in an interaction (cf. [27]).

In order to show the changes of the participants' emotional state during HCI, Huang et al. compared consecutive segments of speech to identify remarkable differences in the affective state of a speaker [61]. For this, they proposed the Emotion Pair Likelihood Ratios allowing an analytical investigation of speech segments based on the generalised likelihood ratio without any need of prior knowledge of emotions. In their experiments, the current speech segment was compared to the previous part given a pre-defined segment length and shift. This approach enables the prediction of disposition changes [61]. Böck et al. tested a different procedure using a classification approach instead of analytically comparing segments of interactions, and applied Gaussian Mixture Models to detect disposition changes. Their investigation was based on emotions which were induced in the beginning of an interaction and then have not been refreshed, vanishing over time. This vanishing effect can be automatically detected by classification [17].

The methods described in [17, 61] both rely on the given ground truth, marking possible points of change in the timing of an interaction. Therefore, a crucial issue in the assessment of disposition changes is the identification or prediction of change positions. To detect such change points, usually a dual-sliding window approach is applied (cf. e.g. [3, 85]). Additionally, Huang and Epps proposed a prediction method using a Martingale framework based on the idea of testing the exchangeability of two segments statistically. Applying such a framework, it was possible to derive regions for an expected change in the user's disposition. This lead to an improvement of the change detection, since a particular predictor was working on preselected segments, reducing the false alarm probability remarkably [60].

Given the current literature on detection, classification, and prediction of disposition changes from speech, we can conclude that this research area is still developing. The interpretation of personalised dynamics of disposition changes is a crucial issue that is highly related to HCI improvement. Therefore, the relation to personality traits (cf. Sect. 11.4.2) should not be left unattended. Another aspect is the timing of dispositions and their evolution over time. First hints are already given in [17] but more interdisciplinary research is still necessary. Moreover, the aspect of disposition evolution is linked to investigations of engagement in HHI and HCI during a communication (cf. [19, 138]). Especially, the interplay of dispositions and engagement provides additional perspective on an interaction, but also further research is necessary.

11.6 Recent Approaches for Disposition Recognition from Speech

As already discussed in Sects. 11.1 and 11.3, the need of enriched data is highly important in the area of disposition recognition from speech. Such data has to provide both, a suitable amount of speech material rich in disposition variety, and a well-prepared information on the "real disposition" included in the recordings, the so-called ground truth. We already presented ideas and methods how such enriched material can be obtained in Sect. 11.3 and how this data can be analysed in Sects. 11.4 and 11.5. But still, even with enriched data, the wish for more suitable material is omnipresent—we need more data to implement automatic systems being able to cover more disposition characteristics as well as to generalise across various user types. Therefore, we discuss approaches which help to increase the amount of material in either a cross-corpus or an intra-corpus way based on transfer learning and semi-supervised learning. Finally, we present three recent approaches based on neural networks that are usable for disposition recognition from speech on smaller datasets, two of which are able to also cover temporal aspects in a biologically inspired way.

Table 11.4 Recognition rates comparing six most similar datasets with all nine corpora as well as the amount of available material are given, together with standard deviation (STD). The table is adapted from [115]

#Corpora	Arousal	Valence	
	UAR (STD)	UAR (STD)	HH:MM
SIX	0.7172 (0.1020)	0.6043 (0.1173)	12:27
NINE	0.4586 (0.1474)	0.4953 (0.2551)	15:48

11.6.1 Corpus-Independent Analysis of Features

Since the effort in generating an enriched large corpus is high, a suitable way of increasing the amount of usable material is to combine various small data sets. Here we are faced with two issues: (1) How can we deal with different recording conditions of different corpora?; (2) Which corpora should be joined to avoid recognition performance decrease caused by unfavourable combinations?

Considering the first question, Böck et al. provided an extensive study on normalisation approaches in a cross-corpus analysis. The authors tested different normalisation strategies on nine well-known benchmark corpora, namely ABC, AVIC, DES, emoDB, eNTERFACE, SAL, SmartKom, SUSAS, and VAM (cf. e.g. [102]), applying several distance- and non-distance-based classification methods. In particular, standardisation, range normalisation, centering, and respective versions estimated on neutral speech only, were investigated. They concluded that standardisation provides the best opportunity to handle different recording conditions considering distance-based classifiers and thus increasing the recognition performance [13]. This gives insights to an alignment of various corpora in a possible set of combinations.

Using a statistical approach, Siegert et al. proposed a framework for the identification of possible corpora to be combined to increase both, the amount of training material and the recognition performance. The authors used the idea of PCA to identify the most informative features per corpus given a shared feature set—in this case, "emobase". Given the PCA-listed features, a ranking across corpora was calculated assuming that similar corpora shared the same most informative features. This method provides hints which corpora should be joined still avoiding unfavourable combinations. Applying this approach to the aforementioned nine benchmark corpora in a subsequent investigation, Siegert et al. identified a set of six quite similar datasets, namely ABC, emoDB, eNTERFACE, SAL, SmartKom, and SUSAS, covering almost 75% of available material in terms of recorded time (cf. Table 11.4) [115].

11.6.2 *Transfer Learning and Active Learning in Disposition Recognition*

As already discussed, various feature-based approaches have been observed checking the discriminative power for disposition recognition from speech. This also includes the opportunity to use these features in cross-corpus experiments transferring the knowledge obtained on one corpus to another. This idea can also be investigated on the classifier level, resulting in the *transfer learning* approach. For this, the knowledge obtained during training on one corpus is transferred to other corpora which are currently not annotated or used. Applying this technique, the particular information on a specific task or corpus can be used for different situations allowing a reduction of annotation and labelling effort, since the applied classifier provides an estimation that has just to be cross-checked manually. This idea was tested by Song et al., who proposed to apply transfer learning to emotion recognition from speech. They utilised an SVM-based approach trained on emoDB for annotation of a Chinese corpus [123]. Huang et al. show that transfer learning is also a feasible method to be used in combination with Deep Neural Networks (DNNs) [59]. Egorow et al. implemented a similar method to semi-automatically annotate FPs on human-human interviews and speeches using a recognition model trained on HCI data [31].

An approach related to transfer learning is *active learning*, which also allows a reduction of annotation and labelling effort (cf. e.g. [66]). Here the main idea is to use a small, manually annotated part of a corpus to train a classifier which provides afterwards a pre-classification of novel parts of the same corpus. The classification result is then cross-checked and corrected manually for other small corpus segments, resulting in an active classifier adaptation during the annotation process. Using this idea, Thiam et al. showed that with a small number of iterations (less than 35), active learning allows to achieve the detection of up to 82% of relevant dispositions [128].

Both approaches are able to reduce the effort usually necessary to edit a corpus—which is especially important in the case of big data with its many thousands of samples. Furthermore, in our case of enriched data, we have the option to increase the available amount of material by transferring given knowledge and combining different speech sources across corpora (cf. [112]).

11.6.3 *Selected Neural Methods for Speech-Based Disposition Classification*

Currently, biologically inspired neural approaches are widely used in the community to handle big data as well as to tackle disposition recognition. We discuss two methods which are biologically inspired in terms of processing the input information and the classifier design, namely Segmented-Memory Recurrent Neural Network (SMRNNs) and Extreme Learning Machine (ELMs). Furthermore, we present ideas how to apply Convolutional Neural Networks (CNNs) for disposition recognition from speech.

Fig. 11.5 SMRNN structure
representing a stacked set of
SRNs grouped by the level
of progress. The figure is
taken from [12]

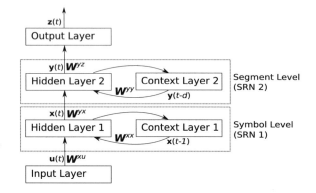

SMRNNs, developed by Chen and Chaudhari, reflect the human way of information processing, since in the first hidden layer only symbols are processed which will be aggregated in time to segments being analysed in the second hidden layer (cf. Fig. 11.5) [23]. Given the recurrent connection in the hidden layers, the network is able to handle dynamics instantaneously. This is also of interest in detection of dispositions from speech and provides furthermore the option to analyse the temporal evolution in the user's characteristics as discussed in Sect. 11.5.2. SMRNNs were already applied to the task of disposition recognition, for instance in [43], achieving benchmark results comparable to common approaches like Hidden Markov Models or SVMs (cf. [102]), but using only one third of the input features. In [43], the authors report classification results on the emoDB corpus achieving an unweighted average accuracy of 73.5% as well as an weighted average accuracy of 71.0% on testing material.

The ELM, introduced by Huang et al. allows a one-step training of the network based on the calculation of the inverse matrix of all connection weights [58]. Glüge et al. propose an extension of ELMs, namely deep ELMs (cf. [133]), to classify dispositions. Applied to benchmark corpora, a remarkable improvement in performance of 7.0 and 5.2% absolute is achieved compared to both SVMs and other neural approaches [42]. Given these achievements and the easy training process, (deep) ELMs should be further investigated as a benchmark classifier for this task.

Besides these highly biologically inspired neural approaches, the current research related to big data concentrates on various deep learning methods such as DNNs and CNNs. This trend can be also seen in the disposition recognition community with deep learning architectures (cf. e.g. [59, 124]) trained on large corpora (cf. e.g. [70]). CNNs are also in the focus of current investigation—recently, three research groups presented state-of-the-art results applying CNNs to different tasks (cf. [5, 77, 148]). They used spectrograms as inputs extracted from speech samples applying window techniques and Fast Fourier Transform. Furthermore, the classification (cf. [77, 148]) and the prediction (cf. [5]) relied on small corpora providing less than 4000 samples. They achieved promising results comparable to common techniques applied in the community (cf. [102]). Additionally, Weißkirchen et al. discussed "sketchy results" utilising the deep dreaming approach to highlight parts of the spectrograms potentially meaningful during classification [148].

11.7   Conclusion and Future Trends

Big data is a hot topic in various fields of computer science—and this trend can be also seen in the affective computing community (cf. [105]), by longing for data which can be utilised to generate systems able to detect human affective states. In particular, considering disposition recognition from speech, today big data is rather a dream, especially in terms of reasonably annotated corpora. Furthermore, the use of big data often involves the risk to lose focus since the enormous amount of data can obscure subtle yet important effects that shall be recognised (cf. Sect. 11.1). Another important issue is that the way of communication and thus the presentation of dispositions is influenced by various social constraints and personal habits—in both, HHI and HCI. Therefore, we argued for a sophisticated consideration of data and corpora called *enriched data*. By introducing this term we wanted to emphasize the importance of additional information obtained by questionnaires, social and demographic characteristics as well as well-elaborated study designs in order to provide detailed insights to the analysed subject and the related task. We encourage the community to select material based on the qualitative content and various hints provided for a certain task or situation, rather than for the mere number of samples comprised by a corpus. Analysing the subject is a complex process and needs well-established and well-informed material to achieve significant results, especially in natural interactions (cf. [70]). This issue can be interpreted as a follow-up discussion on the question which features are meaningful for speech affect recognition (cf. [125] and Sect. 11.6.1).

In this chapter, we discussed our vision of enriched data with the aid of two corpora, namely the LMC [93] and the iGF [132], providing close-to-real-life HCI. Both datasets cover a broad variety of conditions and factors which can be used for investigations related to various research communities.

For the first corpus, LMC, it was shown that acoustic features of the users' voices differ significantly between the different interaction stages (cf. [14, 15]). Furthermore, the material can be used to automatically recognise the interaction stage alongside physiological features (cf. [33]). The usage of Discourse Particles as "interaction patterns" (cf. Sect. 11.5.1) and their relation to psychometric parameters was investigated in [117]. Additionally, there are transdisciplinary analyses of given material that can highlight and dig novel insights which might not be found in "isolated" investigations. For instance, Rösner et al. combined methods applied in text analysis, analytical physiology, and neurobiology to investigate the relations of affective reaction cross-modally. For this, they used the LMC, where in addition to the HCI-related experiments, some of the participants also contributed to a follow-up experiment including neurobiological investigation. The authors analysed the relation between HCI-induced dispositions and the underlying reactions in the human brain using Functional Magnetic Resonance Imaging (FMRI). Especially, the impact of the system's reaction delays was under consideration [95]. In the recorded interactions, an acoustic feedback on a button hit was connected with the physiological brain behaviour (cf. [68, 69]). Complementary, the influence of prosodically modulated acoustic feedback in HCI was analysed considering brain reactions using FMRI (cf. [151]).

Considering the second corpus, iGF, the editing of the material is currently still ongoing. Therefore, we present initial results achieved on two particular topics: (1) Thiers et al. investigated user profiles related to the planning of physical gait exercises [130]. This was linked to different gait styles to be observed for types of falls (cf. [49]). (2) Siegert et al. suggested acoustic markers for intention recognition using an automatic device for planning of gait training courses [119]. In particular, the authors discussed reasonable acoustic markers and features for cognitive underload and overload detection, relating both states to uttered DPs during the interaction (cf. [132]).

Given the broad variety of possible investigations on the presented corpora (cf. Sect. 11.3), and also considering the socio-demographic aspects discussed in Sect. 11.4, as well as recent approaches in disposition recognition (cf. Sect. 11.6), we identified four possible trends for future research.

Socio-Demographic Features: As already discussed, socio-demographic information on an interaction participant can improve the recognition performance in both HHI and HCI (cf. Sect. 11.4). The interpretation of different characteristics is inherent for humans, since we are trained to adapt our expectations regarding communication style, wording, pronunciation, etc. based on various social and cultural features. Currently, for technical systems such observations are often neglected or obscured by massive use of training material. In contrast, considering these characteristics, it is possible to establish classification systems gaining high performance utilising well-selected material. Figure 11.3 shows recognition results for classifiers trained on material pre-selected by demographic characteristics. Although the number of available samples was reduced, the achieved results show a significant improvement compared to the mere use of the full material. Linking these achievements to personality information and the way special speech events such as DPs are used will further increase the performance of disposition recognition systems.

Multimodality: Connecting several sources of information extracted from different inputs is a powerful way to improve the disposition recognition performance. This linking of different sources was already used in the community (cf. e.g. [74, 89]), although mainly in a "restricted" manner. Usually, audio and video inputs are processed together to enable advanced insights into the user's state. We highly argue for a broader consolidation of various modalities, for instance by including textual information, demographic and social characteristics of a subject, gestures, poses, and facial expressions. Additionally, further intrinsic and extrinsic influencing factors can be observed. Especially, the current communication situation provides further reasonable information sources for the interpretation of dispositions. This is also related to the context of an interaction which is often assessed by psychologists relying on context cues to match the situation and the communication (cf. e.g. [4]). Moreover, extrinsic triggers are given by the ongoing interaction itself, as known from the investigations of Schulz von Thun and Watzlawick (cf. [106, 146]). They argued that each interaction is influencing the communication partners since each interlocutor is already communicating by mere presence. In the current research, these ideas are not considered and reflected, yet. We definitely need a solid interpretation of the communication's history and situation, otherwise a well-informed interpretation of dispositions and affects will be not possible. To obtain a broad assessment of the interaction partners, we also need an intrinsic view, that is, the personality as well as the bio-physiological parameters providing pure information on their characteristics.

Bio-Physiological Data: In particular, the investigation of bio-physiological data of a subject during an interaction allows insights on his or her internal characteristics Therefore, studies, as presented in [68, 151], showed the relation of brain activities and the interaction course. On the one hand, given such investigations on brain level, we can obtain information on the underlying concepts of affects and interactions. This helps to assess and model communication flows also in cases where bio-physiological parameters cannot be measured. On the other hand, various physiological parameters like heart rate can be obtained already applying image-based approaches. Therefore, we should incorporate this important information in current frameworks for disposition recognition.

Bio-Inspired Fusion: We highlight the concept of a bio-inspired fusion of multiple inputs. This includes the common fusion of audio, video, or physiological inputs (cf. e.g. [89, 92]) but extends these techniques towards a model-based approach. In particular, the analysis of brain activities can provide models and concepts of certain courses during an interaction which are considered in the fusion process. Especially, the weighting of certain inputs can be controlled by a-priori information with no need of sample-intensive training. Therefore, the approach helps to generate classifiers on smaller amounts of training data.

Based on the presented studies and results as well as the discussed trends, we encourage the research community to work with well-selected enriched data.

Acknowledgements We acknowledge support by the project "Intention-based Anticipatory Interactive Systems" (IAIS) funded by the European Funds for Regional Development (EFRE) and by the Federal State of Sachsen-Anhalt, Germany, under the grant number ZS/2017/10/88785. Further, we thank the projects "Mova3D" (grant number: 03ZZ0431H) and "Mod3D" (grant number: 03ZZ0414) funded by 3Dsensation within the Zwanzig20 funding program by the German Federal Ministry of Education and Research (BMBF). Moreover, the project has received funding from the European Union's Horizon 2020 research and innovation programme under the ADAS and ME consortium, grant agreement No 688900.

References

1. Abraham, W.: Multilingua. J. Cross-Cult. and Interlang. Commun. **10**(1/2) (1991). s.p
2. Allwood, J., Nivre, J., Ahlsn, E.: On the semantics and pragmatics of linguistic feedback. J. Semant. **9**(1), 1–26 (1992)
3. Anguera, X., Bozonnet, S., Evans, N., Fredouille, C., Friedland, G., Vinyals, O.: Speaker diarization: a review of recent research. IEEE Trans. Audio Speech Lang. Process. **20**(2), 356–370 (2012)
4. Bachorowski, J.A., Owren, M.J.: Vocal expression of emotion: acoustic properties of speech are associated with emotional intensity and context. Psycholog. Sci. **6**(4), 219–224 (1995)

5. Badshah, A.M., Ahmad, J., Rahim, N., Baik, S.W.: Speech emotion recognition from spectrograms with deep convolutional neural network. In: Proceedings of 2017 International Conference on Platform Technology and Service, pp. 1–5. IEEE, Busan, South Korea (2017)
6. Baimbetov, Y., Khalil, I., Steinbauer, M., Anderst-Kotsis, G.: Using big data for emotionally intelligent mobile services through multi-modal emotion recognition. In: Proceedings of 13th International Conference on Smart Homes and Health Telematics, pp. 127–138. Springer, Geneva, Switzerland (2015)
7. Batliner, A., Fischer, K., Huber, R., Spiker, J., Nöth, E.: Desperately seeking emotions: actors, wizards and human beings. In: Proceedings of the ISCA Workshop on Speech and Emotion: A Conceptual Framework for Research, pp. 195–200. Textflow, Belfast, UK (2000)
8. Batliner, A., Nöth, E., Buckow, J., Huber, R., Warnke, V., Niemann, H.: Whence and whither prosody in automatic speech understanding: A case study. In: Proceedings of the Workshop on Prosody and Speech Recognition 2001, pp. 3–12. ISCA, Red Bank, USA (2001)
9. Bazzanella, C.: Phatic connectives as interactional cues in contemporary spoken italian. J. Pragmat. **14**(4), 629–647 (1990)
10. Biundo, S., Wendemuth, A.: Companion-technology for cognitive technical systems. KI-Künstliche Intell. **30**(1), 71–75 (2016)
11. Biundo, S., Wendemuth, A. (eds.): Companion Technology—A Paradigm Shift in Human-Technology Interaction. Springer, Cham, Switzerland (2017)
12. Böck, R.: Multimodal automatic user disposition recognition in human-machine interaction. Ph.D. thesis, Otto von Guericke University Magdeburg (2013)
13. Böck, R., Egorow, O., Siegert, I., Wendemuth, A.: Comparative study on normalisation in emotion recognition from speech. In: Intelligent Human Computer Interaction, pp. 189–201. Springer, Cham, Switzerland (2017)
14. Böck, R., Egorow, O., Wendemuth, A.: Speaker-group specific acoustic differences in consecutive stages of spoken interaction. In: Proceedings of the 28. Konferenz Elektronische Sprachsignalverarbeitung, pp. 211–218. TUDpress (2017)
15. Böck, R., Egorow, O., Wendemuth, A.: Acoustic detection of consecutive stages of spoken interaction based on speaker-group specific features. In: Proceedings of the 28. Konferenz Elektronische Sprachsignalverarbeitung of the 29. Konferenz Elektronische Sprachsignalverarbeitung, pp. 247–254. TUDpress (2018)
16. Böck, R., Hübner, D., Wendemuth, A.: Determining optimal signal features and parameters for hmm-based emotion classification. In: Proceedings of the 28. Konferenz Elektronische Sprachsignalverarbeitung of the 15th IEEE Mediterranean Electrotechnical Conference, pp. 1586–1590. IEEE, Valletta, Malta (2010)
17. Böck, R., Siegert, I.: Recognising emotional evolution from speech. In: Proceedings of the 28. Konferenz Elektronische Sprachsignalverarbeitung of the International Workshop on Emotion Representations and Modelling for Companion Technologies. pp. 13–18. ACM, Seattle, USA (2015)
18. Bolinger, D.: Intonation and its uses: Melody in Grammar and Discourse. Stanford University Press, Stanford, CA (1989)
19. Bonin, F.: Content and context in conversations : the role of social and situational signals in conversation structure. Ph.D. thesis, Trinity College Dublin (2016)
20. Butler, L.D., Nolen-Hoeksema, S.: Gender differences in responses to depressed mood in a college sample. Sex Roles **30**, 331–346 (1994)
21. Byrne, C., Foulkes, P.: The mobile phone effect on vowel formants. Int. J. Speech Lang. Law **11**, 83–102 (2004)
22. Carroll, J.M.: Human computer interaction—brief intro. The Interaction Design Foundation, Aarhus, Denmark, 2nd edn. (2013). s.p
23. Chen, J., Chaudhari, N.: Segmented-memory recurrent neural networks. IEEE Trans. Neural Netw. **20**(8), 1267–1280 (2009)
24. Chowdhury, S.A., Riccardi, G.: A deep learning approach to modeling competitiveness in spoken conversations. In: 2017 IEEE International Conference on Acoustics, Speech and Signal Processing (ICASSP), pp. 5680–5684. IEEE (2017)

25. Costa, P., McCrae, R.: NEO-PI-R Professional manual. Revised NEO Personality Inventory (NEO-PI-R) and NEO Five Factor Inventory (NEO-FFI). Psychological Assessment Resources, Odessa, USA (1992)
26. Cowie, R.: Perceiving emotion: towards a realistic understanding of the task. Philos. Trans. R. Soc. Lond. B: Biol. Sci. **364**(1535), 3515–3525 (2009)
27. Cowie, R., Douglas-Cowie, E., Savvidou, S., McMahon, E., Sawey, M., Schröder, M.: 'feeltrace': An instrument for recording perceived emotion in real time. In: Proceedings of the ISCA Workshop on Speech and Emotion: A Conceptual Framework for Research, pp. 19–24. Textflow, Belfast, UK (2000)
28. Crispim-Junior, C.F., Ma, Q., Fosty, B., Romdhane, R., Bremond, F., Thonnat, M.: Combining multiple sensors for event recognition of older people. In: Proceedings of the 1st Workshop on Multimedia Indexing and information Retrieval for Healthcare, pp. 15–22. ACM, Barcelona, Spain (2013)
29. Cuperman, R., Ickes, W.: Big five predictors of behavior and perceptions in initial dyadic interactions: personality similarity helps extraverts and introverts, but hurts 'disagreeables'. J. Personal. Soc. Psychol. **97**, 667–684 (2009)
30. Dobrišek, S., Gajšek, R., Mihelič, F., Pavešić, N., Štruc, V.: Towards efficient multi-modal emotion recognition. Int. J. Adv. Robot Syst. **10** (2013). s.p
31. Egorow, O., Lotz, A., Siegert, I., Böck, R., Krüger, J., Wendemuth, A.: Accelerating manual annotation of filled pauses by automatic pre-selection. In: Proceedings of the 2017 International Conference on Companion Technology (ICCT), pp. 1–6 (2017)
32. Egorow, O., Siegert, I., Andreas, W.: Prediction of user satisfaction in naturalistic human-computer interaction. Kognitive Syst. **2017**(1) (2017). s.p
33. Egorow, O., Wendemuth, A.: Detection of challenging dialogue stages using acoustic signals and biosignals. In: Proceedings of the WSCG 2016, pp. 137–143. Springer, Plzen, Chech Republic (2016)
34. Egorow, O., Wendemuth, A.: Emotional features for speech overlaps classification. In: INTERSPEECH 2017, pp. 2356–2360. ISCA, Stockholm, Sweden (2017)
35. Etemadpour, R., Murray, P., Forbes, A.G.: Evaluating density-based motion for big data visual analytics. In: IEEE International Conference on Big Data, pp. 451–460. IEEE, Washington, USA (2014)
36. Eyben, F., Scherer, K.R., Schuller, B.W., Sundberg, J., André, E., Busso, C., Devillers, L.Y., Epps, J., Laukka, P., Narayanan, S.S., et al.: The Geneva minimalistic acoustic parameter set (GeMAPS) for voice research and affective computing. IEEE Trans. Affect. Comput. **7**(2), 190–202 (2016)
37. Eyben, F., Weninger, F., Gross, F., Schuller, B.: Recent developments in openSMILE, the Munich open-source multimedia feature extractor. In: Proceedings of the 21st ACM International Conference on Multimedia, pp. 835–838. ACM, Barcelona, Spain (2013)
38. Eyben, F., Wöllmer, M., Schuller, B.: Openearintroducing the munich open-source emotion and affect recognition toolkit. In: Proceedings of the 2009th ACII, pp. 1–6. IEEE, Amsterdam, Netherlands (Sept 2009)
39. Forgas, J.P.: Feeling and doing: affective influences on interpersonal behavior. Psychol. Inq. **13**, 1–28 (2002)
40. Frommer, J., Rösner, D., Haase, M., Lange, J., Friesen, R., Otto, M.: Detection and Avoidance of Failures in Dialogues–Wizard of Oz Experiment Operator's Manual. Pabst Science Publishers (2012)
41. Gill, A., French, R.: Level of representation and semantic distance: Rating author personality from texts. In: Proceedings of the Second European Cognitive Science Conference. Taylor & Francis, Delphi, Greece (2007). s.p
42. Glüge, S., Böck, R., Ott, T.: Emotion recognition from speech using representation learning in extreme learning machines. In: Proceedings of the 9th IJCCI, pp. 1–6. INSTICC, Funchal, Madeira, Portugal (2017)
43. Glüge, S., Böck, R., Wendemuth, A.: Segmented-Memory recurrent neural networks versus hidden markov models in emotion recognition from speech. In: Proceedings of the 3rd IJCCI, pp. 308–315. SCITEPRESS, Paris, France (2011)

44. Goldberg, J.A.: Interrupting the discourse on interruptions: an analysis in terms of relationally neutral, power-and rapport-oriented acts. J. Pragmat. **14**(6), 883–903 (1990)
45. Goldberg, L.R.: The development of markers for the Big-five factor structure. J. Pers. Soc. Psychol. **59**(6), 1216–1229 (1992)
46. Gosztolya, G.: Optimized time series filters for detecting laughter and filler events. Proc. Interspeech **2017**, 2376–2380 (2017)
47. Goto, M., Itou, K., Hayamizu, S.: A real-time filled pause detection system for spontaneous speech recognition. In: EUROSPEECH 1999, pp. 227–230. ISCA, Budapest, Hungary (1999)
48. Gross, J.J., Carstensen, L.L., Pasupathi, M., Tsai, J., Skorpen, C.G., Hsu, A.Y.: Emotion and aging: experience, expression, and control. Psychol. Aging **12**, 590–599 (1997)
49. Hamacher, D., Hamacher, D., Müller, R., Schega, L., Zech, A.: Exploring phase dependent functional gait variability. Hum. Mov. Sci. **52**(Supplement C), 191–196 (2017)
50. Hamzah, R., Jamil, N., Seman, N., Ardi, N., Doraisamy, S.C.: Impact of acoustical voice activity detection on spontaneous filled pause classification. In: Proceedings of the IEEE ICOS-2014, pp. 1–6. IEEE, Subang, Malaysia (2014)
51. Hattie, J.: Visible Learning. A Bradford Book, Routledge, London, UK (2009)
52. Hölker, K.: Zur Analyse von Markern: Korrektur- und Schlußmarker des Französischen. Steiner, Stuttgart, Germany (1988)
53. Hölker, K.: Französisch: Partikelforschung. Lexikon der Romanistischen Linguistik **5**, 77–88 (1991)
54. Honold, F., Bercher, P., Richter, F., Nothdurft, F., Geier, T., Barth, R., Hoernle, T., Schüssel, F., Reuter, S., Rau, M., Bertrand, G., Seegebarth, B., Kurzok, P., Schattenberg, B., Minker, W., Weber, M., Biundo-Stephan, S.: Companion-technology: towards user- and situation-adaptive functionality of technical systems. In: 2014 International Conference on Intelligent Environments, pp. 378–381. IEEE, Shanghai, China (2014)
55. Honold, F., Schüssel, F., Weber, M.: The automated interplay of multimodal fission and fusion in adaptive HCI. In: 2014 International Conference on Intelligent Environments, pp. 170–177. IEEE, Shanghai, China (2014)
56. Horowitz, L., Alden, L., Wiggins, J., Pincus, A.: Inventory of Interpersonal Problems Manual. The Psychological Corporation, Odessa, USA (2000)
57. Hossain, M.S., Muhammad, G., Alhamid, M.F., Song, B., Al-Mutib, K.: Audio-visual emotion recognition using big data towards 5g. Mob. Netw. Appl. **21**(5), 753–763 (2016)
58. Huang, G.B., Zhou, H., Ding, X., Zhang, R.: Extreme learning machine for regression and multiclass classification. IEEE Trans. Syst. Man, Cybern Part B (Cybernetics) **42**(2), 513–529 (2012)
59. Huang, Y., Hu, M., Yu, X., Wang, T., Yang, C.: Transfer learning of deep neural network for speech emotion recognition. In: Pattern Recognition—Part 2, pp. 721–729. Springer, Singapore (2016)
60. Huang, Z., Epps, J.: Detecting the instant of emotion change from speech using a martingale framework. In: 2016 IEEE International Conference on Acoustics. Speech and Signal Processing, pp. 5195–5199. IEEE, Shanghai, China (2016)
61. Huang, Z., Epps, J., Ambikairajah, E.: An investigation of emotion change detection from speech. In: INTERSPEECH 2015, pp. 1329–1333. ISCA, Dresden, Germany (2015)
62. Izard, C.E., Libero, D.Z., Putnam, P., Haynes, O.M.: Stability of emotion experiences and their relations to traits of personality. J. Person. Soc. Psychol. **64**, 847–860 (1993)
63. Jahnke, W., Erdmann, G., Kallus, K.: Stressverarbeitungsfragebogen mit SVF 120 und SVF 78, 3rd edn. Hogrefe, Göttingen, Germany (2002)
64. Jiang, A., Yang, J., Yang, Y.: General Change Detection Explains the Early Emotion Effect in Implicit Speech Perception, pp. 66–74. Springer, Heidelberg, Germany (2013)
65. Jucker, A.H., Ziv, Y.: Discourse Markers: Introduction, pp. 1–12. John Benjamins Publishing Company, Amsterdam, The Netherlands (1998)
66. Kächele, M., Schels, M., Meudt, S., Kessler, V., Glodek, M., Thiam, P., Tschechne, S., Palm, G., Schwenker, F.: On annotation and evaluation of multi-modal corpora in affective human-computer interaction. In: Multimodal Analyses Enabling Artificial Agents in Human-Machine Interaction, pp. 35–44. Springer, Cham (2015)

67. Kindsvater, D., Meudt, S., Schwenker, F.: Fusion architectures for multimodal cognitive load recognition. In: Schwenker, F., Scherer, S. (eds.) Multimodal Pattern Recognition of Social Signals in Human-Computer-Interaction, pp. 36–47. Springer, Cham (2017)
68. Kohrs, C., Angenstein, N., Brechmann, A.: Delays in human-computer interaction and their effects on brain activity. PLOS One 11(1), 1–14 (2016)
69. Kohrs, C., Hrabal, D., Angenstein, N., Brechmann, A.: Delayed system response times affect immediate physiology and the dynamics of subsequent button press behavior. Psychophysiology 51(11), 1178–1184 (2014)
70. Kollias, D., Nicolaou, M.A., Kotsia, I., Zhao, G., Zafeiriou, S.: Recognition of affect in the wild using deep neural networks. In: 2017 IEEE Conference on Computer Vision and Pattern Recognition Workshops, pp. 1972–1979. IEEE (2017)
71. Krüger, J., Wahl, M., Frommer, J.: Making the system a relational partner: Users ascrip-tions in individualization-focused interactions with companion-systems. In: Proceedings of the 8th CENTRIC 2015, pp. 48–54. Barcelona, Spain (2015)
72. Lange, J., Frommer, J.: Subjektives Erleben und intentionale Einstellung in Interviews zur Nutzer-Companion-Interaktion. In: proceedings der 41. GI-Jahrestagung, pp. 240–254. Bonner Köllen Verlag, Berlin, Germany (2011)
73. Laukka, P., Neiberg, D., Forsell, M., Karlsson, I., Elenius, K.: Expression of affect in spontaneous speech: acoustic correlates and automatic detection of irritation and resignation. Comput. Speech Lang. 25(1), 84–104 (2011)
74. Lee, C.C., Lee, S., Narayanan, S.S.: An analysis of multimodal cues of interruption in dyadic spoken interactions. In: INTERSPEECH 2008, pp. 1678–1681. ISCA, Brisbane, Australia (2008)
75. Lee, C.M., Narayanan, S.S.: Toward detecting emotions in spoken dialogs. IEEE Trans. Speech Audio Proc. 13(2), 293–303 (2005)
76. Lefter, I., Jonker, C.M.: Aggression recognition using overlapping speech. In: Proceedings of the 2017th ACII, pp. 299–304 (2017)
77. Lim, W., Jang, D., Lee, T.: Speech emotion recognition using convolutional and recurrent neural networks. In: Proceedings of 2016 Asia-Pacific Signal and Information Processing Association Annual Summit and Conference, pp. 1–4. IEEE, Jeju, South Korea (2016)
78. Linville, S.E.: Vocal Aging. Singular Publishing Group, San Diego, USA (2001)
79. Lotz, A.F., Siegert, I., Wendemuth, A.: Automatic differentiation of form-function-relations of the discourse particle "hm" in a naturalistic human-computer interaction. In: Proceedings of the 26. Konferenz Elektronische Sprachsignalverarbeitung. vol. 78, pp. 172–179. TUDpress, Eichstätt, Germany (2015)
80. Lotz, A.F., Siegert, I., Wendemuth, A.: Classification of functional meanings of non-isolated discourse particles in human-human-interaction. In: Human-Computer Interaction. Theory, Design, Development and Practice, pp. 53–64. Springer (2016)
81. Lotz, A.F., Siegert, I., Wendemuth, A.: Comparison of different modeling techniques for robust prototype matching of speech pitch-contours. Kognitive Syst. 2016(1) (2016). s.p
82. Luengo, I., Navas, E., Hernáez, I.: Feature analysis and evaluation for automatic emotion identification in speech. IEEE Trans. Multimed. 12(6), 490–501 (2010)
83. Mairesse, F., Walker, M.A., Mehl, M.R., Moore, R.K.: Using linguistic cues for the automatic recognition of personality in conversation and text. J. Artif. Intell. Res. 30, 457–500 (2007)
84. Matsumoto, D., LeRoux, J., Wilson-Cohn, C., Raroque, J., Kooken, K., Ekman, P., Yrizarry, N., Loewinger, S., Uchida, H., Yee, A., Amo, L., Goh, A.: A new test to measure emotion recognition ability: matsumoto and ekman's Japanese and caucasian brief affect recognition test (JACBART). J. Nonverbal Behav. 24(3), 179–209 (2000)
85. Moattar, M., Homayounpour, M.: A review on speaker diarization systems and approaches. Speech Commun. 54(10), 1065–1103 (2012)
86. Murino, V., Gong, S., Loy, C.C., Bazzani, L.: Image and video understanding in big data. Comput. Vis. Image Underst. 156, 1–3 (2017)
87. Murray, I.R., Arnott, J.L.: Toward the simulation of emotion in synthetic speech: a review of the literature on human vocal emotion. J. Acoust. Soc. Am. 93(2), 1097–1108 (1993)

88. Pantic, M., Cowie, R., D'Errico, F., Heylen, D., Mehu, M., Pelachaud, C., Poggi, I., Schroeder, M., Vinciarelli, A.: Social signal processing: the research agenda. In: Visual Analysis of Humans: Looking at People, pp. 511–538. Springer, London, UK (2011)

89. Poria, S., Cambria, E., Bajpai, R., Hussain, A.: A review of affective computing: from unimodal analysis to multimodal fusion. Inf. Fusion 37(Supplement C), 98–125 (2017)

90. Prylipko, D., Egorow, O., Siegert, I., Wendemuth, A.: Application of image processing methods to filled pauses detection from spontaneous speech. In: INTERSPEECH 2014, pp. 1816–1820. ISCA, Singapore (2014)

91. Resseguier, B., Léger, P.M., Sénécal, S., Bastarache-Roberge, M.C., Courtemanche, F.: The influence of personality on users' emotional reactions. In: Proceedings of Third International Conference on the HCI in Business, Government, and Organizations: Information Systems, pp. 91–98. Springer, Toronto, Canada (2016)

92. Ringeval, F., Amiriparian, S., Eyben, F., Scherer, K., Schuller, B.: Emotion recognition in the wild: incorporating voice and lip activity in multimodal decision-level fusion. In: Proceedings of the 16th ICMI, pp. 473–480. ACM, Istanbul, Turkey (2014)

93. Rösner, D., Frommer, J., Andrich, R., Friesen, R., Haase, M., Kunze, M., Lange, J., Otto, M.: Last minute: a novel corpus to support emotion, sentiment and social signal processing. In: Proceedings of the Eigth LREC, pp. 82–89. ELRA, Istanbul, Turkey (2012)

94. Rösner, D., Haase, M., Bauer, T., Günther, S., Krüger, J., Frommer, J.: Desiderata for the design of companion systems. KI - Künstliche Intell. 30(1), 53–61 (2016)

95. Rösner, D., Hazer-Rau, D., Kohrs, C., Bauer, T., Günther, S., Hoffmann, H., Zhang, L., Brechmann, A.: Is there a biological basis for success in human companion interaction? In: Proceedings of the 18th International Conference on Human-Computer Interaction, pp. 77–88. Springer, Toronto, Canada (2016)

96. Sani, A., Lestari, D.P., Purwarianti, A.: Filled pause detection in indonesian spontaneous speech. In: Proceedings of the PACLING-2016, pp. 54–64. Springer, Bali, Indonesia (2016)

97. Schels, M., Kächele, M., Glodek, M., Hrabal, D., Walter, S., Schwenker, F.: Using unlabeled data to improve classification of emotional states in human computer interaction. J. Multimodal User Interfaces 8(1), 5–16 (2014)

98. Scherer, K.R.: Vocal affect expression: a review and a model for future research. Psychol. Bull. 99(2), 143 (1986)

99. Schmidt, J.E.: Bausteine der Intonation. In: Neue Wege der Intonationsforschung, Germanistische Linguistik, vol. 157–158, pp. 9–32. Georg Olms Verlag (2001)

100. Schneider, T.R., Rench, T.A., Lyons, J.B., Riffle, R.: The influence of neuroticism, extraversion and openness on stress responses. Stress Health: J. Int. Soc. Investig. Stress 28, 102–110 (2012)

101. Schuller, B., Steidl, S., Batliner, A., Nöth, E., Vinciarelli, A., Burkhardt, F., Son, van, V., Weninger, F., Eyben, F., Bocklet, T., Mohammadi, G., Weiss, B.: The INTERSPEECH 2012 Speaker Trait Challenge. In: INTERSPEECH2012. ISCA, Portland, USA (2012). s.p

102. Schuller, B., Vlasenko, B., Eyben, F., Rigoll, G., Wendemuth, A.: Acoustic emotion recognition: a benchmark comparison of performances. In: Proceedings of the IEEE Automatic Speech Recognition and Understanding Workshop, pp. 552–557. IEEE, Merano, Italy (2009)

103. Schuller, B., Batliner, A., Steidl, S., Seppi, D.: Recognising realistic emotions and affect in speech: state of the art and lessons learnt from the first challenge. Speech Commun. 53(9–10), 1062–1087 (2011)

104. Schuller, B., Steidl, S., Batliner, A., Vinciarelli, A., Scherer, K., Ringeval, F., Chetouani, M., Weninger, F., Eyben, F., Marchi, E., et al.: The INTERSPEECH 2013 computational paralinguistics challenge: social signals, conflict, emotion, autism. In: INTERSPEECH 2013. ISCA, Lyon, France (2013). s.p

105. Schuller, B.W.: Speech analysis in the big data era. In: Proceedings of the 18th International Conference Text, Speech, and Dialogue, pp. 3–11. Springer, Plzen, Czech Republic (2015)

106. Schulz von Thun, F.: Miteinander reden 1 - Störungen und Klärungen. Rowohlt, Reinbek, Germany (1981)

107. Shahin, I.M.A.: Gender-dependent emotion recognition based on HMMs and SPHMMs. Int. J. Speech Technol. 16, 133–141 (2013)

108. Shriberg, E., Stolcke, A., Baron, D.: Observations on overlap: findings and implications for automatic processing of multi-party conversation. In: INTERSPEECH, pp. 1359–1362 (2001)
109. Sidorov, M., Brester, C., Minker, W., Semenkin, E.: Speech-based emotion recognition: feature selection by self-adaptive multi-criteria genetic algorithm. In: Proceedings of the Ninth LREC. ELRA, Reykjavik, Iceland (2014)
110. Sidorov, M., Schmitt, A., Semenkin, E., Minker, W.: Could speaker, gender or age awareness be beneficial in speech-based emotion recognition? In: Proceedings of the Tenth LREC, pp. 61–68. ELRA, Portorož, Slovenia (2016)
111. Siegert, I., Böck, R., Vlasenko, B., Ohnemus, K., Wendemuth, A.: Overlapping speech, utterance duration and affective content in HHI and HCI—an comparison. In: Proceedings of 6th Conference on Cognitive Infocommunications, pp. 83–88. IEEE, Györ, Hungary (2015)
112. Siegert, I., Böck, R., Vlasenko, B., Wendemuth, A.: Exploring dataset similarities using PCA-based feature selection. In: Proceedings of the 2015th ACII, pp. 387–393. IEEE, Xi'an, China (2015)
113. Siegert, I., Böck, R., Wendemuth, A.: Modeling users' mood state to improve human-machine-interaction. In: Cognitive Behavioural Systems, pp. 273–279. Springer (2012)
114. Siegert, I., Böck, R., Wendemuth, A.: Inter-Rater reliability for emotion annotation in human-computer interaction—comparison and methodological improvements. J. Multimodal User Interfaces 8, 17–28 (2014)
115. Siegert, I., Böck, R., Wendemuth, A.: Using the PCA-based dataset similarity measure to improve cross-corpus emotion recogniton. Comput. Speech Lang. 1–12 (2018)
116. Siegert, I., Hartmann, K., Philippou-Hübner, D., Wendemuth, A.: Human behaviour in HCI: complex emotion detection through sparse speech features. In: Human Behavior Understanding, Lecture Notes in Computer Science, vol. 8212, pp. 246–257. Springer (2013)
117. Siegert, I., Krüger, J., Haase, M., Lotz, A.F., Günther, S., Frommer, J., Rösner, D., Wendemuth, A.: Discourse particles in human-human and human-computer interaction—analysis and evaluation. In: Proceedings of the 18th International Conference on Human-Computer Interaction, pp. 105–117. Springer, Toronto, Canada (2016)
118. Siegert, I., Lotz, A.F., Duong, L.L., Wendemuth, A.: Measuring the impact of audio compression on the spectral quality of speech data. In: Proceedings of the 27. Konferenz Elektronische Sprachsignalverarbeitung, pp. 229–236 (2016)
119. Siegert, I., Lotz, A.F., Egorow, O., Böck, R., Schega, L., Tornow, M., Thiers, A., Wendemuth, A.: Akustische Marker für eine verbesserte Situations- und Intentionserkennung von technischen Assistenzsystemen. In: Proceedings of the Zweite transdisziplinäre Konferenz. Technische Unterstützungssysteme, die die Menschen wirklich wollen, pp. 465–474. University Hamburg, Hamburg, Germany (2016)
120. Siegert, I., Philippou-Hübner, D., Hartmann, K., Böck, R., Wendemuth, A.: Investigation of speaker group-dependent modelling for recognition of affective states from speech. Cogn. Comput. 6(4), 892–913 (2014)
121. Siegert, I., Philippou-Hübner, D., Tornow, M., Heinemann, R., Wendemuth, A., Ohnemus, K., Fischer, S., Schreiber, G.: Ein Datenset zur Untersuchung emotionaler Sprache in Kundenbindungsdialogen. In: Proceedings of the 26. Konferenz Elektronische Sprachsignalverarbeitung, pp. 180–187. TUDpress, Eichstätt, Germany (2015)
122. Siegert, I., Prylipko, D., Hartmann, K., Böck, R., Wendemuth, A.: Investigating the form-function-relation of the discourse particle "hm" in a naturalistic human-computer interaction. In: Recent Advances of Neural Network Models and Applications. Smart Innovation, Systems and Technologies, vol. 26, pp. 387–394. Springer, Berlin (2014)
123. Song, P., Jin, Y., Zhao, L., Xin, M.: Speech emotion recognition using transfer learning. IEICE Trans. Inf. Syst. E97.D(9), 2530–2532 (2014)
124. Stuhlsatz, A., Meyer, C., Eyben, F., Zielke, T., Meier, H.G., Schuller, B.W.: Deep neural networks for acoustic emotion recognition: raising the benchmarks. In: Proceedings of the ICASSP, pp. 5688–5691. IEEE (2011)
125. Tahon, M., Devillers, L.: Towards a small set of robust acoustic features for emotion recognition: challenges. IEEE/ACM Trans. Audio Speech Lang. Process. 24(1), 16–28 (2016)

126. Tamir, M.: Differential preferences for happiness: extraversion and trait-consistent emotion regulation. J. Pers. **77**, 447–470 (2009)
127. Terracciano, A., Merritt, M., Zonderman, A.B., Evans, M.K.: Personality traits and sex differences in emotions recognition among african americans and caucasians. Ann. New York Acad. Sci. **1000**, 309–312 (2003)
128. Thiam, P., Meudt, S., Kächele, M., Palm, G., Schwenker, F.: Detection of emotional events utilizing support vector methods in an active learning HCI scenario. In: Proceedings of the 2014 Workshop on Emotion Representation and Modelling in Human-Computer-Interaction-Systems, pp. 31–36. ACM, Istanbul, Turkey (2014)
129. Thiam, P., Meudt, S., Schwenker, F., Palm, G.: Active Learning for Speech Event Detection in HCI. In: Proceedings of the 7th IAPR TC3 Workshop on Artificial Neural Networks in Pattern Recognition, pp. 285–297. Springer, Ulm, Germany (2016)
130. Thiers, A., Hamacher, D., Tornow, M., Heinemann, R., Siegert, I., Wendemuth, A., Schega, L.: Kennzeichnung von Nutzerprofilen zur Interaktionssteuerung beim Gehen. In: Proceedings of the Zweite transdisziplinäre Konferenz. Technische Unterstützungssysteme, die die Menschen wirklich wollen, pp. 475–484. University Hamburg, Hamburg, Germany (2016)
131. Tighe, H.: Emotion recognition and personality traits: a pilot study. Summer Res. (2012). s.p
132. Tornow, M., Krippl, M., Bade, S., Thiers, A., Siegert, I., Handrich, S., Krüger, J., Schega, L., Wendemuth, A.: Integrated health and fitness (iGF)-Corpus - ten-Modal highly synchronized subject dispositional and emotional human machine interactions. In: Proceedings of Multimodal Corpora: Computer vision and language processing, pp. 21–24. ELRA, Portorož, Slovenia (2016)
133. Uzair, M., Shafait, F., Ghanem, B., Mian, A.: Representation learning with deep extreme learning machines for efficient image set classification. Neural Comput. Appl. pp. 1–13 (2016)
134. Valente, F., Kim, S., Motlicek, P.: Annotation and recognition of personality traits in spoken conversations from the ami meetings corpus. In: INTERSPEECH 2012. ISCA, Portland, USA (2012). s.p
135. Valli, A.: The design of natural interaction. Multimed. Tools Appl. **38**(3), 295–305 (2008)
136. van der Veer, G.C., Tauber, M.J., Waem, Y., van Muylwijk, B.: On the interaction between system and user characteristics. Behav. Inf. Technol. **4**, 289–308 (1985)
137. Verkhodanova, V., Shapranov, V.: Multi-factor method for detection of filled pauses and lengthenings in russian spontaneous speech. In: Proceedings of the SPECOM-2015, pp. 285–292. Springer, Athens, Greece (2015)
138. Vinciarelli, A., Esposito, A., Andre, E., Bonin, F., Chetouani, M., Cohn, J.F., Cristani, M., Fuhrmann, F., Gilmartin, E., Hammal, Z., Heylen, D., Kaiser, R., Koutsombogera, M., Potamianos, A., Renals, S., Riccardi, G., Salah, A.A.: Open challenges in modelling, analysis and synthesis of human behaviour in human-human and human-machine interactions. Cogn. Comput. **7**(4), 397–413 (2015)
139. Vinciarelli, A., Pantic, M., Boulard, H.: Social signal processing: survey of an emerging domain. Image Vis. Comput. **12**(27), 1743–1759 (2009)
140. Vlasenko, B., Philippou-Hübner, D., Prylipko, D., Böck, R., Siegert, I., Wendemuth, A.: Vowels formants analysis allows straightforward detection of high arousal emotions. In: Proceedings of the ICME. IEEE, Barcelona, Spain (2011). s.p
141. Vlasenko, B., Prylipko, D., Böck, R., Wendemuth, A.: Modeling phonetic pattern variability in favor of the creation of robust emotion classifiers for real-life applications. Comput. Speech Lang. **28**(2), 483–500 (2014)
142. Vogt, T., André, E.: Comparing feature sets for acted and spontaneous speech in view of automatic emotion recognition. In: Proceedings of the ICME, pp. 474–477. IEEE, Amsterdam, The Netherlands (2005)
143. Vogt, T., André, E.: Improving automatic emotion recognition from speech via gender differentiation. In: Proceedings of the Fiveth LREC. ELRA, Genoa, Italy (2006). s.p
144. Walter, S., Kim, J., Hrabal, D., Crawcour, S.C., Kessler, H., Traue, H.C.: Transsituational individual-specific biopsychological classification of emotions. IEEE Trans. Syst. Man Cybern.: Syst. **43**(4), 988–995 (2013)

145. Walter, S., Scherer, S., Schels, M., Glodek, M., Hrabal, D., Schmidt, M., Böck, R., Limbrecht, K., Traue, H., Schwenker, F.: Multimodal emotion classification in naturalistic user behavior. In: Human-Computer Interaction. Towards Mobile and Intelligent Interaction Environments, pp. 603–611. Springer (2011)
146. Watzlawick, P., Beavin, J.H., Jackson, D.D.: Menschliche Kommunikation: Formen, Störungen, Paradoxien. Verlag Hans Huber, Bern, Switzerland (2007)
147. Weinberg, G.M.: The Psychology of Computer Programming. Van Nostrand Reinhold, New York, USA (1971)
148. Weißkirchen, N., Böck, R., Wendemuth, A.: Recognition of emotional speech with convolutional neural networks by means of spectral estimates. In: Proceedings of the 2017th ACII, pp. 1–6. IEEE, San Antonio, USA (2017)
149. White, S.: Backchannels across cultures: a study of americans and japanese. Lang. Soc. **18**(1), 59–76 (1989)
150. Wilks, Y.: Artificial companions. Interdiscip. Sci. Rev. **30**(2), 145–152 (2005)
151. Wolff, S., Brechmann, A.: Carrot and stick 2.0: the benefits of natural and motivational prosody in computer-assisted learning. Comput. Hum. Behav. **43**(Supplement C), 76–84 (2015)
152. Yang, L.C.: Visualizing spoken discourse: prosodic form and discourse functions of interruptions. In: Proceedings of the Second SIGdial Workshop on Discourse and Dialogue, pp. 1–10. Association for Computational Linguistics, Aalborg, Denmark (2001)

Chapter 12
Humans Inside: Cooperative Big Multimedia Data Mining

**Shahin Amiriparian, Maximilian Schmitt, Simone Hantke,
Vedhas Pandit and Björn Schuller**

Abstract Deep learning techniques such as convolutional neural networks, autoencoders, and deep belief networks require a big amount of training data to achieve
an optimal performance. Multimedia resources available on social media represent
a wealth of data to satisfy this need. However, a prohibitive amount of effort is
required to acquire and label such data and to process them. In this book chapter,
we offer a threefold approach to tackle these issues: (1) we introduce a complex
network analyser system for large-scale big data collection from online social media
platforms, (2) we show the suitability of intelligent crowdsourcing and active learning approaches for effective labelling of large-scale data, and (3) we apply machine
learning algorithms for extracting and learning meaningful representations from the
collected data. From YouTube—the world's largest video sharing website we have
collected three databases containing a total number of 25 classes for which we have
iterated thousands videos from a range of acoustic environments and human speech
and vocalisation types. We show that, using the unique combination of our big data
extraction and annotation systems with machine learning techniques, it is possible to
create new real-world databases from social multimedia in a short amount of time.

12.1 Introduction

We live in perhaps the most interesting times of the information era, where each
one of us—knowingly or unknowingly—is responsible for generating data. Information technology companies such as Google, Microsoft, and Facebook collect and

S. Amiriparian (✉) · M. Schmitt · S. Hantke · V. Pandit · B. Schuller
ZD.B Chair of Embedded Intelligence for Health Care and Wellbeing,
Augsburg, Germany
e-mail: shahin.amiriparian@tum.de

S. Amiriparian · S. Hantke
Machine Intelligence & Signal Processing Group, Technische Universität München,
Munich, Germany

B. Schuller
Group on Language Audio and Music, Imperial College London, London, UK

© Springer Nature Switzerland AG 2019   235
A. Esposito et al. (eds.), *Innovations in Big Data Mining and Embedded Knowledge*,
Intelligent Systems Reference Library 159,
https://doi.org/10.1007/978-3-030-15939-9_12

maintain data in exabyte proportions [40]. Importantly, a large proportion of data has been made publicly available with very few restrictions limiting its collection. For example, it was estimated that in July 2015, more than 400 h of video data were uploaded to the popular video sharing website YouTube every minute [1]. This growing amount of multimedia material publicly available online, including produced content and non-acted personal home videos, represents an untapped wealth of data for research purposes.

Equally important as available data resources is the advent of machine learning paradigms which facilitate the labelling of large datasets with minimal human assistance [12, 70, 71]. Strategies like semi-supervised active learning use confidence values from a machine learner to decide whether to keep the automatically assigned label or to request a human annotator to label the instance in question [21, 71]. Such paradigms have been used in a range of tasks, like emotion recognition [64, 68] and sound classification [21, 66, 68, 70].

Moreover, thanks to recent advances in crowdsourcing, when human annotation is required, the labels can be obtained efficiently with low financial overhead. Crowdsourcing is the utilisation of a large group of non-experts to perform a common task under the assumption that the collective opinion of this group is quicker and less challenging to collect than that of a smaller group of trained experts who are familiar with the annotation task. Further, this collective opinion has been shown to be, in terms of quality, as good as annotations determined by a small groups of experts, at a fraction of the cost [10, 22, 29]. Crowdsourcing is gaining popularity in the industry, e.g., as alternative workforce model for software development [33] and has already been successfully applied in a range of computing applications [13, 39, 58].

The aim of this chapter is to propose and assess the feasibility of a system which combines state-of-the-art big data knowledge extraction and annotation systems with machine learning algorithms for generation of real-world databases from multimedia data available on the Internet.

12.2 System Structure

While online multimedia archives contain a wealth of data, its practical application in training machine learning systems is hindered by three obstacles: (1) finding the relevant data, (2) segmenting these into meaningful and coherent segments, e.g., image segments which represent the audio content of a video or vice versa, and (3) reliably labelling the segments, so that they can be useful for machine learning. In the following subsections, we describe our solutions to these problems. A high-level overview of our system for big data extraction, annotation and processing is given in Fig. 12.1. First, we use our CAS^2T toolkit for large-scale data collection from online social media platforms [5] (the first block in Fig. 12.1). This approach applies complex systems theory (cf. Sect. 12.3) to identify most related videos clips on YouTube based on a provided source video. Subsequently, in order to annotate the most relevant clips collected from YouTube, we use our intelligent crowdsourcing platform

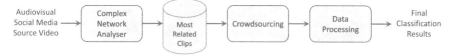

Fig. 12.1 Block diagram of our system for big data extraction and processing. A detailed account of the procedure is given in Sect. 12.2

(cf. Sect. 12.4) to obtain labelled databases (cf. Sect. 12.5). Finally, we process the annotated data with our machine learning approaches (cf. Sect. 12.6) in order to check the feasibility of our proposed system.

12.3 Complex Network Analyser

The main role of CAS^2T—the complex network analyser component—is to enable rapid identification of related multimedia data from online resources. It is currently developed to work with YouTube which has one of the world's largest and most professional video recommendation systems [15]. Whilst the exact mechanism of YouTube is not publicly available, it is known to be based on factors, such as the number of video clicks, video title, description and associated metadata, search query tokens, viewer demographics, and the number of likes and dislikes [15]. CAS^2T operates under the assumption that the content of recommended videos is related to the original video and exploits the complex networks of interconnections generated by YouTube's recommendation system by modelling them as graphs with small-world properties [63]. Watts and Strogatz have shown that many social and technological networks have small path lengths [63] and call them a 'small-world'. These properties can also be found in human brain networks [11, 49]. Given an initial source video link (cf. 'Source Video' in Fig. 12.1), CAS^2T generates an undirected graph G of the videos most highly recommended to the viewer. The *vertices* of G represent videos that are considered to be potentially related to the topic of interest (source video), and the *edges* correspond to the recommendations between videos.

The graph G can therefore be considered as a mapping of YouTube's recommendation space in relation to the source video. To the best of the authors' knowledge, such a mapping has been published for the first time in [5]. This mapping is essential for revealing the extent of mutual relationships between related videos of interest. Within G, the videos (vertices) with a greater relation to the topic of interest have a higher number of connections (edges) with other vertices. Respectively, we apply the *Local Clustering Coefficient* (LCC) algorithm to identify highly related videos [41, 56, 63].

The LCC of a video v_i in G quantifies how close its neighbours are to being a clique, i.e., a complete graph, and how likely they are part of larger highly connected video groups. The LCC can be determined via

$$C_{v_i} = \frac{2n}{k_{v_i}(k_{v_i} - 1)}, \qquad (12.1)$$

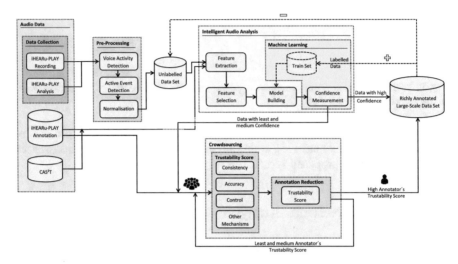

Fig. 12.2 iHEARu-PLAY's interaction between the intelligent audio analysis, the machine learning, and the data quality management components, including the annotator trustability calculation and the annotation reduction components. Adapted from [22, 25]

where C_{v_i} is the LCC for v_i, n is the number of edges that actually pass between the neighbours of v_i, and k_{v_i} is the number of neighbours of v_i [63]. Highly related videos in G, i.e., videos with a high number of edges, will have a high LCC value and conversely, unrelated videos will have a low LCC. CAS^{2}T uses the LCC to locate highly mutually related content groups in G, which can be downloaded and then sent to the crowdsourcing platform for annotation and further processing (cf. Sect. 12.4).

12.4 Intelligent Crowdsourcing for Data Annotation

The YouTube clips which were collected with CAS^{2}T (cf. Sect. 12.2) are sent to our gamified intelligent crowdsourcing platform iHEARu-PLAY[1] [22, 25] for labelling purposes. The platform offers audio, video, and image labelling for a diverse range of annotation tasks as well as audio-visual data collections. In addition, iHEARu-PLAY ensures a high quality of annotations through an optimised data quality management, while the gamification aspect aims to reduce the boredom of the annotators [26]. Making use of the data quality management, annotator trustability-based machine learning algorithms are integrated resulting in an intelligent way of gathering annotations and at the same time reducing the manual annotation workload [25, 27].

An overview of the iHEARu-PLAY components is depicted in Fig. 12.2 and will be described in the following sections in more detail.

---

[1] https://www.ihearu-play.eu.

12.4.1 Intelligent Audio Analysis

Data owners, i.e., researchers, upload their audio data to iHEARu-PLAY, which will then automatically run through the *Intelligent Audio Analysis* component. After having chosen a feature set, the acoustic features are automatically extracted by using the integrated openSMILE toolkit [18]. Then, a classifier is automatically trained with the on iHEARu-PLAY pre-labelled small amount of training data and the results are automatically transferred to the *Trustability-based Machine Learning* component.

12.4.2 Active Learning Algorithms

The proposed intelligent crowdsourcing approach combines different state-of-the-art Active Learning (AL) algorithms with the crowdsourcing platform iHEARu-PLAY aiming to combine the accuracy of manual labelling with the speed and cost-effectiveness of machine learning classifiers. The main concept of an AL algorithm is based on the idea that the algorithm can improve the classification accuracy with as little training data as possible by actively choosing the data the algorithm is most certain about [14, 34]. The major drawback of conventional (static) AL algorithms though is that they still rely predominantly on annotators to provide the correct label for each instance, meaning they wait until a predefined number of manual annotations are gathered for an instance before determining the final label [25, 69]. Alternately, a dynamic learning process tries to further reduce the annotation workload by applying an adaptive query strategy without sacrificing performance [65].

The integrated trustability-based Dynamic Active Learning (TDAL) [25], as shown in Algorithm 1, describes the machine learning algorithms applied in this work which are based on the *least certainty* query strategy and on the *medium certainty* query strategy, which have been introduced in [24, 67]. The main improvement of this technique compared to the static AL method is that for every instance in a subset $\mathscr{S}_u$, the TDAL approach first requests only l annotations and after that asks for one more label each time until the predefined agreement level of an annotator's trustability score T_a is reached for one class. For each query strategy, the algorithm starts by classifying all instances of the unlabelled data pool U using a model trained on a small set of data L, previously labelled using iHEARu-PLAY. Upon the posterior probability, the confidence values assigned to each instance are ranked and stored in a queue Q in descending order. Finally, a subset S_u of the unlabelled data set U corresponding to those instances predicted with least or medium confidence values are sent to manual annotation within iHEARu-PLAY. Thenceforth, these instances are added to the labelled set L and removed from U. This sequential process is repeated until a predefined number of instances are selected or until a predefined stopping criterion is met [24, 25, 67].

Algorithm 1 Trustability-based Dynamic Active Learning (TDAL) with least and medium certainty query strategy [25].

Given:

- $\mathscr{L}$: a small set of labelled data
- $\mathscr{U}$: a large pool of unlabelled data
- $\mathscr{S}_u$: a subset of the unlabelled data set $\mathscr{U}$
- $\mathscr{M}$: an initial model trained by $\mathscr{L}$
- C: classifier's confidence value
- T_a: calculated trustability score of an annotator a

repeat

- (Optional) Upsample labelled set $\mathscr{L}$ to obtain even class distribution $\mathscr{L}_D$.
- Use $\mathscr{L}/\mathscr{L}_D$ to train enhancing classifier $\mathscr{M}$, then classify unlabelled set $\mathscr{U}$.
- Calculate corresponding classifier's confidence value C.
- Rank data based on the prediction confidence values C and store them in a queue Q.
- **Choose a query strategy:**
    - Least certainty query strategy: Select subset $\mathscr{S}_u$ of unlabelled data set $\mathscr{U}$ whose elements are 'at the bottom' of the ranking queue Q.
    - Medium certainty query strategy: Select subset $\mathscr{S}_u$ of unlabelled data set $\mathscr{U}$ whose elements are 'in the middle' of the ranking queue Q.
- Submit selected instances $\mathscr{S}_u$ to manual annotation.

    repeat

    - Compute aggregated manual labels l and assess annotators' confidence using the trustability score T_a calculation.
    - Add instances with high annotators' trustability scores $S_{u(high)}$ and their aggregated labels to labelled set $\mathscr{L}$, $\mathscr{L} = \mathscr{L} \cup S_{u(high)}$.
    - Remove $S_{u(high)}$ from unlabelled set U, $U = U - S_{u(high)}$.
    - Keep instances with low $S_{u(low)}$ or medium $S_{u(medium)}$ annotators' trustability scores for next iteration in order to obtain more manual labels, $\mathscr{S}_u = S_{u(low)} \cup S_{u(medium)}$.

    until manual annotation is not possible OR a predefined number of iterations is met.

until there is no data in the unlabelled data set predicted as belonging to the target class OR model training converges OR manual annotation is not possible OR a predefined number of iterations is met.

12.4.3 Annotator Trustability Score and Annotation Reduction

Whilst (dynamic) active learning is mainly applied to expedite the learning process and aims to reduce the labelling efforts, on the other hand, crowdsourcing can potentially reduce the cost or workload per annotation in a fast and efficient way. A major problem of gathering annotations, especially when using crowdsourcing, is that the quality of the annotations can be lower. This can result in training the model using wrongly labelled data, and cause reductions in the accuracy of a classifier trained using this data. Therefore, iHEARu-PLAY aims to obtain annotations from non-expert annotators that are qualitatively close to gold standard annotations created by experts.

Within many well-known crowdsourcing platforms such as Amazon Mechanical Turk[2] or CrowdFlower,[3] the annotator's reliability and the annotation quality is usually assessed through a pretest comprising different questions to determine if the annotator is performing the given task correctly. Inspired by this Quality Management System (QMS), we introduced a more detailed QMS throughout the learning algorithm to assess the novel quality mechanism called trustability score [25]. For calculating this trustability score, several quality measuring features were implemented, including consistency, accuracy, and control questions to compute the intra-annotator and inter-annotator agreement:

(i) *Control Questions*, which contain definitely wrong answers are mixed into the shown answer options and are used to detect annotators who do not read the question and/or just select a random answer, (ii) *Consistency Questions*, which are repeated certain questions from within the data files currently being labelled by a particular annotator, and then comparing the answer with the previous answer(s), and (iii) *Accuracy Questions*, which focus on the relation of the given answer to other annotators' answers towards the same data instance. Giving 'inconsistent' or 'wrong' answers decreases an annotator's trustability score whilst a 'correct' answer increases or maintains it.

Besides these three main components for calculating the trustability score, iHEARu-PLAY checks an annotator's behaviour and further QMS features are integrated such as gamification [26], as opposed to monetary motivation [62], and pre-annotation listening checks. Our integrated QMS and the calculated trustability score have been shown to be a promising approach to obtain annotations from non-expert annotators that are qualitatively close to gold standard annotations created by experts [24, 25, 27].

All these methods can be used as a measurement to weight the gathered annotations, for example with the help of the *Weighted Trustability Evaluator* [23]. In addition, under the assumption that a 'good' annotator with a high trustability score creates better annotations than a 'bad' annotator with a low trustability score, the number of redundant annotations can be minimised, since this information can be used to define a quality threshold for the number of needed annotations. Annotations from trusted annotators will then be added to the final labelled dataset. If an annotator is not trusted, the data instance will be sent back to a new round of labelling until the given minimum number of required annotations is met. This procedure is iterated until a defined criterion (cf. Algorithm 1) is met and all annotations are given.

---

[2]https://www.mturk.com.

[3]https://www.crowdflower.com.

12.5 Databases

After collecting the initial sets of YouTube clips with CAS^{2}T (cf. Sect. 12.3) and annotating them using iHEARu-PLAY (cf. Sect. 12.4), we have obtained a range of real-world databases: the SPEECH AND VOCALISATION database, with 6 different human speech and vocalisation types [5], the VEHICLES database, including 10 various transportation systems, and the BALL SPORTS database, containing 9 types of ball sports. In Table 12.1 we give an overview on details of all data sets which we then use as training and evaluation sets for our machine learning experiments.

12.5.1 SPEECH AND VOCALISATION *Database*

The SPEECH AND VOCALISATION corpus contains 7851 unique audio recordings. All tasks in this database are based on the concept of acoustic surveillance [28]. The *Freezing, Intoxication, Screaming,* and *Threatening* topics are related to audio-based surveillance for security purposes in noisy public places. The *Coughing* and *Sneezing* topics are related to the monitoring of everyday activity—in terms of, e.g., personal health—in common, relatively quiet environments such as home or office [28]. All tasks offer a two-class classification problem, i.e., they have a target class, e.g., *freezing* and a 'normal speech' class which contains audio samples that are not affected by the target class.

12.5.2 VEHICLES *Database*

Classification of various acoustic scenes and environmental sounds is gaining popularity in recent years [37, 47]. *Detection and Classification of Acoustic Scenes and Events* challenges provide a very good benchmark for this recognition task [36, 37]. The VEHICLES corpus contains YouTube recordings from acoustic environments in which an audio activity related to a transportation system is present. It offers a 10-class classification problem and contains 1 158 audio instances. Such a corpus can be used, e.g., for devices requiring environmental awareness [9, 59].

12.5.3 BALL SPORTS *Database*

The BALL SPORTS database contains 1 269 YouTube clips that are recorded in a statium or a place in which there is an audio activity related to a ball sport. This corpus offers a 9-class classification problem and can be considered as an extension for the acoustic scene data sets.

Table 12.1 Specifications of classes available in the data sets. l_{total}: the total length of the clips; l_{min} and l_{max}: the minimum and maximum lengths of the recording; σ: standard deviation; n: the number of all recordings in each class

Tasks	Train				Evaluation			
	l_{total} (s)	l_{min}/l_{max} (s)	σ (s)	n	l_{total} (s)	l_{min}/l_{max} (s)	σ (s)	n
SPEECH AND VOCALISATION Database								
Coughing	5658	0.5/28.8	3.5	2088	3834	0.5/23.2	2.7	1571
Freezing	4554	2.0/29.4	5.8	614	1344	2.0/28.6	5.9	171
Intoxication	8382	2.0/29.9	6.5	1069	1002	2.0/24.8	5.3	152
Screaming	3216	2.0/29.9	7.6	375	1320	2.1/29.9	5.5	189
Sneezing	402	0.5/8.0	1.3	238	552	0.5/9.3	1.4	291
Threatening	6396	2.0/29.8	7.4	652	2748	2.0/29.2	5.2	441
All	28608	0.5/29.9	5.4	5036	10800	0.5/29.9	3.1	2815
VEHICLES Database								
Bicycle	1254	6.3/28.7	5.0	75	441	7.6/27.6	5.0	29
Bus	640	6.2/20.8	3.2	45	324	8.7/19.6	2.8	22
Car	926	8.2/24.5	3.5	63	234	11.3/21.2	2.7	15
Metro	634	5.1/10.0	1.1	101	349	5.1/10.9	1.2	57
Motorcycle	1081	7.4/28.7	4.5	69	504	6.2/28.7	6.1	32
S-Bahn	2449	6.8/34.2	6.3	113	1253	8.9/33.9	6.2	58
Scooter	1152	5.5/29.4	4.8	67	515	6.9/27.1	5.7	33
Train	562	5.3/36.0	7.3	66	501	5.3/9.7	0.7	84
Tram	694	5.3/11.1	0.9	108	271	5.6/11.3	1.1	42
Truck	1583	5.1/505.4	72.7	56	289	7.0/16.5	2.5	23
All	10975	5.1/505.4	21.1	763	4681	5.1/33.9	6.9	395
BALL SPORTS Database								
Am. Football	377	5.1/9.4	0.9	64	217	5.1/7.0	0.5	38
Badminton	742	5.4/16.1	2.1	91	1698	5.0/21.9	3.0	227
Basketball	1053	9.7/25.1	3.6	63	754	11.6/260.2	43.9	31
Handball	1118	7.5/27.8	4.3	72	560	9.7/33.9	5.6	33
Soccer	1222	7.2/26.4	3.5	74	466	9.3/24.2	3.6	30
Table Tennis	1098	8.2/25.2	4.3	71	483	7.7/28.3	4.5	32
Tennis	1221	5.2/30.4	6.4	72	539	6.2/29.0	5.2	33
Volleyball	1440	10.5/141.0	14.3	80	557	9.7/52.3	7.3	31
Water Polo	1788	5.1/23.4	3.4	198	229	5.3/12.4	1.8	29
All	10058	5.1/141.0	7.2	785	5503	5.0/260.2	12.8	484

12.6 Machine Learning Approaches

Big data analytics and machine learning are two spotlights of data science. The major focus of machine learning approaches is the representation of the training data and generalisation of the learnt patterns for use on unseen (evaluation) data.

Good quality data with meaningful contents will lead to good representations while noisy data presents a big challenge for the machine learner and can lead to suboptimal prediction accuracies. Especially for training a classifier from YouTube data which is mostly recorded under noisy and real-world conditions, sophisticated machine learning algorithms are needed. For learning robust representations from the YouTube data we apply three state-of-the-art unsupervised feature extraction systems: (1) AUDEEP, a recurrent sequence-to-sequence autoencoder (cf. Sect. 12.6.1), (2) a bag-of-deep-features (BoDF) approach (cf. Sect. 12.6.2), which quantises representations extracted via pre-trained image CNNs descriptors, and (3) a crossmodal-bag-of-words system for representing media of variable length in a fixed-dimensional feature vector (cf. Sect. 12.6.3). All algorithms can be applied for processing of the segmented audio and videos from the collected YouTube clips. However, for the classification experiments, we use the audio modality, which—in contrast to video modality—compromises most relevant features for the SPEECH AND VOCALISATION database [5] and it is the more challenging modality to recognise the target classes in the VEHICLES and BALL SPORTS databases.

In order to predict the class labels for the audio instances in each of the corpora, we train a linear support vector machine (SVM) classifier. The evaluation metric is unweighted average recall (UAR), which is defined as the sum of the class-wise recalls for each task divided by the number of classes. We use this measure, as our corpora have a slightly imbalanced distribution of instances (cf. Table 12.1). For the classifier, we use the open-source linear SVM implementation provided in the *scikit-learn* machine learning library [45]. For the extracted features except for the BOAW features, we do not apply standardisation, i.e., subtracting the mean and dividing by the standard deviation as they have been found to negatively impact classifier performance.

12.6.1 Recurrent Sequence-to-Sequence Autoencoders

In this approach, feature sets are obtained through unsupervised representation learning with recurrent sequence-to-sequence autoencoders, using the AUDEEP toolkit[4] [19]. The recurrent sequence-to-sequence autoencoders which are employed by AUDEEP, in particular, explicitly model the inherently sequential nature of audio with recurrent neural networks (RNNs) within the encoder and decoder networks [3, 19]. A high-level overview of this approach is given in Fig. 12.3. In the AUDEEP approach, Mel-scale spectrograms are first extracted from the raw waveforms in a data set (cf. Fig. 12.3a). In order to eliminate some background noise available in the YouTube data, power levels are clipped below two given thresholds (-30 and -60 dB) in these spectrograms, which results in two separate sets of spectrograms per data set. Subsequently, a distinct recurrent sequence to sequence autoencoder is trained on each of these sets of spectrograms in an unsupervised way, i.e., without any

---

[4]https://github.com/auDeep/auDeep.

Fig. 12.3 A high-level overview of the sequence to sequence representation learning system. The autoencoder training is completely unsupervised. A detailed description of the procedure is given in Sect. 12.6.1

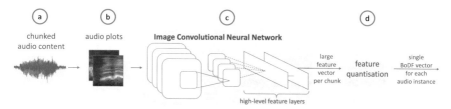

Fig. 12.4 Illustration of the BoDF system for quantising DEEP SPECTRUM features. A detailed account of the procedure is given in Sect. 12.6.2. Figure adapted from [6]

label information. The learnt representations of a spectrogram are then extracted as feature vectors for the corresponding instance (cf. Fig. 12.3b). Afterwards, these feature vectors are concatenated to obtain the final feature vector (cf. Fig. 12.3c). Finally, we train a classifier (cf. Fig. 12.3d) on the feature vectors to predict the labels of the recordings. For further details on AUDEEP, the interested reader is referred to [3, 19].

12.6.2 Bag-of-Deep-Features

An overview of our bag-of-deep-features (BoDF) system [6] is given in Fig. 12.4. First, we segment each YouTube clip into multiple chunks with equal length (cf. Fig. 12.4a). We then generate DEEP SPECTRUM features [4][5] by forwarding the visual representations of the audio recordings, in particular Mel-spectrograms (cf. Fig. 12.4b) through pre-trained image convolutional neural networks (CNNs) such as AlexNet [32], VGG16 and VGG19 [54], and GoogLeNet [57] (cf. Fig. 12.4c). Afterwards, by using openXBOW [52],[6] we quantise all extracted DEEP SPECTRUM features from the chunked audio content in order to create a single BoDF vector for each YouTube clip (cf. Fig. 12.4d). Finally, we train a SVM classifier to predict the class labels based on the quantised feature sets. We apply SVMs as they are able to deal with imbalanced class ratio.

---

[5]https://github.com/DeepSpectrum/DeepSpectrum.

[6]https://github.com/openXBOW/openXBOW.

Fig. 12.5 A sample
Mel-spectrogram taken from
the *sneezing* database using
two different colour maps,
magma and *viridis*

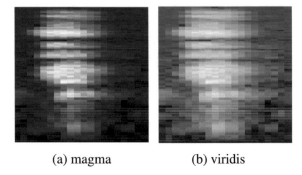

(a) magma   (b) viridis

12.6.2.1   Creating Mel-Spectrograms

The Mel-Spectrograms are computed with a window size of 2 048 and an overlap of
1 024 from the log-magnitude spectrum by dimensionality reduction using a Mel-
filter. We use 128 filter banks equally spaced on the Mel-scale defined in Eq. (12.2):

$$f_{mel} = 2595 \cdot \log_{10}\left(1 + \frac{f_{Hz}}{700}\right), \tag{12.2}$$

where f_{mel} is the resulting frequency on the mel-scale calculated in mels and f_{Hz} is the
normal frequency measured in Hz. The Mel-scale is based on the frequency response
of the human ear, which has higher resolution at lower frequencies. For the Mel-
spectrogram plots, we use two different colour mappings: *viridis*, and *magma*. These
colour mappings are shown in Fig. 12.5. It has been demonstrated that pre-trained
CNNs performing strongly different on spectra with various colours maps [2, 4].
In Fig. 12.6, we highlight the audio similarities and differences that potentially exist
between different classes in the VOCALISATION AND SPEECH corpus by showing an
example Mel-spectrogram from each target class and from a 'normal speech' which
was not affected by the target classes.

12.6.2.2   Deep Feature Extractors

We use four different image CNN architectures of pre-trained networks to extract
deep representations from the Mel-spectrograms described in Sect. 12.6.2.1. All four
networks have been trained for the task of object recognition on the large ImageNet [16] corpus which provides more than 1 million images labelled with over 1 000
object classes. The architectural differences and similarities of AlexNet, VGG16,
and VGG19 are given in Table 12.2. The GoogLeNet's architecture is depicted
in Fig. 12.7.

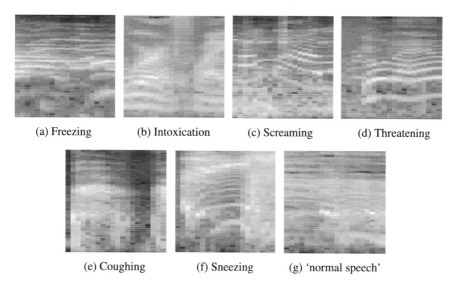

(a) Freezing   (b) Intoxication   (c) Screaming   (d) Threatening

(e) Coughing   (f) Sneezing   (g) 'normal speech'

Fig. 12.6 Example Mel-spectrograms (**a–f**) extracted from the target classes contained in the VOCALISATION AND SPEECH databases. The last example Mel-spectrogram is from an audio sample considered to be a 'normal speech' utterance which was not affected by the target classes. The range of the time (horizontal) and frequency (vertical) axes are [0–0.45] s and [0–4 096] Hz

AlexNet

AlexNet has 5 convolutional layers, in cascade with 3 fully connected layers [32]. The feature maps generated by the first, the second, and the third convolutional layers are downsampled using an overlapping maxpooling operation. The network makes use of a rectified linear unit (ReLU) non-linearity, as this non-saturating function regularises the training, while also improving the network's generalisation capabilities. The 4 096 activations of the network's seventh layer (commonly denoted as *fc7*) are used as features in our experiments.

VGG16/VGG19

While the filter sizes vary across the layers in AlexNet, both VGG16 and VGG19 utilise a constant 3×3-sized receptive field in all of their convolutional layers [54]. The two architectures consist of 2 additional maxpooling layers in comparison with AlexNet, and have deeper fully connected layers in cascade. Both VGG16 and VGG19 employ ReLUs for response normalisation, just like AlexNet. We employ the 4 096 activations of the second fully connected layer as the feature vectors.

Table 12.2 Overview of the architectural similarities and differences between the three of the NNs used for the extraction of DEEP SPECTRUM features, AlexNet, VGG16, and VGG19. *conv* denotes convolutional layers and *ch* stands for channels. The table is adapted from [4]

AlexNet	VGG16	VGG19
input: RGB image		
1×conv	2×conv	
size: 11; ch: 96; stride: 4	size: 3; ch: 64; stride: 1	
maxpooling		
1×conv	2×conv	
size: 5; ch: 256	size: 3; ch: 128	
maxpooling		
1×conv	3×conv	4×conv
size: 3; ch: 384	size: 3; ch: 256	size: 3; ch: 256
	maxpooling	
1×conv	3×conv	4×conv
size: 3; ch: 384	size: 3; ch: 512	size: 3; ch: 512
	maxpooling	
1×conv	3×conv	4×conv
size: 3; ch: 256	size: 3; ch: 512	size: 3; ch: 512
maxpooling		
fully connected *fc6*, 4 096 neurons		
fully connected *fc7*, 4 096 neurons		
fully connected, 1 000 neurons		
output: soft-max of probabilities for 1 000 object classes		

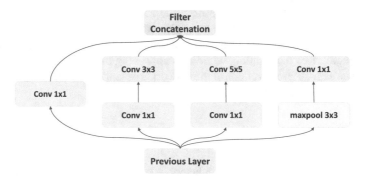

Fig. 12.7 An inception module applied in the GoogLeNet architecture. To reduce the dimensionality small 1×1 convolutions are used. Filters of various path sizes are concatenated to combine information found at different scales. Figure adapted from [6]

GoogLeNet

Instead of the most typical layers that the previously discussed three networks use, GoogleNet employs the so-called inception modules (cf. Fig. 12.7) in succession. The module consists of a set of parallel convolution layers and a maxpooling layer, outputs of all of which get concatenated to produce a single output. This module thus aggregates multi-level features from every input on different scales. The activations of the last pooling layer are employed as the features.

12.6.2.3 DEEP SPECTRUM Features

We use a state-of-the-art system based on the introduced CNN image descriptors (cf. Sect. 12.6.2.2). The basic system architecture (before quantisation) is shown in the left part of Fig. 12.4. We extract the DEEP SPECTRUM features as follows. First, Mel-spectrograms are created from the chunked (each 0.5 s) audio recordings using the audio and music analysis library *librosa* [35]. We choose Mel-spectrograms since they have been successfully applied for a wide range of audio recognition tasks [7, 44, 50, 61]. The Mel-spectrograms are then transformed to images by creating colour mapped plots. The second step consists of feeding the created plots to the pre-trained CNNs and extracting the activations of a specific layer from each CNN as large feature vectors. These features are a high-level representation of the plots generated from low-level audio features.

12.6.2.4 Bag-of-Deep-Features

The last important component of our BoDF system is the feature quantisation block (cf. Fig. 12.4d). In this stage, we bag the extracted DEEP SPECTRUM features which we described in Sect. 12.6.2.3 to analyse the denoising effect of the deep feature quantisation. In order to achieve this, we generate a fixed length histogram representation of each audio recording. This is done by first identifying a set of 'deep audio words' from some given training data, and then quantising the original feature space, with respect to the generated codebook, to form the histogram representation. The histogram shows the frequency of each identified deep audio word in a given audio instance [42, 43, 51].

We normalise the features to [0, 1] and random sample a codebook with fixed size from the training partition. Afterwards, each input feature vector (from the training and evaluation partitions) is applied a fixed number of its closest vectors from the codebook. We then apply logarithmic term-frequency weighting to the generated histograms.

The size of the codebook and the number of assigned codebook words (cw) are optimised with $size \in \{50, 100, 200, 500, 1\,000\}$, $cw \in \{25, 50, 100, 200, 500\}$ and evaluated on the evaluation partition.

12.6.3 Crossmodal-Bag-of-Words

The BoDF representation is a special case of the well-known *bag-of-words*. It is an efficient and convenient way to represent media of variable length in a fixed-dimensional vector and has been generalised to arbitrary domains of symbolic or numeric features. In the original approach, developed in the field of *natural language processing (NLP)*, a sparse histrogram vector is generated, counting the frequencies of each word from a predefined dictionary occuring in a given document and so representing a document of arbitrary length in a fixed-dimensional vector [30]. The size of this vector directly corresponds to the dictionary size and can be used with any machine learning model.

In the audio and video domains, the method is known as *bag-of-audio-words (BoAW)* or *bag-of-visual-words*, respectively. Here, low-level descriptors (LLDs) describing temporal and/or local content of the audio, images, or video are extracted first. As acoustic LLDs, very often, short-time spectral band energy, or speech-related features, e.g., *Mel-Frequency Cepstral Coefficients (MFCCs)* or pitch [17], are used. As visual features, as an example, *histogram of oriented gradients* [46] are employed or combined detection and description algorithms such as *Scale-Invariant Feature Transform (SIFT)* [60]. If the images or video contain faces, facial landmark trackers are very often used to provide relevant information on, e.g., the emotional expression of the subject's face [51]. Next, the LLDs are quantised based on a codebook of templates ('audio/visual words') previously learnt from a given amount of training samples. Usually, the codebook is generated with a standard clustering method, such as *k-means* or *k-means++* [8]. The codebook size is an important parameter in this context, as it controls the amount of sparsity [53] and needs to be adapted to the actual task and the employed LLDs. Each LLD vector is then assigned to the template in the codebook which is closest in terms of the Euclidean distance.

Finally, normalisation techniques introduced in NLP are usually exploited, such as logarithmic term-frequency weighting of the histogram or *inverse document-frequency weighting*. These techniques are all integrated in the *crossmodal bag-of-words toolkit* OPENXBOW [51], illustrated in Fig. 12.8. In the shown processing chain, four different modalities are combined: video, audio, text, and metadata. While the video and audio modality provide *numeric* descriptors each, the text and metadata modalities are *symbolic* and processed by methods from NLP, e.g., *stop words* to exclude words from the dictionary that are very rare or function words, such as articles and conjunctions. If the text, i.e., the transcription of the speech in the audio is not given directly, it can be transcribed using *automatic speech recognition* software. Metadata can be, e.g., tags given by the users of a social media platform.

The bags-of-words from all modalities are fused to create the final feature vector, which is of constant length, independent from the size or length of the input media. Generally, fusing different modalities usually improves the accuracy in multimedia retrieval tasks [51], but also in other kinds of multimodal analysis, e.g., health care [31].

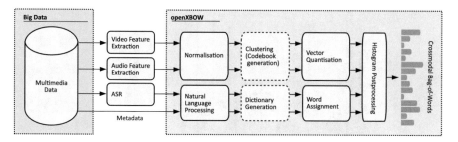

Fig. 12.8 Overview of a crossmodal bag-of-words (XBOW) processing chain using the toolkit OPENXBOW. Both *numeric* (e.g., video or audio features) and *symbolic* features (e.g., transcriptions of the speech present or metadata) can be used as inputs to generate a fixed dimensional XBOW feature vector for input multimedia instances of arbitrary length. ASR: Automatic speech recognition (speech-to-text)

For the experiments (cf. Sect. 12.7), MFCCs and their temporal derivatives of 1st (deltas) and 2nd order (double-deltas) were extracted from the audio. As mentioned above, MFCCs are one of the most commonly used acoustic LLDs. They are computed in a similar way as the above mention Mel-Spectrogram features (cf. Sect. 12.6.2.1), but with a final decorrelation using *Discrete Cosine Transform*, ending up in a vector of a dimension of 13. For each, the coefficients, the deltas, and the double-deltas, a codebook was trained on the training partition, where its size was optimised between 100 and 1 000. The BoAW representations of the three descriptor types were computed and concatenated by OPENXBOW and a logarithmic term-frequency weighting was applied to compress their dynamic range.

12.7 Results

We have conducted an extensive series of experiments to evaluate the quality of the collected data (cf. Sect. 12.5) and analyse the performance of our machine learning approaches for feature extraction (cf. Sect. 12.6).

Table 12.3 shows the feasibility of applied machine learning approaches for all recognition tasks. The results imply that the collected and annotated data sets contain relevant information for each target class. For the paralinguistic task we have compared the results obtained by the BoDF system [6] and AUDEEP with the baseline results (cf. BOAW column in the Table 12.3) introduced in [5]. BOAW, which is trained using a 39-dimensional MFCC feature representation, achieves the best classification results for the *Coughing* (97.6 % UAR), *Sneezing* (79.8 % UAR), *Vehicles*, and *Ball Sports* tasks while BoDF shows the best performance for the acoustic tasks *Freezing* (72.9 % UAR), *Intoxication* (73.6 % UAR), *Screaming* (98.5 % UAR), and *Threatening* (77.3 % UAR). For the non-binary classification tasks GOOGLENET performed weaker than other CNN descriptors. We assume that representations extracted by GOOGLENET do not contain enough discriminative information for the classes

Table 12.3 Classification results of each paralinguistic task from the baseline paper [5] by Bag-of-Audio-Words (BoAW) compared with our results from the bag-of-deep-features (BoDF) and AUDEEP. The best result for each corpus is highlighted with a light grey shading. The chance level for each task in the SPEECH AND VOCALISATION database is 50.0 % UAR, for the VEHICLES database 10.0 % UAR, and for the BALL SPORTS database 11.1 % UAR

% UAR	MFCCs BoAW	AlexNet BoDF	GG16 BoDF	VGG19 BoDF	GoogLeNet BoDF	auDeep
Speech and Vocalisation Database (binary-class classification)						
Coughing	**97.6**	95.3	95.3	95.2	92.0	91.5
Freezing	65.6	70.4	**72.9**	69.1	71.6	64.5
Intoxication	66.7	61.9	64.7	71.3	**73.6**	67.1
Screaming	94.0	**98.5**	96.7	98.2	94.3	94.0
Sneezing	**79.8**	74.6	74.9	79.4	71.8	73.6
Threatening	67.0	76.4	73.9	70.3	**77.3**	66.8
Vehicles Database (10-class classification)						
Vehicles	**59.6**	50.0	47.9	51.9	47.8	53.5
Ball Sports Database (9-class classification)						
Ball Sports	**85.1**	70.0	72.0	71.3	69.0	72.4

in both VEHICLES and BALL SPORTS databases. The results obtained with AUDEEP, which showed its strength in a range of acoustic and environmental sound classification tasks [3, 19] are comparable with the BoDF results but not as strong as the BoAW results. AUDEEP contains a wide range of adjustable hyperparameters that prohibit an exhaustive exploration of the parameter space. By fine-tuning the autoencoder hyperparameters and testing more parameters for the pre-processing step (extraction of the mel-spectrograms), it should be possible to learn stronger representations from the databases and achieve higher classification results.

12.8   Closing Remarks

Our proposed system in Sect. 12.2 is highly efficient at rapidly constructing new databases by exploiting the data available on YouTube. It is a unique combination of large-scale multimedia data extraction (cf. Sect. 12.3) and annotation systems (cf. Sect. 12.4) with state-of-the-art machine learning approaches (cf. Sect. 12.6). With our system, we have searched thousands of videos on YouTube and collected recordings from a range of acoustic sounds and human speech and vocalisation types (cf. Sect. 12.5). We achieved strong classification results (cf. Sect. 12.7) on the collected data by applying two deep learning approaches and a crossmodal bag-of-words system. All three toolkits are open source and publicly available (cf. footnotes in Sect. 12.6 for the corresponding GitHub links).

The inherent large *variability* as well as the sheer *volume* of online multimedia data will further enable us to develop robust systems for real-life environments by ensuring that the system evaluations are not too optimistic and that the systems will be capable of working under realistic, noisy, and unpredictable conditions.

Potential future work includes extending the current system fusion to operate on a range of social media platforms, such as *Wikimedia*[7] which offers a collection of more than 45 millions freely usable media files. The efficacy of the classification system may be increased by extracting and combining both audio and visual information from YouTube clips. It is also interesting to explore the benefits of fine-tuning the pre-trained CNNs on larger real-world databases like *AudioSet* [20] or data sets for Acoustic Scene Classification and Sound Event Detection challenges [36–38]. Finally, the autonomy of the system can be further increased by exploring natural language processing and deep zero-resource learning techniques [48, 55] to enable self-gathering and self-labelling of truly large in-the-wild datasets, setting the stage for the next generation of intelligent big data analytics systems.

Acknowledgements This work was supported by the European Unions's seventh Framework Programme under grant agreement No. 338164 (ERC StG iHEARu).

---

[7]https://commons.wikimedia.org/.

References

1. (2018) Hours of video uploaded to YouTube every minute as of July 2015. https://www.statista.com/topics/2019/youtube. Accessed 5 Mar 2018
2. Amiriparian, S., Cummins, N., Ottl, S., Gerczuk, M., Schuller, B.: Sentiment analysis using image-based deep spectrum features. In: Proceedings of the Biannual Conference on Affective Computing and Intelligent Interaction (ACII), San Antonio, TX, pp. 26–29 (2017)
3. Amiriparian, S., Freitag, M., Cummins, N., Schuller, B.: Sequence to sequence autoencoders for unsupervised representation learning from audio. In: Proceedings of the Detection and Classification of Acoustic Scenes and Events Challenge Workshop (DCASE), Munich, Germany, pp. 17–21 (2017)
4. Amiriparian, S., Gerczuk, M., Ottl, S., Cummins, N., Freitag, M., Pugachevskiy, S., Schuller, B.: Snore sound classification using image-based deep spectrum features. proceedings of INTER-SPEECH, pp. 3512–3516. Stockholm, Sweden (2017)
5. Amiriparian, S., Pugachevskiy, S., Cummins, N., Hantke, S., Pohjalainen, J., Keren, G., Schuller, B.: CAST a database: Rapid targeted large-scale big data acquisition via small-world modelling of social media platforms. In: Proceedings of the Biannual Conference on Affective Computing and Intelligent Interaction (ACII), San Antonio, TX, pp. 340–345 (2017)
6. Amiriparian, S., Gerczuk, M., Ottl, S., Cummins, N., Pugachevskiy, S., Schuller, B.: Bag-of-Deep-Features: Noise-Robust Deep Feature Representations for Audio Analysis. In: Proceedings of 31st International Joint Conference on Neural Networks (IJCNN), IEEE, IEEE, Rio de Janeiro, Brazil, pp. 2419–2425 (2018)
7. Amiriparian, S., Cummins, N., Gerczuk, M., Pugachevskiy, S., Ottl, S., Schuller, B.: Are you playing a shooter again?!" deep representation learning for audio-based video game genre recognition. In: IEEE Transactions on Computational Intelligence and AI in Games PP, submitted, 10 p. (2018)
8. Arthur, D., Vassilvitskii, S.: k-means++: The advantages of careful seeding. In: Proceedings of 18th Annual ACM-SIAM Symposium on Discrete Algorithms, pp. 1027–1035. SIAM, New Orleans, USA (2007)
9. Battaglino, D., Lepauloux, L., Pilati, L., Evans, N.: Acoustic context recognition using local binary pattern codebooks. In: 2015 IEEE Workshop on Applications of Signal Processing to Audio and Acoustics (WASPAA), pp. 1–5. IEEE (2015)
10. Brabham, D.C.: Crowdsourcing. Wiley Online Library (2013)
11. Braun, U., Muldoon, S.F., Bassett, D.S.: On Human Brain Networks in Health and Disease. eLS (2015)
12. Burmania, A., Abdelwahab M, Busso C (2016) Tradeoff between quality and quantity of emotional annotations to characterize expressive behaviors. In: Proceedings of ICASSP, Shanghai, China, pp. 5190–5194
13. Burmania, A., Parthasarathy, S., Busso, C.: Increasing the reliability of crowdsourcing evaluations using online quality assessment. IEEE Trans. Affect. Comput. 7(4), 374–388 (2016b)
14. Costa, J., Silva, C., Antunes, M., Ribeiro, B.: On using crowdsourcing and active learning to improve classification performance. In: Proceedings of International Conference on Intelligent Systems Design and Applications, pp. 469–474. IEEE, Cordoba, Spain (2011)
15. Covington, P., Adams, J., Sargin, E.: Deep neural networks for youtube recommendations. In: Proceedings of the ACM Conference on Recommender Systems, pp. 191–198. ACM, New York, USA (2016)
16. Deng, J., Dong, W., Socher, R., Li, L.J., Li, K., Fei-Fei, L.: Imagenet: A large-scale hierarchical image database. In: IEEE Conference on Computer Vision and Pattern Recognition, 2009. CVPR 2009, pp. 248–255. IEEE (2009)
17. Eyben, F.: Real-time Speech and Music Classification by Large Audio Feature Space Extraction. Springer (2015)
18. Eyben, F., Wöllmer, M., Schuller, B.: openSMILE—the munich versatile and fast open-source audio feature extractor. In: Proceedings of ACM International Conference on Multimedia (ACMMM), Florence, Italy, pp. 1459–1462 (2010)

19. Freitag, M., Amiriparian, S., Pugachevskiy, S., Cummins, N., Schuller, B.: audeep: Unsupervised learning of representations from audio with deep recurrent neural networks. J. Mach. Learn. Res. **18**(1), 6340–6344 (2017)
20. Gemmeke, J.F., Ellis, D.P.W., Freedman, D., Jansen, A., Lawrence, W., Moore, R.C., Plakal, M., Ritter, M.: Audio set: An ontology and human-labeled dataset for audio events. In: 2017 IEEE International Conference on Acoustics, Speech and Signal Processing (ICASSP), pp. 776–780 (2017)
21. Han, W., Coutinho, E., Ruan, H., Li, H., Schuller, B., Yu, X., Zhu, X.: Semi-Supervised active learning for sound classification in hybrid learning environments. PLoS One **11**(9) (2016)
22. Hantke, S., Eyben, F., Appel, T., et al.: iHEARu-PLAY: introducing a game for crowdsourced data collection for affective computing. In: Proceedings of the 1st International Workshop on Automatic Sentiment Analysis in the Wild (WASA 2015) Held Conjunction with 6th Biannual Conference on Affective Computing and Intelligent Interaction (ACII 2015), pp. 891–897. IEEE, Xi'an, PR China (2015)
23. Hantke, S., Marchi, E., Schuller, B.: Introducing the weighted trustability evaluator for crowdsourcing exemplified by speaker likability classification. In: Proceedings of the International Conference on Language Resources and Evaluation, Portoroz, Slovenia, pp. 2156–2161 (2016)
24. Hantke, S., Zhang, Z., Schuller, B.,: Towards intelligent crowdsourcing for audio data annotation: integrating active learning in the real world. In: Proceedings of Interspeech 18th Annual Conference of the International Speech Communication Association, pp. 3951–3955. ISCA, Stockholm, Sweden (2017)
25. Hantke, S., Abstreiter, A., Cummins, N., Schuller, B.: Trustability-based Dynamic Active Learning for Crowdsourced Labelling of Emotional Audio Data. IEEE Access **6**, p. 12 (2018). to appear
26. Hantke, S., Appel, T., Schuller, B.: The inclusion of gamification solutions to enhance user enjoyment on crowdsourcing platforms. In: Proceedings of the 1st Asian Conference on Affective Computing and Intelligent Interaction (ACII Asia 2018), p. 6. IEEE, Beijing, P.R. China (2018)
27. Hantke, S., Stemp, C., Schuller, B.: Annotator Trustability-based Cooperative Learning Solutions for Intelligent Audio Analysis. In: Proceedings Interspeech 2018, 19th Annual Conference of the International Speech Communication Association, p. 5. ISCA, Hyderabad, India (2018). to appear
28. Härmä, A., McKinney, M.F., Skowronek. J.: Automatic surveillance of the acoustic activity in our living environment. In: Proceedings of the International Conference on Multimedia and Expo. IEEE, Amsterdam, The Netherlands (2005). no pagination
29. Hsueh, P., Melville, P., Sindhwani, V.: Data quality from crowdsourcing: a study of annotation selection criteria. In: Proceedings of the NAACL HLT Workshop on Active Learning for Natural Language Processing, pp. 27–35. ACL, Boulder, USA (2009)
30. Joachims, T.: Text categorization with support vector machines: learning with many relevant features. European Conference on Machine Learning, pp. 137–142. Springer, Chemnitz, Germany (1998)
31. Joshi, J., Goecke, R., Alghowinem, S., Dhall, A., Wagner, M., Epps, J., Parker, G., Breakspear, M.: Multimodal assistive technologies for depression diagnosis and monitoring. J. MultiModal User Interfaces **7**(3), 217–228 (2013)
32. Krizhevsky, A., Sutskever, I., Hinton, G.E.: ImageNet classification with deep convolutional neural networks. In: Advances in Neural Information Processing Systems, vol. 25, pp. 1097–1105. Curran Associates, Inc. (2012)
33. Lakhani, K., Garvin, D., Lonstein, E.: Topcoder (a): Developing Software Through Crowdsourcing (2010)
34. McCallumzy, A.K., Nigamy K.: Employing em and pool-based active learning for text classification. In: Proceedings of Conference on Machine Learning, Madison, Wisconsin, pp. 359–367 (1998)
35. McFee, B., McVicar, M., Nieto, O., Balke, S., Thome, C., Liang, D., Battenberg, E., Moore, J., Bittner, R. Yamamoto, R., Ellis, D., Stoter, F.R., Repetto, D., Waloschek, S., Carr, C., Kranzler, S., Choi, K., Viktorin, P., Santos, J.F., Holovaty, A., Pimenta, W., Lee, H.: librosa 0.5.0 (2017)

36. Mesaros, A., Heittola, T., Virtanen, T.: TUT database for acoustic scene classification and sound event detection. 24th European Signal Processing Conference (EUSIPCO 2016), pp. 1128–1132. IEEE, Budapest, Hungary (2016)
37. Mesaros, A., Heittola, T., Diment, A., Elizalde, B., Shah, A., Vincent, E., Raj, B., Virtanen, T.: DCASE 2017 challenge setup: tasks, datasets and baseline system. In: Proceedings of the Detection and Classification of Acoustic Scenes and Events Workshop (DCASE2017). Munich, Germany (2017)
38. Mesaros, A., Heittola, T., Benetos, E., Foster, P., Lagrange, M., Virtanen, T., Plumbley, M.D.: Detection and classification of acoustic scenes and events: outcome of the DCASE 2016 challenge. IEEE/ACM Transactions on Audio, Speech, and Language Processing **26**(2), 379–393 (2018)
39. Morris, R., McDuff, D., Calvo, R.: Crowdsourcing techniques for affective computing. In: Calvo, R.A., D'Mello, S., Gratch, J., Kappas, A. (eds.) Handbook of Affective Computing, pp. 384–394. Oxford University Press, Oxford Library of Psychology (2015)
40. Najafabadi, M., Villanustre, F., Khoshgoftaar, T.M., Seliya, N., Wald, R., Muharemagic, E.: Deep learning applications and challenges in big data analytics. J. Big Data **2**(1), 1 (2015)
41. Newman, M., Watts, D., Strogatz, S.: Random graph models of social networks. Proc. Natl. Acad. Sci. **99**(suppl 1), 2566–2572 (2002)
42. Pancoast, S., Akbacak, M.: Bag-of-audio-words approach for multimedia event classification. In: Proceedings of Interspeech: 13th Annual Conference of the International Speech Communication Association, pp. 2105–2108. ISCA, Portland, OR, USA (2012)
43. Pancoast, S., Akbacak, M.: Softening quantization in bag-of-audio-words. In: 2014 IEEE International Conference on Acoustics, Speech and Signal Processing (ICASSP), pp. 1370–1374. IEEE (2014)
44. Panwar, S., Das, A., Roopaei, M., Rad, P.: A deep learning approach for mapping music genres. In: 2017 12th System of Systems Engineering Conference (SoSE), pp. 1–5. IEEE (2017)
45. Pedregosa, F., Varoquaux, G., Gramfort, A., Michel, V., Thirion, B., Grisel, O., Blondel, M., Prettenhofer, P., Weiss, R., Dubourg, V., Vanderplas, J., Passos, A., Cournapeau, D., Brucher, M., Perrot, M., Duchesnay, E.: Scikit-learn: machine learning in python. J. Mach. Learn. Res. **12**, 2825–2830 (2011)
46. Peng, X., Wang, L., Wang, X., Qiao, Y.: Bag of visual words and fusion methods for action recognition: comprehensive study and good practice. Comput. Vis. Image Underst. **150**, 109–125 (2016)
47. Piczak, K.J.: ESC: Dataset for environmental sound classification. In: Proceedings of the 23rd ACM International Conference on Multimedia, pp. 1015–1018. ACM (2015)
48. Romera-Paredes, B., Torr, P.: An embarrassingly simple approach to zero-shot learning. In: International Conference on Machine Learning, pp. 2152–2161 (2015)
49. Rubinov, M., Knock, S.A., Stam, C.J., Micheloyannis, S., Harris, A.W., Williams, L.M., Breakspear, M.: Small-world properties of nonlinear brain activity in schizophrenia. Hum. Brain Mapp. **30**(2), 403–416 (2009)
50. Salamon, J., Bello, J.P.: Unsupervised feature learning for urban sound classification. In: 2015 IEEE International Conference on Acoustics, Speech and Signal Processing (ICASSP), pp. 171–175. IEEE (2015)
51. Schmitt, M., Schuller, B.: openXBOW—introducing the passau open-source crossmodal bag-of-words toolkit. J. Mach. Learn. Res. **18**(96), 1–5 (2017a)
52. Schmitt, M., Schuller, B.: openxbow introducing the passau open-source crossmodal bag-of-words toolkit. J. Mach. Learn. Res. **18**(96), 1–5 (2017b)
53. Schmitt, M., Ringeval, F., Schuller, B.: At the border of acoustics and linguistics: bag-of-audio-words for the recognition of emotions in speech. In: Proceedings of Interspeech, San Francisco, CA, pp. 495–499 (2016)
54. Simonyan, K., Zisserman, A.: Very deep convolutional networks for large-scale image recognition. Comput. Res. Repos. (CoRR) (2014). arXiv:1409.1556
55. Socher, R., Ganjoo, M., Manning, C.D., Ng, A.: Zero-shot learning through cross-modal transfer. In: Advances in Neural Information Processing Systems, pp. 935–943 (2013)

56. Strogatz, S.: Exploring complex networks. Nature **410**(6825), 268–276 (2001)
57. Szegedy, C., Liu, W., Jia, Y., Sermanet, P., Reed, S., Anguelov, D., Erhan, D., Vanhoucke, V., Rabinovich, A.: Going deeper with convolutions. In: Proceedings of the IEEE Conference on Computer Vision and Pattern Recognition, pp. 1–9 (2015). IEEE, Boston, MA, USA (2015)
58. Tarasov, A., Delany, S.J., Cullen, C.: Using crowdsourcing for labelling emotional speech assets. In: Proceedings of the W3C Workshop on Emotion Markup Language (EmotionML), pp. 1–5. Springer, Paris, France (2010)
59. Tchorz, J., Wollermann, S., Husstedt, H.: Classification of environmental sounds for future hearing aid applications. In: Proceedings of the 28th Conference on Electronic Speech Signal Processing (ESSV 2017), Saarbrücken, Germany, pp. 294–299 (2017)
60. Tirilly, P., Claveau, V., Gros, P.: Language modeling for bag-of-visual words image categorization. In: Proceedings of the International Conference on Content-based Image and Video Retrieval, pp. 249–258. ACM, Niagara Falls, Canada (2008)
61. Valenti, M., Diment, A., Parascandolo, G., Squartini, S., Virtanen, T.: Dcase 2016 acoustic scene classification using convolutional neural networks. In: Proceedings of the Detection and Classification of Acoustic Scenes and Events 2016 Workshop (DCASE2016), pp. 95–99 (2016)
62. Von Ahn, L.: Games with a purpose. Computer **39**, 92–94 (2006)
63. Watts, D., Strogatz, S.: Collective dynamics of small-world networks. Nature **393**(6684), 440–442 (1998)
64. Zhang, Z., Schuller, B.: Active learning by sparse instance tracking and classifier confidence in acoustic emotion recognition. In: Proceedings of Interspeech, pp. 362–365. ISCA, Portland, OR, USA (2012)
65. Zhang, Y., Coutinho, E., Zhang, Z., Quan, C., Schuller, B.: Dynamic active learning based on agreement and applied to emotion recognition in spoken interactions. In: Proceedings of International Conference on Multimodal Interaction, pp. 275–278. ACM, Seattle, USA (2015a)
66. Zhang, Z., Coutinho, E., Deng, J., Schuller, B.: Cooperative learning and its application to emotion recognition from speech. IEEE Trans. Audio Speech Lang. Process. **23**(1), 115–126 (2015b)
67. Zhang, Z., Coutinho, E., Deng, J., Schuller, B.: Cooperative learning and its application to emotion recognition from speech. IEEE Trans. Audio Speech Lang. Process. **23**, 115–126 (2015c)
68. Zhang, Z., Ringeval, F., Dong, B., Coutinho, E., Marchi, E., Schuller, B.: Enhanced semi-supervised learning for multimodal emotion recognition. In: Proceedings of ICASSP, pp 5185–5189. IEEE, Shanghai, P.R. China (2016a)
69. Zhang, Y., Zhou, Y., Shen, J., Schuller, B.: Semi-autonomous data enrichment based on cross-task labelling of missing targets for holistic speech analysis. In: Proceedings of 41st IEEE International Conference on Acoustics, Speech, and Signal Processing, ICASSP 2016, pp. 6090–6094. IEEE, Shanghai, P.R. China (2016b)
70. Zhang, Z., Cummins, N., Schuller, B.: Advanced data exploitation in speech analysis—an overview. IEEE Signal Process. Mag. **34**, 24 (2017)
71. Zhu, X.: Semi-supervised learning literature survey. Technical report. TR 1530, Department of Computer Sciences, University of Wisconsin at Madison, Madison, WI (2006)

Chapter 13
Conversational Agents and Negative Lessons from Behaviourism

Milan Gnjatović

Abstract This chapter addresses the question of whether it is enough to extract from data the knowledge needed to implement socially believable conversational agents. Contrary to the popular views, the answer is negative. In this respect, the chapter points to some shortcomings of fully data-driven approaches to dialogue management, including the lack of external criteria for the selection of dialogue corpora, and the misconception of dialogue structure and dialogue context. To point to these shortcomings is not to undervalue data-driven approaches, but to emphasize the message that big data provide only a partial account of human-machine dialogue, and thus must not remain wedded to small linguistic theory, as it is currently the case.

13.1 Introduction

Since its inception in the mid-twentieth century, the field of artificial intelligence is characterized by a strong intellectual divide between the two research traditions: representationalism and behaviourism. The research question of human-machine dialogue modelling was not an exception in this respect. Related to this particular research question, the first tradition (i.e., representationalism) assumes that rules to manage human-machine dialogue should be represented in logical or at least human-interpretable forms, and that the functionality of conversational agents should be analytically tractable and with explanatory power. The second tradition (i.e., behaviourism) assumes that rules to manage human-machine dialogue can be automatically derived from dialogue corpora, e.g., by means of statistical and neural network-based methods, without the need to understand underlying dialogue phenomena [38, 39].

M. Gnjatović (✉)
Faculty of Technical Sciences, University of Novi Sad, Trg Dositeja Obradovića 6, 21000 Novi Sad, Serbia
e-mail: milangnjatovic@uns.ac.rs

© Springer Nature Switzerland AG 2019
A. Esposito et al. (eds.), *Innovations in Big Data Mining and Embedded Knowledge*, Intelligent Systems Reference Library 159, https://doi.org/10.1007/978-3-030-15939-9_13

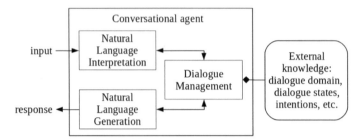

Fig. 13.1 The basic conversational agent architecture

The standard textbook architecture of a conversational agent is depicted in Fig. 13.1 [13, 14, 40]. The number of its components may vary,[1] and Fig. 13.1 shows only those components that are important for human-machine dialogue management. The task of the natural language interpreter is to recover communicative intent of a user from her dialogue act and available information about dialogue domain, syntactic expectations, etc. Based on the obtained interpretation of a user's dialogue act, the dialogue manager updates its representation of dialogue context, and determines the agent's response according to the underlying dialogue control model. The natural language generator maps the determined response onto a dialogue act.

The traditional approaches to dialogue management can be categorised in dialogue grammars, frame-based, plan-based and collaborative. Although they differ significantly among themselves, their common characteristic is that they are grounded, i.e., based on presupposed knowledge on dialogue domain, dialogue structure, users' intentions, etc. These approaches—that may be referred to as representational—have been criticised on account of the fact that this external knowledge is often hand-coded, highly task-dependent, and not elaborate enough to capture all dialogue phenomena relevant for natural dialogue, which causes conversational agents to lack flexibility and portability [40].

The current trend in research on human-machine dialogue management is aimed at overcoming these drawbacks. Thus, at the specification level, it is aimed at developing open-domain end-to-end conversational agents. At the methodological level, the current trend is strongly influenced by the success of data-driven approaches in pattern recognition (e.g., in machine translation and speech recognition). This methodological paradigm assumes that it is possible to conduct a fully data-driven training of end-to-end conversational agents, as illustrated in Fig. 13.2 (cf. [28]).

One of the most obvious advantages of automated machine learning analyses of large-scale dialogue corpora is that they may reveal rules to manage human-machine dialogue that are otherwise neither intuitive nor evident to researchers. However, such analyses provide a third person view of an observer on how dialogues unfold, but not facts that represent first person awareness of an interlocutor in a conversation.

---

[1]A typical architecture often found in literature contains two additional components for speech recognition and synthesis. They are omitted in Fig. 13.1, since they are not in the focus of this chapter.

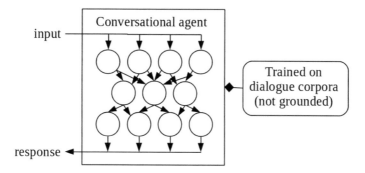

Fig. 13.2   End-to-end trainable conversational agent

In other words, they do not reveal facts of what interlocutors know, believe or intend in conversation (cf. [37]). The currently popular data-driven approaches to dialogue management are based on a strong belief that by analysing surface manifestations of dialogues we can recover *all* rules relevant for dialogue management.

In this chapter, I argue that this principal approach alone cannot, in general, provide the knowledge needed to implement socially believable[2] conversational agents whose functionalities go substantially beyond chatbots. This criticism is not intended to provide a complete account of methodological shortcomings present in data-driven approaches, but to emphasize selected conceptual and methodological points that may induce a critical reconsideration of the unduly optimism toward fully data-driven approaches to human-machine dialogue modelling.

The chapter is organized as follows. In Sect. 13.2, I discuss the fact that automatic production of large-scale dialogue corpora lacks external criteria according to which dialogues are selected, introducing thus a bias into corpora. Sections 13.3 and 13.4 include criticism of the hidden Markov model-based approach and the neural network-based approach to human-machine dialogue management, respectively. The criticism of the hidden Markov model-based approach is (mainly) levelled against the misconception of dialogue structure, while the criticism of the neural network-based approach is (mainly) levelled against the misconception of dialogue context. The distinction between these criticisms is not firm, and the reader is advised to consider them as two facets of the same problem. Section 13.5 concludes the chapter.

13.2   Lack of External Criteria for Dialogue Corpora Production

The machine learning approaches to dialogue management are based on the assumption that all dialogue phenomena show as patterns within the data. It may very well

---

[2]The notion of a socially believable conversational agent is discussed in [6] in more detail. Here I rely on the reader's intuitive understanding of this notion.

be so. But it does not imply that we can derive a systemic account of these patterns from dialogue corpora. On the one hand, the notions of representativeness and balance of a dialogue corpus are quite vague, and there are still no definite answers to fundamental questions of what sort of dialogues should be collected, and how to identify dialogue instances that can serve as models for the population, etc. (cf. [32, 35, 38]). On the other hand, criteria used to collect dialogue corpora in actual research in the field of data-driven conversational agents do not even seem to be designed to ensure, or at least to increase, representativeness and balance. This is no surprise. Fully data-driven approaches require large-scale dialogue corpora that can be practically collected only by means of automatic extraction—almost by rule— from web sources. Two prominent examples are the Twitter corpus, containing 1.3 million conversations extracted from the microblogging service Twitter [23], and the Ubuntu Dialogue Corpus, containing almost one million conversations scraped from the Ubuntu technical support chat logs [16].

What usually remains unnoticed is that such datasets are called *corpora* only nominally, but actually they are just *collections* or *archives*. The difference is not just terminological, but rather essential. To produce a dialogue *corpus*, one needs to first define *external criteria*, based on an examination of the communicative function of a dialogue, according to which dialogue instances should be selected to represent dialogue variety as a source of data for research (cf. [32]). In contrast to this, the automatic production of large-scale dialogue *collections* is never conducted according to some well-elaborated external criteria, but rather by *internal criteria* that reflect details of selected dialogues.

This can be exemplified for the Twitter corpus [23]. Twitter was crawled for a two month period, according to the following protocol. First, a limited set of most recently active Twitter users was obtained. This set was expanded by a limited number of additional users who had engaged in conversation with the seed users. Then, the chain of replies was recursively followed to recover entire conversations between the selected users. However, the described selection of Twitter users is biased— more active users are more probably selected, and that introduces the bias into the dialogue corpus. The user selection bias could have been corrected, e.g., by applying a Metropolis-filtered random walk on the Twitter graph [15], but even that would not necessarily imply that the recovered conversations are representative, because their properties and communicative functions were never taken into account in the first place.

Another problem can be exemplified for the Ubuntu Dialogue Corpus [16]. In contrast to the Twitter corpus, it comprises dialogues from a task-specific domain. Dyadic dialogues were extracted from the chat room multi-party conversations scraped from the Ubuntu technical support chat logs. Conversations with less than three dialogue turns, or longer than five sentences where one user produces more than eighty percent of the sentences, were disregarded as not representative. However, these constraints do not ensure the representativeness of the corpus. The extracted conversations evolve between two interlocutors, where one of them asks for a solution to a given Ubuntu-related problem, and the other one supposedly tries to provide a solution. But the latter interlocutor may omit to provide a solution, or provide an incorrect solution.

The inclusion of such conversations affects the representativeness of the corpus, because a (useful) conversational agent should not be trained on conversations that do not satisfy the informational need of the first interlocutor.

In general, the research question of determining appropriate external criteria for the automatic production of large-scale dialogue corpora is still open, and not likely to be answered any time soon (cf. [32, 35]). This fact is too often neglected by researchers in the field of data-driven machine learning.

13.3 Criticism of the Hidden Markov Model-Based Approach

Traditionally, in the field of human-machine dialogue modelling, the simplest and most often applied dialogue architecture relates to dialogue grammars [14]. The dialogue grammar approach tends to capture surface patterns of dialogue. In the most strict version of this approach, i.e., the finite state approach, the dialogue structure is defined beforehand. Less strict versions of this approach are focused on identifying local surface patterns, such as sequencing regularities in linguistic exchange [40], e.g., so-called *adjacency pairs*: question followed by answer, proposal followed by acceptance or rejection, etc. [26].

This approach has been criticised on the grounds that it is too restrictive, highly domain-dependent and unnatural, as the user has no control of the conversation, but is constrained to follow a predefined script. In order to overcome these drawbacks, the notion of probabilistic discourse grammar was proposed [14, 22, 34]. The motivation underlying this dialogue modelling framework is that dialogue structure can and should be automatically detected.

At the methodological level, this approach is based on well known techniques applied for speech recognition and machine translation, such as n-grams and hidden Markov models [14, 22, 34]. The core idea is briefly presented here. Let $D = d_1, d_2, \ldots, d_k$ be a sequence of dialogue acts representing hidden states, and $O = o_1, o_2, \ldots, o_k$ a set of lexical, collocational, and prosodic features that represent observations, where observation o_i is related to dialogue act d_i, for $1 \leq i \leq k$ (cf. Fig. 13.3).

The problem of dialogue management is reduced to the problem of determining a sequence of dialogue acts $\hat{D}$ that has the highest posterior probability $P(\hat{D}|O)$, i.e., rewritten by applying the Bayes' rule:

Fig. 13.3 Hidden Markov modelling of dialogue

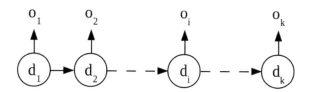

$$\hat{D} = \text{argmax } P(D|O) = \text{argmax } \frac{P(O|D)P(D)}{P(O)} = \text{argmax } P(O|D)P(D) . \quad (13.1)$$

The observation likelihood $P(O|D)$ is estimated as:

$$P(O|D) \approx \prod_{i=1}^{k} P(o_i|d_i) , \quad (13.2)$$

based on the assumption that the probabilities $P(o_i|d_i)$ are mutually independent.

The prior probability of a dialogue sequence $P(D)$ is estimated by using a dialogue act n-gram. The use of n-grams is based on a very specific conceptualization of dialogue. The most obvious characteristic of this conceptualization is that dialogue is a linear sequence of dialogue acts. Another, less obvious characteristic is related to the Markovian assumption underlying n-gram models that it is possible to predict a dialogue act by taking into account only several immediately preceding dialogue acts, i.e.:

$$P(d_k|d_1, d_2, \ldots, d_{k-1}) \approx P(d_k|d_{k-n+1}, \ldots, d_{k-1}) , \quad (13.3)$$

where n is the order of n-gram. In other words, n-grams can capture relations between dialogue acts whose distance in a linear sequence is smaller than n. This further implies that only local relations between dialogue acts are important for the purpose of dialogue modelling.[3]

However, this conceptualization of dialogue is too restrictive—neither dialogue has a sequential structure, nor are all relevant relations between dialogue acts local, nor are probabilities $P(o_i|d_i)$ independent. This is demonstrated by the following examples emphasizing the hierarchical structure of dialogue and selected dialogue cohesive agencies (i.e., ellipsis-substitutions and references). Without loss of generality, let us assume that a trigram model (i.e., n-gram model of order three) is used to capture structural relations between dialogue acts, i.e.:

$$P(d_k|d_1, d_2, \ldots, d_{k-1}) \approx P(d_k|d_{k-2}, d_{k-1}) . \quad (13.4)$$

(i) Hierarchical structure. In linguistics, it has been already widely acknowledged that dialogue structure is essentially hierarchical, and not sequential [11, 24, 27]. This implies that two dialogue acts that form a so-called *adjacency pair* do not necessarily need to be adjacent, but may have intervening dialogue acts between them. This is illustrated by the fragment of dialogue between two participants given in Fig. 13.4.

The command stated in d_1 is ambiguous in the given interaction context, so participant B cannot provide the expected response in her turn (and thus close the linguistic exchange). Instead, she asks for a clarification in d_2, opening a subordinate linguistic exchange. Participant A provides additional information in d_3. Only after the successful closure of the subordinate exchange, participant B provides, in d_4, the expected

---

[3]It should be kept in mind that the order of n-gram models in practical applications is typically not very high, due to the requirement of efficiency.

Fig. 13.4 Dialogue
fragment that illustrates
recursive development of
dialogue structure

Participant	Dialogue act
A	d_1: Move that blue object.
B	d_2: Do you mean blue cylinder?
A	d_3: Yes.
B	d_4: Here it is.

Fig. 13.5 Structure of the
dialogue fragment given in
Fig. 13.4 (cf. [24])

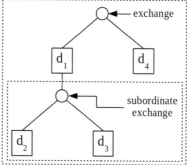

Fig. 13.6 Dialogue
fragment that illustrates
ellipsis-substitutions

Participant	Dialogue act
A	d_5: Move the blue cylinder.
B	d_6: I think that you need the red *one*.
A	d_7: *Do* what I say.
B	d_8: Here it is.

response to the command specified in d_1 (cf. Fig. 13.5). In this fragment, dialogue acts d_1 and d_4 are related, but the relation between them is not captured within the trigram (d_2, d_3, d_4).

The next two examples show that even in such cases when a short dialogue fragment can be represented as a sequence of dialogue acts, the Markovian assumption does not necessarily hold. This will be illustrated by observing two lexicogrammatical relationships between dialogue acts: ellipsis-substitutions and references.

(ii) Ellipsis-substitutions. The speaker may use ellipsis-substitutions to leave out or substitute parts of a dialogue act, in order to emphasize its contrastive parts in a given interaction context. Thus, ellipsis-substitutions are usually limited to immediately preceding dialogue acts, and are particularly characteristic of adjacency pairs (for more details on ellipsis-substitutions, the reader may consult [12]). Let us consider the dialogue fragment given in Fig. 13.6.

This fragment contains two ellipsis-substitutions. In d_6, the *one* substitutes for *cylinder*. The speaker omits *cylinder*, in order to emphasize the prominence status of the adjective *red*. In d_7, the verb *do* substitutes for the action required in d_5. Participant A does not explicitly provide information in d_7 on the required action, assuming that it can be recovered from the context, and gives more prominence to the expression of his attitude.

Similarly as in the previous example, dialogue acts d_5 and d_8 are related. However, since the in-between dialogue acts d_6 and d_7 do not contain information that could

Fig. 13.7 Dialogue
fragment that illustrates
anaphoric references

Participant	Dialogue act
A	d_9: This is a cylinder.
B	d_{10}: Nice colour.
A	d_{11}: Yes, *it* is.
B	d_{12}: And that object next to *it*?

help in recovering the intent encoded in d_5 (i.e., this information is omitted), the relation between d_5 and d_8 is not captured within the trigram (d_6, d_7, d_8).

(iii) References. References are cohesive agencies that establish relations with elements that were either explicitly mentioned in the preceding dialogue (e.g., anaphoric references), or are recoverable from the context (e.g., exophoric references) [12]. Anaphoric references[4] are illustrated by the dialogue fragment given in Fig. 13.7.

Dialogue acts d_{11} and d_{12} contains two anaphoric references (cf. pronoun *it*). Although these two references appear same at the surface level, the elements that are anaphorically pointed differ. The pronoun *it* in d_{11} refers to the color mentioned in d_{10}, while the pronoun *it* in d_{12} refers to the object mentioned in d_9. Again, the relation between the anaphora *it* in d_{12} and its antecedent *cylinder* in d_9 is not captured within the trigram (d_{10}, d_{11}, d_{12}). In contrast to ellipsis-substitutions, anaphoric references are not restricted to immediately preceding dialogue acts, but may extend back into the dialogue history—which deviates even more from the assumption that all relevant relations between dialogue acts are local.

In addition, it should be noted that the cohesive agencies such as ellipsis-substitutions and anaphora inherently transcend dialogue act boundaries. This implies that the assumption—underlying Eq. (13.2)—that the probabilities $P(o_i|d_i)$ are independent does not hold.

In general, it can be said that the hidden Markov model-based approach does not capture contextual information necessary for nontrivial dialogue management. This limitation is reflected at the practical level. This approach to dialogue modelling has been recently intensively criticised for lacking contextual information and producing too generic and dull systems' responses (cf. [17] and references therein). However, the popular criticism differs from the criticism stated in this section. While I argue here that the conceptualization of dialogue structure and dialogue cohesion in this dialogue modelling framework is inappropriate (as shown in above examples), the popular criticism is based on the view that the training objective (i.e., the maximum likelihood estimation) is over-simplified. This view is discussed in the next section.

---

[4]I consider here anaphoric references simply because they can be resolved within linguistic context. Exophoric references point outward from the dialogue, e.g., they may be recoverable from a spatial context. They are not included in the example in order to show that even when an element that is pointed to anaphorically (i.e., an antecedent) is explicitly stated in the dialogue, the established relation is not necessarily local.

13.4 Criticism of the Neural Network-Based Approach

When compared against the hidden Markov model-based approach discussed in the previous section, neural network-based approaches have recently been demonstrated as being more successful in generating *conversationally appropriate* dialogue acts [5, 17, 29, 30, 33, 36]. One of the main underlying reasons is that neural networks can practically capture a significantly broader span of dialogue history. The core idea of the neural network-based approach to context-sensitive response generation is briefly presented here.

This approach to dialogue management is related to the notion of word embeddings (cf. [2, 18, 19]). Let V be a vocabulary, and each word $w \in V$ be encoded by its one-hot representation, i.e., a vector of dimensionality $|V|$ in which the element corresponding to word w is set to 1, while all other elements are set to 0. To reduce the dimensionality of one-hot representations, each word w is assigned a learned distributed word feature vector in $\mathfrak{R}^k$, i.e., a real-valued vector of dimensionality k, where $k \ll |V|$. This vector is called *embedding*.

The joint probability function of a word sequence is then expressed in terms of embeddings of words contained in the given sequence. However, word embeddings and the probability function are learned simultaneously, by applying a recurrent neural network language model. This model contains three layers: an input layer $x_t \in \mathfrak{R}^{|V|}$ representing one-hot representation of input word at time t, a hidden layer $s_t \in \mathfrak{R}^k$ representing sentence history at time t, and an output layer $y_t \in \mathfrak{R}^{|V|}$ representing probability distribution over words at time t. At more practical level, this recurrent neural network is described by three matrices (cf. [33]):

- M_x—input matrix of dimension $|V| \times k$. Each word from vocabulary V is assigned a row in M_x containing its k-dimensional embedding, i.e., the embedding of word x_t is equal to $x_t^T \times M_x$, where x_t^T is the transposed one-hot representation of x_t.
- M_s—recurrent matrix of dimension $k \times k$ representing sentence history. Recurrent connections allow for cycling of information inside the network for arbitrary long time (as illustrated in Fig. 13.8), which is used to overcome the problem of fixed length dialogue context mentioned in the previous section.
- M_y—output matrix of dimension $k \times |V|$ that maps the hidden state on a probability distribution over words.

Fig. 13.8 Cycling of sentence history information inside the recurrent neural network model (cf. [18])

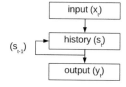

These matrices are initialized randomly, and then calculated as follows:

$$s_t = \begin{cases} f(x_t^T M_x + s_{t-1}^T M_s), & \text{if } 1 \leq t \leq T , \\ 0, & \text{if } t = 0 , \end{cases} \tag{13.5}$$

$$y_t = s_t^T M_y , \tag{13.6}$$

where f is the sigmoid activation function (i.e., logistic activation function):

$$f(z) = \frac{1}{1 + e^{-z}} . \tag{13.7}$$

The probability distribution of next word w, given the previous words is estimated by the softmax activation function (i.e., multiple logistic function):

$$P(w_t = w | w_1, w_2, \ldots, w_{t-1}) = \frac{e^{y_{tw}}}{\sum_{v \in V} e^{y_{tv}}} . \tag{13.8}$$

The probability of a word sequence $w_1, w_2, \ldots, w_n$ is then estimated as:

$$P(w_1, w_2, \ldots, w_n) = \prod_{i=1}^{n} P(w_i | w_1, \ldots, w_{i-1}) . \tag{13.9}$$

This estimation of the probability of a word sequence allows for accounting of much longer dialogue history than was practically achievable by n-grams. Let $H = h_1, h_2, \ldots, h_n$ be a word sequence of previous dialogue exchanges, immediately followed by a word sequence $D_A = a_1, a_2, \ldots, a_p$ representing a dialogue act of one interlocutor. The probability of a word sequence $D_B = b_1, b_2, \ldots, b_q$ representing the ensuing dialogue act of the other interlocutor is estimated as:

$$P(D_B | H, D_A) = \prod_{i=1}^{q} P(b_i | \underbrace{h_1, h_2, \ldots, h_n}_{\substack{\text{history} \\ \text{sequence}}}, \underbrace{a_1, a_2, \ldots, a_p}_{\substack{\text{participant A's} \\ \text{dialogue act}}}, \underbrace{b_1, b_2, \ldots, b_{i-1}}_{\substack{\text{partial response} \\ \text{of participant B}}}) ,$$

$$\tag{13.10}$$

which is illustrated in Fig. 13.9.

In contrast to the hidden Markov model-based approach, dialogue history in the recurrent neural network model is not restricted to only local relations between dialogue acts, but may be arbitrarily long. This is an obvious advantage. However, it is important to note that the neural network-based approach does not give up the assumption of sequential dialogue structure, which means that the notion of dialogue context is reduced to the notion of dialogue history. This conceptualization of dialogue context is inappropriate, as is discussed below.

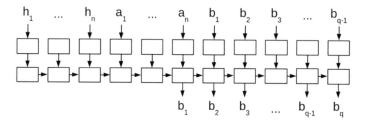

Fig. 13.9 Illustration of the recurrent neural network model for dialogue management

Dialogue segment Participant Dialogue act

		A	d_{13}: How many visits were made to our site during the last week?
		B	d_{14}: How can I tell?
		A	d_{15}: Check the control panel.
S_1	DS_2	B	d_{16}: Can I use your phone to log-in?
		A	d_{17}: Sure, you can.
		B	d_{18}: OK. I am logged-in.
	DS_3	B	d_{19}: Which option should I choose?
		A	d_{20}: See in the main menu.

Fig. 13.10 Dialogue fragment that illustrates dominance relationships between discourse segments

(iv) Dialogue history and dialogue context. The notions of dialogue history and dialogue context differ. While dialogue history may be conceptualized as a sequence of dialogue acts, the notion of dialogue context is closely related to the notions of attention and intention [11]. Attentional state is one of the components of a dialogue structure that contains information that are most salient at any given moment. Information may become salient when explicitly stated in dialogue acts, or when it is retrieved as external knowledge relevant for the purpose of dialogue acts processing. I consider here the former case, when salient information is explicitly stated in dialogue history, while the latter case (i.e., the grounding) is discussed latter.

The relations between dialogue history, attentional state and intentional structure are elaborated in Grosz and Sidner's theory of discourse structure. In their words, attentional state "serves to summarize information from previous utterances crucial for processing subsequent ones, thus obviating the need for keeping a complete history of the discourse" [11, p. 177]. In addition, "attentional state is parasitic upon the intentional structure" [11, p. 180]. We clarify these relations in an example.

According to Grosz and Sidner, there are two structural relations underlying discourse structure (including dialogue structure): dominance and satisfaction-precedence. The former is of interest for this discussion. In general, discourse segment DS_A dominates discourse segment DS_B when the purpose of DS_B contributes to the purpose of DS_A. This is illustrated in the dialogue fragment given in Fig. 13.10. This fragment has three discourse segments: DS_1, DS_2, and DS_3, whose purposes are expressed by dialogue acts d_{13}, d_{16}, and d_{19}, respectively. Discourse segment DS_1 is initiated by participant A, while discourse segments DS_2 and DS_3 are initiated by

Fig. 13.11 Illustration of the process of manipulating focus spaces

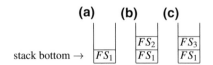

participant B. Since the purposes of DS_2 and DS_3 contribute to the purpose of DS_1, it may be concluded that DS_1 dominates DS_2 and DS_3.

To each of these segments is associated a focus space containing those semantic entities that are salient. Thus, focus space FS_1 associated with DS_1 may contain the following semantic entities: *visits*, *site*, *last week*, and *control panel*. Focus space FS_2 associated with DS_2 may contain *phone* and *log-in*, and focus space FS_3 associated with DS_3 may contain *option* and *main menu*.

The process of manipulating focus spaces is modelled by a stack [9, 11]. Thus, the processing of discourse segment DS_1 pushes focus space FS_1 on the stack (cf. Fig. 13.11a). When discourse segment DS_2 is being processed, focus space FS_2 is also pushed on the stack (cf. Fig. 13.11b). After DS_2 has been processed, FS_2 is popped from the stack, and the processing of DS_3 pushes FS_3 on the stack (cf. Fig. 13.11c). At this moment, for the purpose of dialogue act processing, semantic entities from FS_3 are most salient and available, semantic entities from FS_1 are less salient but still available, while semantic entities from FS_2 are not available at all.

This is an important difference in modelling dialogue context between the theory of discourse structure and the recurrent neural network model. In the theory of discourse structure, the contextual information relevant for the processing of a dialogue act belonging to discourse segment DS_3 contains semantic entities encoded in discourse segments DS_1 and DS_3. In contrast to this, the neural network model assumes that the relevant contextual information includes sentences from all three segments DS_1, DS_2, and DS_3—although segment DS_2 is not relevant for the purpose of processing dialogue act belonging to DS_3.

In general, the problem of the recurrent neural network model is that it equalizes dialogue history and dialogue context, which is not appropriate. More precisely, this model assumes that all relations between dialogue segments are of linear precedence,[5] while the dominance relations are not recognized. However, this assumption is not correct. Dominance relations between dialogue segments are inherently present, determining the hierarchical structure of dialogue [11, 24, 27]. Thus, the contextual information relevant for the processing of ensuing dialogue acts should not encapsulate all dialogue acts from dialogue history, but only those that belong to the current dialogue segment and dominating dialogue segments. In the neural network-based approach, the conversational agent undergoes unselective training on all preceding dialogue acts, which basically means that training data include irrelevant dialogue acts.

---

[5]This is actually the relation of satisfaction-precedence (cf. [11]), but the phrase "linear precedence" is used to emphasize the assumption that dialogue structure is sequential.

(v) Grounding and dialogue context. It was mentioned above that dialogue context contains salient information that were either explicitly stated during the interaction or retrieved as relevant external knowledge. The described recurrent neural network model is focused only on linguistic context, and not grounded in external knowledge [33], so generated responses seldom include "factual" content [5]. Researchers are aware of this, and efforts to ground data-driven models in the real world are starting to emerge. The goal of these efforts is to ground dialogue models, without giving up data-driven scalability.

The approach described in [5] extends the recurrent neural network model by conditioning a conversational agent's responses not only on (sequential) dialogue history, but also on external knowledge selected as relevant to the current dialogue context. The main idea may be represented as follows. It is assumed that the required external knowledge is dominantly stored in collections of unstructured, or loosely structured, documents available on the Web (e.g., in collections such as Wikipedia, Goodreads, Foursquare, etc.) that can be indexed with named entities. Given a dialogue history (i.e., a sequence of dialogue acts), attentional state is conceptualized as a text span that contains a set of named entities. They can be determined by means of simple keyword matching or other more advanced techniques, which is not important for this discussion. These entities are then used to retrieve external knowledge that is relevant to the current attentional state. Both dialogue history and the retrieved knowledge are fed into two distinct encoders within the neural network architecture.

At the performance level, two problems related to this approach have been already recognized: the system's responses may include irrelevant and self-contradictory facts, as reported in [5]. However, the causes of these problems have not been explained, except by the remark that data contained in underlying collections of documents may be subjective and inaccurate. It is important to note that even if data in a document collection were correct and consistent, these problems would still be present. To explain this, I turn again to what I believe to be one of the main causes of these problems. The conceptualization of attentional state as a text span is (still) too restrictive, since it does not account for dominance relationships between dialogue segments, as already discussed above. Therefore, not all determined entities in the observed text span are necessarily relevant for the processing of an ensuing dialogue act. Since these entities are used to retrieve external knowledge, it can cause that retrieved knowledge is also not really relevant to the current dialogue context.

13.5 Discussion and Conclusion

It is fair to say that currently popular data-driven approaches to human-machine dialogue management are behaviouristic. By giving up insights in dialogue structure and cohesion provided by linguistic and cognitive theories, the research in the field becomes widely, if not dogmatically, acknowledged as a science of behaviour. In this respect, it has something in common with psychology during the first half of the twentieth century. It that time, the prevalent view in psychology was that "the only objective evidence available is, and must be, behavioral" [20]. Today, the prevalent view in the field of natural language processing is "that the only real object is a corpus of data and that by automated analysis [...] one can derive everything that's relevant about the language" [4].

This chapter emphasized some of the conceptual and methodological shortcomings of data-driven approaches to dialogue management. To point to these shortcomings is not to undervalue data-driven approaches. It should be kept in mind that traditional representational approaches were prone to similar shortcomings. For example, dialogue structure has been too often misunderstood as a predefined task structure. In conversational agents based on this misconception (for prominent examples cf. Sundial [3] and Verbmobil [1] systems), the processing of a dialogue act was reduced to assigning it to one of the predefined task-specific roles (for more details, cf. [6]). Although traditional representational approaches to dialogue modelling were hand-coded and task-dependent, they still did not result in human-machine dialogue of satisfying level of naturalness. In contrast to them, neural networks have demonstrated a potential to make human-machine interaction more fluid. And that may very well be where the main value of neural network-based approaches to dialogue management lies.

However, big data provide only a partial account of human-machine dialogue and thus must not remain wedded to small linguistic theory, as it is currently the case. The core problem is that practitioners of neural networks in the field of natural language processing too often completely neglect any linguistic theory, and get rid of the responsibility to understand language phenomena. And sometimes this attitude works. A case in point: one does not need to have an understanding of a certain language in order to make an automatic speech recognition system for that language. This is possible because speech can be adequately modelled as having a linear structure [21], but the assumption of linear structure does not hold for dialogue. Consequently, speech recognition technology has significantly advanced in the last three decades, in contrast to the pretty static field of dialogue management. It should not be surprising that we cannot apply the data-driven paradigm to address the research question of dialogue modelling in the same manner as we apply it to the speech recognition problem.

If we aim at developing task-oriented and socially-believable conversational agents, it is necessary to break the grip of behaviourism on the field, i.e., to integrate the principles of dialogue with data-driven approaches (cf. [10, 31]). Although representational approaches to dialogue modelling have been much neglected for a while by the research community, they were not completely abandoned. In the meantime, they have also been further advanced with respect to design-oriented approaches to deploying the dialogue principles in conversational agents, although these developments remained less obvious due to the overwhelming data-driven trend. For examples of such representational approaches that are by no means exhaustive, the reader may consult [7–10, 25].

Another historical analogy may be appropriate at the end. By the mid-nineteen-fifties, in psychology it had become evident that the behaviourist approach alone does not suffice, and that cognitive concepts must be also taken into account to explain the behavioural data [20]. Maybe the field of human-machine interaction should attempt to learn from the negative lessons from behaviourism.

Acknowledgements The presented study was sponsored by the Ministry of Education, Science and Technological Development of the Republic of Serbia (research grants III44008 and TR32035), and by the intergovernmental network EUREKA (research grant E!9944). The responsibility for the content of this article lies with the author.

References

1. Alexandersson, J., Reithinger, N.: Learning dialogue structures from a corpus. In: Proceedings of EuroSpeech-97, Rhodes, pp. 2231–2235 (1997)
2. Bengio, Y., Ducharme, R., Vincent, P., Jauvin, C.: A neural probabilistic language model. J. Mach. Learn. Res. **3**, 1137–1155 (2003)
3. Bilange, E.: An approach to oral dialogue modelling. In: Taylor, M.M., Néel, F., Bouwhuis, D.G. (eds.) The Structure of Multimodal Dialogue II, pp. 189–205. John Benjamins Publishing Company, Philadelphia/Amsterdam (2000)
4. Chomsky, N.: Language and the Cognitive Science Revolution(s) (2011). https://chomsky.info/20110408/. Cited 19 Feb 2018
5. Ghazvininejad, M., Brockett, C., Chang, M.-W., Dolan, B., Gao, J., Yih, W.-t., Galley, M.: A Knowledge-Grounded Neural Conversation Model, Association for the Advancement of Artificial Intelligence (2018)
6. Gnjatović, M., Borovac, B.: Toward conscious-like conversational agents. In: Toward Robotic Socially Believable Behaving Systems, Volume II—Modeling Social Signals. Esposito, A., Jain, L.C. (eds.), volume 106 of the series Intelligent Systems Reference Library, pp. 23–45. Springer (2016)
7. Gnjatović, M.: Therapist-centered design of a robot's dialogue behavior. Cogn. Comput. **6**(4), 775–788 (2014)
8. Gnjatović, M., Delić, V.: Cognitively-inspired representational approach to meaning in machine dialogue. Knowl.-Based Syst. **71**, 25–33 (2014)
9. Gnjatović, M., Janev, M., Delić, V.: Focus tree: modeling attentional information in task-oriented human-machine interaction. Appl. Intell. **37**(3), 305–320 (2012)
10. Grosz, B.: Smart enough to talk with us? Foundations and challenges for dialogue capable AI systems. Comput. Linguist. **44**(1), 1–15 (2018)
11. Grosz, B., Sidner, C.: Attention, intentions, and the structure of discourse. Comput. Linguist. **12**(3), 175–204 (1986)
12. Halliday, M.A.K., Matthiessen, C.M.I.M.: An Introduction to Functional Grammar, 3rd edn. Hodder Arnold (2004)
13. Jokinen, K., McTear, M.: Spoken Dialogue Systems. Synthesis Lectures on Human Language Technologies, vol. 2(1), pp. 1–151. Morgan Claypool (2009)
14. Jurafsky, D., Martin, J.H.: Speech and Language Processing: An Introduction to Natural Language Processing, Speech Recognition, and Computational Linguistics, 2nd edn. Prentice-Hall (2009)
15. Lovász, L.: Random walks on graphs: a survey, combinatorics, Paul Erdos is eighty. Bolyai Society Mathematical Studies, vol. 2, pp. 1–46 (1993)
16. Lowe, R., Pow, N., Serban, I., Pineau, J.: The Ubuntu dialogue corpus: a large dataset for research in unstructured multi-turn dialogue systems. In: Special Interest Group on Discourse and Dialogue, SIGDIAL (2015)
17. Li, J., Monroe, W., Shi, T., Ritter, A., Jurafsky, D.: Adversarial Learning for Neural Dialogue Generation. Empirical Methods in Natural Language Processing (EMNLP) (2017)

18. Mikolov, T., Karafiát, M., Burget, L., Černocký, J.H., Sanjeev Khudanpur, S.: Recurrent neural network based language model. In: Proceedings of INTERSPEECH, pp. 1045–1048 (2010)
19. Mikolov, T., Yih, W.-t., Zweig, G.: Linguistic regularities in continuous space word representations. In: Proceedings of NAACL-HLT 2013, Association for Computational Linguistics, pp. 746–751 (2013)
20. Miller, G.A.: The cognitive revolution: a historical perspective. Trends Cogn. Sci. **7**(3), 141–144 (2003)
21. Rabiner, L.R., Schafer, R.W.: Digital Processing of Speech Signals. Prentice-Hall (1978)
22. Ritter, A., Cherry, C., Dolan, W.B.: Data-driven response generation in social media. In: Proceedings of EMNLP 2011, pp. 583–593 (2011)
23. Ritter, A., Cherry, C., Dolan W.B.: Unsupervised modeling of twitter conversations. In: Human Language Technologies: The 2010 Annual Conference of the North American Chapter of the Association for Computational Linguistics, HLT '10, Morristown, NJ, USA, pp. 172–180 (2010)
24. Roulet, E.: On the structure of conversation as negotiation. In: Parret, H., Verschueren, J. (eds.) (On) Searle on Conversation, pp. 91–99. John Benjamins Publishing Company, Philadelphia/Amsterdam (1992)
25. Savić, S., Gnjatović, M., Mišković, D., Tasevski, J., Maček, N.: Cognitively-inspired symbolic framework for knowledge representation. In: Proceedings of the 8th IEEE International Conference on Cognitive Infocommunications (CogInfoCom), Debrecen, Hungary, pp. 315–320 (2017)
26. Schegloff, E.A.: Sequencing in conversational openings. Am. Anthropol. **70**, 1075–1095 (1968)
27. Searle, J.: Conversation. In: Parret, H., Verschueren, J. (eds.) (On) Searle on Conversation, pp. 7–29. John Benjamins Publishing Company, Philadelphia/Amsterdam (1992)
28. Serban, I.V., Lowe, R., Charlin, L., Pineau, J.: A Survey of Available Corpora for Building Data-Driven Dialogue Systems. arXiv e-prints, arXiv:1512.05742 (2015)
29. Serban, I.V., Sordoni, A., Bengio, Y., Courville, A., Pineau, J: Building end-to-end dialogue systems using generative hierarchical neural network models. In: Proceedings of the Thirtieth AAAI Conference on Artificial Intelligence, AAAI'16, pp. 3776–3783 (2016)
30. Shang, L., Lu, Z., Li, H.: Neural responding machine for short-text conversation. In: Proceedings of the 53rd Annual Meeting of the Association for Computational Linguistics and the 7th International Joint Conference on Natural Language Processing, pp. 1577–1586. Beijing, China (2015)
31. Shoham, Y.: Why knowledge representation matters. Commun. ACM **59**(1), 47–49 (2015)
32. Sinclair, J.: Corpus and text—basic principles. In: Wynne, M. (ed.) Developing Linguistic Corpora: A Guide to Good Practice, pp. 1–16. Oxbow Books, Oxford (2005)
33. Sordoni, A., Galley, M., Auli, M., Brockett, C., Ji, Y., Mitchell, M., Nie, J-Y., Gao, J., Dolan, B.: A neural network approach to context-sensitive generation of conversational responses. In: Proceedings of HLT-NAACL, pp. 196–205 (2015)
34. Stolcke, A., Ries, K., Coccaro, N., Shriberg, E., Bates, R., Jurafsky, D., Taylor, P., Martin, R., Van Ess-Dykema, C., Meteer, M.: Dialogue act modeling for automatic tagging and recognition of conversational speech. Comput. Linguist. **26**(3), 339–373 (2000)
35. Tognini-Bonelli, E.: Corpus Linguistics at Work. John Benjamins, Amsterdam (2001)
36. Vinyals, O., Le, Q.V.: A neural conversational model. In: ICML Deep Learning Workshop, arXiv:1506.05869 [cs.CL] (2015)
37. Widdowson, H.G.: On the limitations of linguistics applied. Appl. Linguist. **21**(1), 3–25 (2000)
38. Wilks, Y.: Is there progress on talking sensibly to machines? Science **318**(5852), 927–928 (2007)
39. Wilks, Y.: IR and AI: traditions of representation and anti-representation in information processing. In: McDonald, S., Tait, J. (eds.) Advances in Information Retrieval. ECIR 2004. Lecture Notes in Computer Science, vol. 2997, pp. 12–26. Springer, Berlin, Heidelberg (2004)
40. Wilks, Y., Catizone, R., Turunen, M.: Dialogue Management. COMPANIONS Consortium: State of the Art Papers (2006) Public report. https://pdfs.semanticscholar.org/fea2/03cd009a66cc232ead6b808f1c9f67c86f3f.pdf. Cited 19 Feb 2018

Author Index

© Springer Nature Switzerland AG 2019
A. Esposito et al. (eds.), *Innovations in Big Data Mining and Embedded Knowledge*,
Intelligent Systems Reference Library 159,
https://doi.org/10.1007/978-3-030-15939-9

Printed in the United States
By Bookmasters